The West in a Nutshell

Foundations, Fragilities, Futures.

Paul Monk

with images by
Jörg Schmeisser

BARRALLIER BOOKS

First published in 2009 by Barrallier Books
Registered Office: 35-37 Gordon Avenue, Wast Geelong, Victoria 3220
Australia
www.barrallierbooks.com
Copyright © Paul Monk 2009
Images © Jörg Schmeisser

National Library of Australia cataloguing-in-publication information:

Monk, Paul M., 1956
The West in a nutshell: foundations, fragilities, futures / Paul Monk.

1st ed.
Reprinted 2014

Includes index.

ISBN 978 0 9585838 9 3

I. Civilisation, Western II. Intellectual life—History

Other authors/Contributors Schmeisser Jörg, 1942–2012

909.09821
Book and cover design by Peter Gamble, Ink Pot Graphic Design, Canberra
Set in Garamond Premier Pro 12/17 and P22 Cezanne
Cover image: *JS 461 Thinking of Ariadne*, Jörg Schmeisser, etching, 1991

Contents

vi

Images

Jörg Schmeisser was an artist with a distinctive style and a remarkable range of interests. The superb etchings he offered to enhance *Sonnets to a Promiscuous Beauty* made a return to his work for the present book irresistible. The delicate conundrum lay in which images to choose from his offerings - truly a remarkably rich array. *Thinking of Ariadne* was chosen for the cover because the underlying motif of the book is that of Ariadne's golden thread guiding Theseus through the labyrinth. *The Goddess (Knossos)* was selected as the frontispiece because the labyrinth was located in Knossos, which was the epicentre of archaic Western civilisation during the Bronze Age. *Shell* greets the reader at the beginning of Part One because the shell is evocative of both the origins of life in the sea and the blackened shell of fire-bombed cities. In this it is iconic of the remote origins and doubtful fate of humanity. *Chania* heralds the beginning of Part Two as an image suggestive of the rise of classical Grecian civilisation. *Now I Remember Jerusalem* is the gateway to Part Three, evoking the great debate over the foundations of Western civilisation: the secular claims of Athens ranged against the religious claims of Jerusalem and of Islam, represented here by the Dome of the Rock. *Sketches Also After Dürer* was chosen because, from first to last, it is richly suggestive of both the biological nature of our kind and the combination of art and science by which we have risen to dominate the planet, both for better and for worse.

Para mi Ariadne, con su hilo de oro
Nunca Naxos, pero siempre amor en Atenas

Preface

Many years ago when I was still a university student, a talented friend who had just been offered a Rhodes Scholarship, confided in me that his secret ambition was to become 'a Bloomsbury essayist'. My own ambitions at that point in time were somewhat more grandiose. In the event, as fate would have it, I became the essayist, while he went on to a serious career in government service. The essays I have written are neither scientific papers nor ideological polemics, but pieces written as the occasion presented itself or the mood took me over the past decade. Each was written for the sheer pleasure of trying to articulate a thoughtful position on the subject in question. They are published now not to serve any particular agenda, but simply as an offering to all those who enjoy such reflection.

The book contains three parts, each consisting of ten essays. None of the parts attempts an exhaustive or systematic coverage of the subject matter. Rather, each consists of a broad theme with the essays approaching it from different angles. The essays are invitations to think about key issues from a provocative point of view and each offers a guide as to where the interested reader may find further good reading on the subject. Part I revolves around the question: 'What are we as human beings and how did we assume a position of such presumed mastery in the world?' Part II centres on the question: 'What is our Western heritage as regards the nature of truth?' It begins with the difference between myth and history and ends with the difference between metaphysical certainty and scientific circumspection. The essays in Part III address a range of pressing questions that have arisen in our time and seek to bring to bear upon them something of the wisdom derived from the sources drawn on in Parts I and II – as well as the best contemporary sources I could find.

The word 'essay' itself derives from the French word to *attempt* or try, whence our now seldom-used verb *assay* – to attempt. That was the precise meaning of *Essais* – the French title given by Michel de Montaigne to his famous sixteenth-century collection of short reflections. That book has been the single most influential book of essays in the Western canon. First published in 1580, it quickly became famous for its rich combination of serious intellectual inquiry and anecdote, couched in a tone both moderate and conversational. It is the original model for this book, just as Montaigne has, for many years, been a kind of model for me, ever since I first read his *Essays* as a young man.

Montaigne (1533-1592) lived in an age of the discovery of new continents and of bitter religious strife. Both drove him to reflect deeply on meaning and truth. He is often credited with having invented the essay as a literary genre in the form to which we have become accustomed. He himself, however, modelled his essays on the short reflective writings of the great classical historian, biographer and moralist Plutarch (46-125). Plutarch's *Parallel Lives* (short character sketches of the great Greek and Roman leaders of the ancient republics) and *Moralia* (dozens of short reflections on diverse topics) had a major influence on Montaigne, as they did on his younger contemporary, Shakespeare. Plutarch is mentioned some four hundred times in Montaigne's *Essays*.

I discovered Plutarch and Montaigne at much the same time, as a young undergraduate fascinated alike by the classical world and the world of the Western Renaissance and Reformation. Simply reading such writers helps to form reflective habits of mind and it is unsurprising, therefore, that both Plutarch and Montaigne were also major influences on the famous nineteenth-century American essayist Ralph Waldo Emerson (1803-1882). His *Essays*, initially published in two volumes, in 1841 and 1844, drew much from Plutarch's *Moralia*. He described Plutarch's *Parallel Lives* as 'a Bible for heroes'. Emerson's writing, like the writing of Montaigne, exhibits a free-spirited break with religious orthodoxy, a non-dualist natural philosophy and an emphasis on manners and character in impressive human beings. All three of these essayists, though especially Montaigne, have left their mark on me and I trust that the reader will find in these essays traces of their freedom of spirit and some of their abiding concerns.

What took Montaigne beyond Plutarch was his realisation of the significance of the discovery of the Americas in the decades just before he was born. What took Emerson beyond Montaigne was his realisation of the significance of the discoveries of early modern science, in the century or so before he wrote. There are few passages in Emerson more indicative of the impression made on him by modern science than that, in his famous essay 'Nature', in which he remarked:

> Geology has initiated us into the secularity of nature, and taught us to disuse our dame-school measures, and exchange our Mosaic and Ptolemaic schemes for her large style. We knew nothing rightly, for want of perspective. Now we learn what patient periods must round themselves before the rock is formed; then before the rock is broken, and the first lichen race has disintegrated the thinnest external plate into soil, and opened the door for the remote Flora, Fauna, Ceres and Pomona to come in. How far off yet is the trilobite! How far the quadruped! How inconceivably remote is man! All duly arrive and then race after race of men. It is a long way from granite to the oyster; farther yet to Plato and the preaching of the immortality of the soul.

He was writing barely a decade after the publication of Charles Lyell's *Principles of Geology*. His awe at the depth of perspective that Lyell had opened up on the immense age of the earth tells us a good deal about his freedom of spirit. He enriched himself by reflecting upon reality, rather than seeking to shield himself from reality in a world of reflection.

I would like to think that these essays, collected and arranged in order to offer them to a wider readership, exhibit the same spirit of openness and engagement with reality. They were written for my own enjoyment of depth of perspective and of a wide-ranging, rather than narrowly specialised understanding of the world. No overall design was conceived in their composition, however, before the preparation of this book. Then, when I looked back over the scores of essays written during the better part of a decade, it became apparent that I had covered a great deal of subject matter in time and space, from the roots of our common humanity, to the riches of the classical Western tradition, to some of the pressing concerns of the early twenty-first century world. A selection of essays was made, therefore, which would take the reader on this journey of exploration.

The book's design is aesthetic, however, not ideological. It *introduces* a thematic coherence that was never foreseen as the essays were composed. Many other essays preceded and accompanied them throughout the years in which they were published. The thirty published here were selected from perhaps three times that number with a view to putting between two covers something like a handbook for the well-read citizen or university student. The intended reader is a person looking for a synoptic perspective on the world of Western civilisation, but not a perspective forced into a single narrative or ideological form and not so encyclopedic as to be exhausting. That such a book might be treasured in this way was inferred from the feedback readers provided on these essays over the years when they were originally published.

The attempt at a synoptic introduction to the West is why the title *The West in a Nutshell* was chosen. That was, originally, the title of the essay on Catullus which appears here as 'On the Fate of Classical Erotic Poetry'. The book's title also captures what I would like the book to represent. Understanding what it means to be a citizen of the West, in the light of our underlying humanity, the classical past and our present dilemmas, is something I aspired to even as a young boy. From the time that I was read J. R. R. Tolkien's *The Lord of the Rings* as a fifth grader, I longed to have the depth of perspective on the world and 'the West' exhibited by 'the Wise' – Gandalf, Elrond, Galadriel. I shall never forget listening to my school teacher reading a pivotal chapter of the book, 'The Council of Elrond', in which, responding to the hobbit Frodo's astonishment that he remembered things that had occurred thousands of years earlier, Elrond remarks, 'I have seen three ages in the West of the world and many defeats and many fruitless victories.'

Gaining such depth of perspective on the real world, as distinct from Tolkien's Middle Earth, depended upon gaining a deep education and that meant reading an immense amount in a few decades since I did not have thousands of years ahead of me. It meant putting aside many a more practical or mundane concern and, in an almost monastic endeavour, working my way right into the labyrinth of the human past – always with the intention of 'slaying the Minotaur' and, like Theseus in the famous Greek legend, emerging again into the light of day. Readers of this volume may judge for themselves whether I have succeeded in doing so. A measure of my success will be whether readers find the book itself to be a kind of Ariadne's thread, helping them in their own venture into the labyrinth – and out again.

The tale of Theseus and Ariadne is one of the oldest and most venerable in the Western canon. It takes us back to archaic Greek times during the high Bronze Age (before 1,300 BCE) when the Minoan Cretan kingdom, centering on Knossos, dominated the Aegean world and pre-classical Athens was tributary to it. Every year as part of its tribute, Athens had to send young men and women to the Cretan King Minos, who would send them into his labyrinth where they would be killed by the monstrous Minotaur. When Theseus was sent to Knossos, however, Ariadne, the daughter of King Minos, fell in love with him and secretly gave him a ball of golden thread with which to find his way around and out of the labyrinth. Using it, Theseus evaded and then slew the Minotaur, escaped from the labyrinth and then eloped with Ariadne - only to abandon her on the island of Naxos and sail back to Athens alone, to a hero's welcome. On Naxos however, Ariadne was found by the god Dionysus who married her and gave her a crown of seven stars, which, after her death, became a constellation. The legend has been the subject of artistic representation and elaboration from ancient times (it was the subject of Catullus's longest poem) to the twentieth century (the opera *Ariadne of Naxos* by Richard Strauss, from a libretto by Hugo von Hofmansthal).

Plutarch, Montaigne and Emerson not only served as models for me in the art of essay writing, they also accompanied me intimately as I prepared this book for publication. I found myself thinking of Plutarch, who studied at Plato's Academy in Athens during the reign of Nero, and would, in his later years, be visited by admirers from all over the Roman Empire, coming to his estate at Chaeronea for the pleasure of conversation with him. *The West in a Nutshell* is an offering to all those who still enjoy such conversation. It is my humble equivalent of the famous marble chair from which Plutarch presided over the dialogues at Chaeronea. If I do not actually preside over dialogues, I hope this book might at least stimulate them.

I had occasion also to reflect that, when I abandoned work as an intelligence analyst more than a decade ago and turned to teaching and essay writing, I had been the very age Montaigne was when he commenced writing his own *Essays*. Montaigne, at 38, was weary of the ways of courtiers and the corruptions of reason in the religious strife of his time. He retreated to the tower of his family estate, which he called his citadel, and almost totally isolated himself from society and even family. Drawing upon his library of 1,500 books, he devoted himself to his *attempts* to find wisdom and dignity and to put down what he found so that it might be shared with others in due course, which it was - almost a decade later. While I did not quite enjoy a chateau, looking back over the decade in which my essay writing took place, I found certain surprising parallels between Montaigne's experience and my own.

As I edited the essays, I also found myself appreciating anew Emerson's serenity, his free personality and his catholicity of interests. Breaking free from his more or less orthodox Christian upbringing, he had become a New World free spirit, always insisting that vitality, energy, the natural world, boldness and largeness of mind should be respected more than rancour, envy, pedantry, theology, or what we would call political correctness. Revising these essays for publication, I found the temperate and magnanimous spirit of Emerson a guide. Emerson's *Essays* are a model for this book, less in their style or content than in their spirit. If readers of my own essays come away from them with something of the free perspective and appreciation for the richness of the world that one imbibes from Emerson, then this book will have accomplished its purpose and justified its existence.

The West in a Nutshell

JS 522, *Shell for Dresden*
Jörg Schmeisser, etching, 1997.

Part One – The Roots of our Humanity

... we can understand neither ourselves nor our world until we have finally understood what language is and what it has done for our species.
- Derek Bickerton (1990) [1]

There is no step more uplifting, more momentous in the history of mind design, than the invention of language. When Homo sapiens became the beneficiary of this invention, the species stepped into a slingshot that has launched it far beyond all other earthly species in the power to look ahead and reflect.
- Daniel Dennett (1996) [2]

Lengthening approach distances, as African herds evolved to become more wary since early [Homo] erectus days, would have gradually required more accurate throws (just double your target distance and it will become about eight times more difficult). This is the Red Queen Principle, as when Alice was told [in Through the Looking Glass*] that you had to run faster just to stay in the same place. The Red Queen may be the patron saint of the Homo lineage.*
- William H. Calvin (2004) [3]

1. On the Origins of Language

To borrow a phrase from Nietzsche, our species has built its cities on the slopes of Mount Vesuvius and sailed its ships into uncharted seas.[4] We 'live dangerously'. We feel amazed at the achievements, appalled by the savageries and stupidities of our kind. We are, like Hamlet, awed at ourselves as the paragon of animals, but disgusted by ourselves, when we contemplate our defects and depravities.[5]

At the root of both our achievements and depravities, is the human mind. It defines what we are. More precisely, it enables us to engage in the act of 'defining' (and redefining) things at all, ourselves included.[6] But what is this thing we call 'mind'? Where did it come from? That question has preoccupied shamans, priests, philosophers and cognitive scientists for millennia.

'The concept of something known as "mind" and regarded as somehow problematic goes back to the dawn of recorded thought', wrote linguist Derek Bickerton in 1990. 'Were there two parallel realities, one physical and one not? Was one set of phenomena unreal? ... Over the centuries, almost every possible answer to these questions has been espoused by someone. No solution seems to have worked; least of all the solution of claiming that there was no problem.'[7]

The longstanding debates about an 'apparent' versus a more 'real' metaphysical world and about the mind (or soul) and body showed certain constant themes, he suggested. Few people, however, 'seem to have suspected that the mind-body problem and the problem of language origins might turn out to be related.'[8] What if the very nature of language was such that it created both 'mind' and the metaphysical puzzles philosophers have wrestled with for thousands of years?

What was needed was a cognitive archaeology – an archaeology of the evolution of primate and hominid cognitive capacities and of the neurological foundations of language itself. No philosopher, from Plato to Heidegger, was equipped to undertake any such task. It has only become possible in recent decades, with rapid advances in both the cognitive and archaeological sciences. Those advances have gradually revealed the origins of the human mind within the natural order of things.

Colin Renfrew's *Archaeology and Language* (1988), Bickerton's *Language and Species* (1990), Merlin Donald's *Origins of the Modern Mind: Three Stages in the Evolution of Culture and Cognition*

(1991), Steven Mithen's *The Prehistory of the Mind: A Search for the Origins of Art, Religion and Science*, Donald Johanson and Blake Edgar's *From Lucy to Language* (1996), Terence Deacon's *The Symbolic Species: The Co-Evolution of Language and the Brain* (1997), and Ian Tattersall's *Becoming Human* (1998) were all excellent syntheses of the ongoing scientific work.

William Calvin is at the forefront of this work. In 2000, he co-authored a book with Bickerton called *Lingua ex Machina: Reconciling Darwin and Chomsky with the Human Brain*.[9] In it, they argued that an adequate theory of what language is could not simply posit an innate language module, as Chomsky does, and leave the matter at that. It had to supply an evolutionary pedigree for such a module. They offered the elements of just such a pedigree.

What they were looking for was the evolutionary origin of the neural circuitry that makes language possible at all. Their finding had to be consistent with several given realities. First, that hominids had evolved over an enormous period of time (some seven million years) compared to that in which *Homo sapiens* had overrun the biosphere (50,000 years). Second, that evolution works by mutation, adaptation (or exaptation) and accumulation, not through linear development or holistic, teleological design.[10] Third, that hominids had had large brains for hundreds of thousands of years before there was any evidence of the kind of thinking we associate with language.

What all this meant, they suggest, is that the syntax which makes language what it is had to be an emergent property of a number of other neural and physiological capacities that had evolved over a very long period of time. 'It might be like adding a capstone to an arch,' they suggest, 'which permits the other stones to support themselves without scaffolding and, as a committee, they can defy gravity.' Their task, as they explain, was to conceive of 'the scaffolding that could have initially put such a stable structure in place.'

In *A Brain For All Seasons* (2003), Calvin argues that abrupt climate flips and selection pressure on pre-sapient hominids to enhance their capacity for accurate throwing may have played crucial roles in the development of human brains and specifically in paving the way for language. The 'scaffolding' consisted of the physical and neurophysiological changes that such pressures induced in large-brained hominids over the past two million years or more.

In a further book, *A Brief History of the Mind* (2004), he sketches out not only these hypotheses about how the brain evolved a sapient mind, but how that mind has since invented human culture and the challenges it faces in the twenty-first century. His project in this little book was extraordinarily ambitious. He described it as having been inspired by Stephen Hawking's *A Brief History of Time* and David Fromkin's *The Way of the World*, which attempted similarly ambitious tasks.

Nothing in the book is more intriguing than Calvin's reiteration of the hypothesis that selection pressures for accurate throwing could do much to explain the emergence of a capacity for syntax in the brain. Of the tasks facing pre-sapient hominids, he argues, 'It is only the *accurate throws* that make a lot of demands on the brain compared with what the great apes can do.' This is because accurate throwing requires elaborate coordination of many muscles and an intentional stance that can produce flexible variations of such coordination at any given instant.

This takes us back before language, looking at real, non-metaphysical processes which might have paved the way for it. In other words, it overcomes the mind-body problem by reconstructing a time when there was a human body without 'mind' as we know it, then showing how this 'mind' can have emerged within the body. Better still, it shows how language itself may have emerged, on an unforeseen basis, within a brain trying to cope with various other challenges.

After all, it is the appearance of syntax – the rules governing the construction of sentences – itself that needs to be explained in naturalistic terms if we are to account for the emergence of language and with it 'mind' from among the primates. More precisely, it is *structured thought* that requires explanation. As Calvin puts it, 'syntax is not the only type of structured thought, and perhaps not even the first to evolve. Other structured aspects of thought are multi-stage planning, games with rules that constrain possible moves, chains of logic, structured music and a fascination with discovering hidden order, with imagining how things hang together.'

It's vital to remember that Calvin is not saying all this was caused by accurate throwing alone. A set of conditions had to be in place before the 'capstone' of structured thought or syntax could be slipped into place. The archaeological evidence clearly indicates that early humans engaged in group hunting, camp-building, tool-making of a kind requiring joint attention and intentional instruction and thus social interactions of a sophistication unknown among apes and monkeys.

All this, plus the physiological changes entailed in upright posture (including the freeing up of the throat and chest muscles and nerves for more subtle vocalisations), the change of diet from vegetarian to high meat content (allowing for a shrinkage of the gut and a diversion of blood resources to the growing brain), and the development of the neocortex (which is all about forming new associations between things) within the brain, plainly were among these conditions.

Even so, how could hundreds of thousands of years of relentless work on accurate throwing have hauled the capstone up the scaffolding and dropped it into place, as it were? Calvin makes three basic claims. First, that just getting set for such a throw 'involves a major amount of reassignment of cerebral resources, quite unlike most cognitive tasks.' Second, that there is considerable neural overlap between hand/arm, oral/facial and linguistic queuing or 'planning' in our brains. Third, that the more accurate casting of projectiles would have had both immediate, tangible pay-offs ('your family eats high calorie, non-toxic food for additional days of the month') and a long growth curve (if you keep improving, you get better returns).

This third claim prompts one of his most colourful analogies – the one I've used as an epigraph. As African herds became more and more wary of canny, biped hunters with projectiles, they would move away at any hint of human approach. The task of the hunters became progressively more demanding. Like Alice, in *Through the Looking Glass*, they had to run harder and harder just to stay in the same place.

More precisely, they had to throw more and more accurately, over greater and greater distances to keep getting their meat. Hence Calvin's quip that Lewis Carroll's Red Queen might well be considered the 'patron saint of the *Homo* lineage' because 'she' put our kind on a treadmill which forced us to develop post-pongid neural processes. Far from keeping us in the same place, these finally launched us, in Daniel Dennett's phrase, as if from a slingshot, into our strange and wondrous sapient condition.

However, we are very much 'through the looking glass' here and one might still ask how Calvin deduced that accurate throwing for two million years gave us art, agriculture, writing, philosophy and science within 50,000 years. He puts it like this: 'The structuring that I have in mind for higher intellectual functions is not just a simple chain of events and intermediate products. It is more like a symphony – and that reminds me of another important symphony that the brain had been producing for a million years before intellect. Accurate throwing ... involves a structured plan. Indeed, planning a throw has some nested stages, *strongly reminiscent of syntax*.'

To achieve the desired effect, the brain must coordinate about a hundred muscles from the spine via the shoulder, elbow and wrist to the fingers, be able to pick exactly the right moment to cast and execute in an eighth of a second 'in just the right way or ... miss the target and go hungry.' Calvin offers the intriguing observations, based on this root, that 'the notion of being *impelled down a path* is very strong in us' and that many of our cognitive processes involve arbitrary frameworks of 'allowed moves, against which possible moves must be checked before acting.'

That linguistic capacity took so long to develop from this root is less problematic than one might intuitively think. There are other cases of capability long antedating development. For example, as Calvin points out, 'we were likely capable of structured music ... just after the big transition (around 50,000 years ago) long before Western music got around to using it about a thousand years ago. Just because novel symphonies of hand-arm movement commands had been getting better and better for a million years doesn't mean that secondary use for spoken language had to happen.'

He does believe, however, that proto-language very likely existed long before the throwing 'syntax' was exapted to create full-blown language. What is unclear is when and to what extent such proto-language emerged – given that chimps, for example, from whom we diverged seven million years ago, use various common sounds to register warnings as specific as 'eagle' and 'snake'. Yet, by this account, (full-blown, syntactical) language did not emerge until relatively *late* in the development even of *Homo sapiens*.

What is clear is that, once it *did* emerge, it made us a formidable competitor for any other creature on the planet – as thousands of species have since found. The slingshot that Dennett understandably celebrates made us to other beasts what the cyborgs and rebellious robots of current science fiction are to us (in our imaginations) – uncanny beings of higher and masterful intelligence, entirely capable of exterminating us at will and even bent on doing so, for reasons of their own.

We are uncanny. We need to remind ourselves of that. Our greatest literature reflects an awareness of how strange we are – strangers in a strange world. The world is full of extraordinary things, the great Greek tragedian Sophocles wrote, 2,400 years ago, but nothing is more uncanny than human beings. Of course, our literature is itself the product of millennia of 'slingshot' momentum, though the oldest myths of pre-literate human beings already exhibit a sense that we are unusual and somehow acquainted with 'gods'.

Calvin's eighth through thirteenth chapters briefly trace the beginnings of this slingshot trajectory including, in one chapter (the eleventh) of just twelve pages, how we 'civilized ourselves' by inventing agriculture, writing and 'mind medicine'. If one was to complain about his book, it might be because right here he would have done better to take a little more space and to cover crucial and fascinating topics that we now need to understand clearly in order to grasp what we are.

I have in mind the origins of writing and mathematics, in particular, and the curious history of their highly uneven development. He does not really address the question of how writing was invented and deals only perfunctorily with its cognitive implications and limitations. He is plainly aware of the findings of the best scholarship, as he indicates by simply remarking that 'writing developed from tax accounting about 5,200 year ago in Sumer.'[11] Given how immense a step this was, however, it merited at least a paragraph or two to spell out what the breakthrough development took.

He does not discuss the rise of mathematics at all. This is a strange omission for a natural scientist. Without mathematics, we could not measure and analyse the trajectory of a stone fired from a slingshot, however well we might be able to fire it and however much we might marvel at the force so achieved. We have accomplished altogether extraordinary things through the apprehension and application of mathematical ideas in the past few thousand years. This requires careful and appreciative explanation.

The twentieth-century physicist Eugene Wigner famously spoke of the 'unreasonable effectiveness of mathematics in the natural sciences', by way of arguing that mathematics could not be simply the invention of human thought. There are those who argue that it is, nonetheless, just such an invention. These are the so-called mathematical humanists, as distinct from the Platonists and formalists. It is disappointing, even in a brief history of the mind, to find no reckoning with this important and profound subject.

This is all the more so because Calvin's derivation of syntax from bodily motion and neural exaptation bears a decided resemblance to the argument of the mathematical humanists George Lakoff and Rafael Nunez that abstract mathematical ideas have arisen as a nested set of conceptual metaphors rooted in bodily experiences of spatial perception, tactile encounter, movement and containment. As they declare, 'Mathematics is deep, fundamental and essential to the human experience. As such, it is crying out to be understood. It has not been.'[12]

What Calvin did, on the other hand, was provide a sensible and accessible, albeit very brief, reflection on the cognitive roots of many of our confusions and failings. Even more importantly, in his final chapter, 'The Future of the Augmented Mind', he argues for a down-to-earth approach to coping with the defects of the mind we have acquired by natural selection – now that millennia of feedback, via our record systems, has enabled us to get some perspective on how it works.

In his plainspoken style, he comments, 'much of the higher intellectual function seems half-baked; what you ordinarily see is a prototype rather than a finished, well-engineered product. Perfect you don't get; not from Darwinian evolution. And the quality controls are spotty. But culture – especially education and medicine – can sometimes patch things up, if society works hard enough at it.'

He offers quite a catalogue of 'bugs' in the 'prototype' – far more and much better documented than are offered in the old Christian concept of 'original sin'. These include our susceptibility to colourful rhetoric, our fateful capacity for working in really large scale operations which overwhelm our emotional autonomy and critical faculties, our 'compulsive search for coherence' often leading us to 'finding hidden patterns where none exist', our susceptibility to being blindsided by our logic to the extent that we act on errors or prejudices with great conviction.

Our memories are selective and unreliable, our decision-making easily swayed by the last thing to have made a vivid impression on us, our intuitions about logic, probability and causation are powerful but flawed in quite numerous ways with consequences that are magnified, not diminished, by our creation of complex social and cognitive ways of life. Our capacities for dissociation and deception, including even self-deception, complicate our affairs and all too often issue in cruelties and moral vacuities that have terrible consequences.

On top of all this, we have, as Carl Sagan expressed it, both created a civilisation enormously dependent on science and technology, but 'arranged things so that almost no-one understands' them – 'a prescription for disaster.' The challenges we now face are such that unless we can invent better means for teaching critical thinking and scientific understanding and applying both to public policy, we are at risk of being overwhelmed by 'crashes' that will occur in part because of the breakneck and accelerating speed with which we are now travelling – on our uncanny (if not monstrous) slingshot trajectory.

Calvin believes we need to augment our cognitive abilities and aim to culturally 'debug' the prototype mind we inherited. His pivotal point is the claim that 'Very little education or training is currently based on scientific knowledge of brain mechanisms … To imagine what a difference [such knowledge] could make, consider the history of medicine … One century ago, medicine was still largely empirical and only maybe a tenth had been modified by science … It is only a slight exaggeration to say that the transition from an empirical to a semi-scientific medicine has doubled life-span and reduced suffering by half.'[13] He believes similar gains can be made in education.

Yet education, at the beginning of the twenty-first century, is 'largely still empirical and only slightly scientific', almost as much as medicine was before 1800. 'We know some empirical truths about education, but we don't know how' good learning and advanced critical thinking skills 'are implemented in the brain, and thus we don't know rational ways of improving them.' New information technologies and even genetic engineering could, he believes, enable us – if used wisely – to make gains in education similar to those made in medicine since 1800.

'Wisely' is, of course, the key word. For as long as we self-described 'sapient' human beings have wondered about the origins of the world and the 'mind-body problem', we have also sought wisdom for dealing with one another and the natural world. Many still seek such wisdom – and find a modicum of it – in the old religions of the Iron Age, or in some version of half-baked primitivism. We need a larger vision now, though. We need a universal cognitive humanism – a wisdom based not on platitude, sentiment or good old-time religion, but on a deep understanding of what makes us what we all are.

What Calvin's little book offers is a doorway into such a larger understanding. If you seek wisdom for our time, this is a good place to start. If, of course, what you really seek is consolation, you may find it elsewhere. But the time has come when, in order to overcome our savageries, our defects, our depravities, we surely need to dig deeper, learn profoundly and find ways of thinking that serve us all better. Otherwise, we run an increasing risk of our hour strutting and fretting upon the stage ending, as *Hamlet* ends: the stage strewn with corpses and the prince's luminous mind snuffed out – after all too brief a history.

1.   *Endnotes*

1   Derek Bickerton, *Language and Species*, University of Chicago Press, 1990, p. 257.

2   Daniel Dennett, *Kinds of Minds*, Basic Books, Science Masters, New York, 1996, p. 147.

3   William H. Calvin, *A Brief History of the Mind: From Apes to Intellect and Beyond*, Oxford University Press, 2004, p. 70.

4   Friedrich Nietzsche, *The Gay Science – With a Prelude in Rhymes and an Appendix of Songs*, #283, trans. with commentary Walter Kaufmann, Vintage Books, New York, 1974, p. 228: '... For believe me the secret for harvesting from existence the greatest fruitfulness and the greatest enjoyment is – to live dangerously! Build your cities on the slopes of Vesuvius! Send your ships into uncharted seas ...'

5   William Shakespeare, *The Tragedy of Hamlet, Prince of Denmark*, act 2, scene 2, lines 304-320.

6   Robert Brandom, *Making It Explicit: Reasoning, Representing and Discursive Commitment*, Harvard University Press, 1994.

7   There is a marvellous passage in Nietzsche, in which he attempted to summarise the history of these philosophical debates from Plato to nineteenth-century materialism: Friedrich Nietzsche, *Twilight of the Idols/The Antichrist*, trans. R. J. Hollingdale, Penguin, 1968 (1982) pp. 40-1. 'How the "real" world at last became a myth.'

8   Bickerton, *Language and Species*, pp. 196-7.

9   William H. Calvin and Derek Bickerton, *Lingua ex Machina: Reconciling Darwin and Chomsky with the Human Brain*, MIT Press, 2000.

10   As Ian Tattersall puts it, '... perhaps the most important lesson we can learn from what we know of our own origins involves the significance of what has in recent years increasingly been termed "exaptation". This is a useful name for characteristics that arise in one context before being exploited in another, or for the process by which such novelties are adopted in populations.' 'How We Came to be Human', *Scientific American*, December 2001, p. 43.

11   Denise Schmandt-Besserat, *How Writing Came About*, University of Texas Press, Austin, 1996.

12   George Lakoff and Rafael Nunez, *Where Mathematics Comes From: How the Embodied Mind Brings Mathematics into Being*, Basic Books, New York, 2000.

13   For a fine introduction to this subject, see Roy Porter, *The Greatest Benefit to Mankind: A Medical History of Humanity from Antiquity to the Present*, HarperCollins, 1997.

Then said Jesus to those Jews which believed in him, If ye continue in my word, then are ye my disciples indeed; and ye shall know the truth and the truth shall make you free.
-St John[1]

Grace does not abolish nature, but completes it.
-Thomas Aquinas[2]

'Man will become better when you show him what he is like', wrote Chekhov, and so the new sciences of human nature can help lead the way to a realistic, biologically informed humanism. They expose the psychological unity of our species beneath the superficial differences of physical appearance and parochial culture.
-Steven Pinker[3]

2. On Religion and Evolution

The story of the creation and of original sin in *Genesis* is true', wrote the great French mystic, Simone Weil, in 1942.[4] She was reflecting on how religion might counter the horrors of totalitarianism and global war. She was representative of many who still seek in the Bible the mental means to cope with the formidable challenges of our time. She was committed to the good, but she was in error.

The story, which stands at the foundation of the three great Biblical religions, Judaism, Christianity and Islam, is *not* true. Nor will any nostalgia for an imagined past of religious wisdom and redemptive beliefs make it so. If we are to come to terms with the realities and prospects of religion in the twenty-first century and truly rise to the challenges of our time, it is vital that we understand this. We need to think through its implications also, for they are profound.

Denying the truth of the creation and fall story in *Genesis* will occasion little more than a shrug on the part of many people, but resistance and even outrage on the part of others. It's like stating candidly that the Exodus never took place,[5] that Jesus did not rise from the dead,[6] or that Mohammed was a huckster.[7] Such claims strike at the heart of deeply held and very ancient systems of belief.

Yet the creation and fall story in *Genesis* is not true. Quite simply, we were not created by a Deity six thousand or so years ago in his own image and did not lapse into sin. More fundamentally, even if *Genesis* is interpreted in some allegorical, rather than literal sense, we are not and never have been a 'fallen' species of being. We never needed redemption, whether by a Jewish Messiah, by Christ crucified and raised on the third day, or by adherence to the teachings of Mohammed.

On the contrary, we are a *risen* species – gifted, voracious, capable of fiendish cruelty, extraordinary compassion and astonishing creativity. We are *Homo sapiens* and there is no other creature quite like us.[8] We arose over millions of years of biological and cognitive evolution and over the past hundred thousand years have colonised the entire biosphere. We are language animals, symbol-using, networked creatures; extraordinarily inventive, imaginative, uncanny and (armed with our inventions) a danger to all else that lives and breathes.[9] In historic time – the past five thousand years – we have created ever more complex societies, technologies and systems for symbolic analysis.

The Bible, starting with *Genesis*, is a set of stories originating in the late Bronze and early Iron Ages, in which priests and sages wrestled with the enigmas of the human condition. It has served countless human beings as a means for trying to comprehend the vertiginous sweep of human affairs for more than two millennia; but we need to acknowledge that it consists of often luminous fables, not of a revealed truth – however that 'truth' is interpreted.

It was put together when books were a relatively new and astonishing phenomenon and it has long been accorded unique status, among believers, as *the* Book. It is a classic, with a formidable history,[10] but it cannot pass muster any longer as a source of knowledge about the origins, nature and destiny of our world. This begins with the creation story in *Genesis*, which can only be invoked responsibly now as a fable pointing to the general problem of human beings falling short of their best possibilities.

Even many educated believers acknowledge this. Yet the Biblical religions continue to lay claim *on dogmatic grounds* to the consciences and imaginations of as many as two billion people. They are in need of an 'upgrade', to a truly universal *kerygma* – one grounded not in ideas of fall and redemption, but in acknowledgement of our animal nature and the imperfect development of our cognitive and moral faculties.

After all, to repudiate the ideas of fall and redemption is not to claim that we are either perfect or perfectible. It is certainly not to endorse the conceit that – in Jean-Jacques Rousseau's famous phrase – 'Man was born free, but is everywhere in chains'.[11] We were born in darkness, not in Eden, and have slowly sought the light since our remote ancestors discovered the uses and sustenance of fire. Our religions have played their part in that search, but none of them is more than a small part of the story.[12]

The unparalleled ways in which we have raised ourselves over the ten millennia since the first beginnings of agriculture and the five millennia since the beginnings of writing systems, have wrought enormous material and cognitive changes in what it means to be human. They also have been tumultuous: marred by upheavals and cruelties that appall the moral consciousness we have developed along the way.

Time and again, our experiments with complex social order have involved gross abuse of one another and of the natural world. Often they have collapsed into anarchy or barbarism. The Bible belongs very much within this long history of violence and change. It has provided an anchor for ever so many. Yet the truth is far larger and deeper than anything recorded in it. Our history is, ultimately, that of life on earth; our story that of the entirety of humanity, including *pre-sapiens* hominids. We need to share that history and that story globally.

Above all, we need to develop a common appreciation for how our *thinking* has made us what we are. *Homo sapiens* is (we are) the ape that thinks. We ponder and re-imagine reality, we spin metaphors out of our brains and share them in conversation, we think in a grammar of past and future tenses, of conditional and subjunctive possibilities, of subjects and objects. We think abstractly and invent hypotheses and experiments, discover rules and laws.

No other animal does these things. Yet this is not a disembodied spirit, fallen from grace, that thinks. It is the evolved brain of our kind and it has created the world of fabrics and machines, alloys and electronics, orchestras and ICBMs that we now inhabit.

Bringing all this to bear on the rethinking of Biblical religion is a challenge that our educational institutions have largely evaded. Our secular schools mostly avoid the subject, our religious schools dance around it. Yet surely the time has come when the dogmas of the Biblical religions must be repudiated for the same basic reasons that polytheistic beliefs have long since been repudiated by monotheists.

In the late second century, the great Christian Neo-Platonist, Clement of Alexandria, in his 'Exhortation to the Greeks', wrote of the death of Zeus. 'Where is Zeus himself?' he asked. 'He has grown old, wings and all ... Search for your Zeus. Scour not heaven, but earth. Callimachus the Cretan, in whose land he lies buried, will tell you in his hymns ... Yes, Zeus is dead (take it not to heart), like Leda, like the swan ....'[13]

'Zeus', of course, is the Greek equivalent to the Latin word 'Deus', which is to say 'God'. Clement was declaring to the Greeks, 'Your God is dead.' Most educated Greeks and Romans agreed, as it happens. The critique of the old pagan gods long antedated Christianity. The key to the critique was abstract thought about the ontological and moral nature of 'deity' as such. The old gods made superb subjects for poetry – as they still do – but belief in their 'existence' was another matter, philosophers saw.

Some 1,700 years after Clement, Friedrich Nietzsche declared to the monotheists that *their* God was dead. More precisely – and this seems to be almost always overlooked – Nietzsche had a 'madman' declare the death of God; not to monotheists in their churches, synagogues or mosques, but in the marketplace to 'many of those who did not believe in God'.

His madman came to the marketplace with a lantern in the bright morning hours, crying 'I seek God! I seek God!' He was laughed at by the unbelievers. So he rounded on them, crying, 'We have killed him, you and I. All of us are his murderers ... Do we hear nothing as yet of the noise of the gravediggers who are burying God? Do we smell nothing as yet of the divine decomposition? Gods, too, decompose. God is dead. God remains dead. And we have killed him.'[14]

Nietzsche was not merely repudiating monotheism, as Clement repudiated the cult of Zeus. He was pointing to its cultural significance and the possible consequences of dispensing with it.[15] He was the son and grandson of Lutheran pastors and had great respect for the riches of the Biblical tradition. He remarked, at one point that reverence for the Bible 'is perhaps the best piece of discipline and refinement of manners that Europe owes to Christianity.'[16] Yet he could not see how Biblical theology could any longer be sustained. Worldwide, religion required a further refinement, and it would not come easily.

Anticipating much of the cultural ferment of the twentieth century, Nietzsche saw the Biblical worldview as doomed by the sheer accumulation of knowledge by humanity. While he foresaw this giving rise to nihilism and cataclysms, he also saw it as portending an unprecedented liberation of the human spirit – provided that human beings had the courage to take hold of the freedom that modern insights made possible.

There was scope for 'the most spiritual Shrovetide laughter and wild spirits, for the transcendental heights of the most absolute nonsense and Aristophanic universal mockery ...'[17] But there would be a need to recover and reshape what we could of our ancient traditions of ceremony and ritual from the drastic inroads of modern knowledge and worldliness. 'What was holiest and mightiest of all that the world has yet owned has bled to death under our knives: who will wipe this blood off us?' his madman cries out. 'What water is there for us to clean ourselves? What festivals of atonement, what sacred games shall we have to invent?'[18]

Strip away the tribal, anthropomorphic, superstitious, hallucinatory and garbled aspects of Biblical religion – from the more arcane prohibitions in Leviticus and Deuteronomy to the conception of God as a Father and the idea of the resurrection of the body – and what is left? Nothing at all? An empty shell? No. A three-millennia-long testimony to human spiritual striving and some of the most sublime poetry in the world. Discard the excrescences and what becomes possible? A major existential house-cleaning and spiritual reformation.

Such is the prevalence of dogmatic and folk religion, even now, that many would despair of this being possible. Many others, of a secular and materialist cast of mind, are likely to regard such a project as quixotic or cranky. Yet it is surely possible, in principle. The old religions would clearly outlast such a reformation, just as the pagan gods have long outlasted belief in their existence, or cults dedicated to them.[19] What would become possible, however, is a common, authentic language of existential and ontological orientation, transcending the dogmatic claims of the old religions.

It seems clear that we are in need of such a language. There are all manner of dangers in the revival of fiercely dogmatic religion around the world, whether Christian, Muslim, Jewish or Hindu. There are cultural disorders entailed in the longstanding schism between arcane religious doctrines and scientific knowledge of life and the cosmos. Yet there are abiding human needs that religion seems to provide more fully than anything secular society has yet created.

I am thinking of needs for meaningful ceremonies to mark and dignify births, comings of age, marriages and deaths; also for an historically resonant poetics of existence, community and moral life. The Bible has been a profoundly rich source of these things for a very long time in much of the world. Not only would it be philistine to deny this, it would be quite simply erroneous.

Yet so much of it no longer works and it no longer works because the cognitive dissonance between our secular lives and our ceremonial ones has deepened relentlessly in recent decades. Preachers who engage in superficial, half-hearted apologetics find their congregations melting away. Those who try to shore up the old religions by suppressing cognitive dissonance are winning far more adherents. At the margins of such suppression, deluded and murderous fanatics look for an apocalyptic overthrow of secular civilisation.

In a book called *The Future of Christianity* published in 2002, Alister McGrath, Professor of Historical Theology and Principal of Wycliffe Hall at Oxford University, asserts that 'science and progress' have been toppled from their 'thrones' by a new, non-intellectual 'spirituality' in recent decades, especially in America. In language worthy more of a mountebank than of an Oxford scholar, he writes approvingly: 'Post-modern culture seems fed up with the rather boring platitudes of scientific progress and longs for something rather more interesting and exciting.'[20] In *madrasas* across the Muslim world, similar doctrines are being taught, with incendiary effect.

In a more responsible and scholarly piece of work, also published last year, James Carroll wrestled with his Catholic heritage and ended by urging that the cross, a symbol of torture and death, be repudiated by the Catholic Church and replaced by images of the face of Jesus.[21] Again and again, I found myself writing in the margins of Carroll's book, 'You are surely correct on this point [and this and this], so remind me, why are you a Catholic?'[22] What, after all, is the face of Jesus? Surely, Carroll was reaching for something of which such a face would be symbolic – compassion and human transfiguration?

I think that we need to go much further than Carroll. The problem is not one of Catholicism; it is one of Bible-centered religion. Islam is in travail for fundamentally similar reasons and in need of at least as radical an upheaval of thought. The Koran is every bit as problematic as the Bible and twenty-first-century religion must convert it from a fountainhead of obscurantism and dogma into one resource among others of existential perspective and reflection.[23]

There are many, many paths into the project I am proposing, but I want to offer a single, provocative one here: the Apostles' Creed, dating back as far as 100 CE. It antedates the Nicene Creed by at least two centuries and is rooted in the earliest Christian communities.[24] I appreciate its roots and its beauties,[25] but I want to juxtapose it with what might be called a World Creed, as a radical thought experiment.

The Apostles' Creed reads as follows:

I believe in God, the Father Almighty,
Creator of Heaven and Earth;
And in Jesus Christ, His only Son, our Lord.
Who was conceived by the Holy Spirit,

Born of the Virgin Mary,

Suffered under Pontius Pilate,

Was crucified, died and was buried.

He descended into Hell.

On the third day, he rose again from the dead.

He ascended into Heaven

And is seated at the right hand of God,

The Father Almighty.

From there, he shall come

To judge the living and the dead.

I believe in the Holy Spirit,

The Holy Catholic Church,

The communion of saints,

The forgiveness of sins,

The resurrection of the body

And life everlasting. Amen.

I was taught this as a child. I am steeped in the tradition that is based on it – the eschatological vision of human transformation and the purging of evil from the world. Yet I cannot utter these words in a church as a creed. I simply do not believe them. It is not a matter of agnosticism, or doubt or confusion. I simply believe that the Apostles' Creed is a cultural heirloom, like the far more ancient Egyptian Book of the Dead. It is not something that expresses what I believe to be true about the world.

Here is a credo, written to correspond in both length and ontological scope with the Apostles' Creed, which I could proclaim with integrity:

I believe that all deities are idols of the mind,

That blood sacrifices to them are an abomination,

That dogmas are obstacles to enlightenment.

I believe in the plurality of worlds,[26]

But know of none that can compare with ours

In its abundance of life;

Of a kind that has arisen,

Through countless changes and catastrophes,

Out of the primal waters of the Earth.

I acknowledge that I am of this world,

Though a brief sojourner in it.

I have sprung from it and will pass back into it.

I recognise that my existence,

Both sentient body and sapient mind,

Is possible only within the natural order of things.

Capable of mimesis, metaphor and music,

Of reason and responsibility,

I believe that I am neither fated nor predestined,

But am able to live for possibilities

And move intentionally toward a horizon that is open.

Less poetic than the Apostles' Creed? Less dramatic? Perhaps. The point is that it is true both to biological realities and to human phenomenological experience – globally.[27] I think it is time we filled our religious structures, which Nietzsche's madman called 'the tombs and sepulchres of God', with proclamations of a creed along these lines. We should consign Biblical eschatology to the museum of history, along with blood sacrifices – whether of lambs, or sons of God.[28]

'Ah!' you might well exclaim. 'This is just not going to happen.' Your reasons for thinking so will vary from the presumption that it cannot, because Christian revelation, as expressed in the Apostles' Creed, is simply true; to the equally dubious presumption that religions are inherently irrational and cannot be reformed in such a manner. I cordially disagree.

Call it, if you like – as medieval mystics already did – the Third Age, the Age of the Spirit, but I imagine a future in which such a transformation has occurred.[29] I wander in it, in my mind, as Goethe wandered Rome in 1786, filled with wonder that all the dreams of my youth have come to life.[30] Am I alone in imagining such things? I don't think I am. So, what would it take? That's the real thought experiment.

2.   *Endnotes*

1   *The Holy Bible: King James Version*, The Church of Jesus Christ of Latter Day Saints, Salt Lake City, Utah, 1979, p. 1341, The Gospel According to St John, 8: 31-32.

2   *Summa Theologica* I: i. 8; quoted in Richard E. Rubinstein, *Aristotle's Children: How Christians, Muslims and Jews Rediscovered Ancient Wisdom and Illuminated the Dark Ages*, Harcourt, 2003, p. 198. Rubinstein remarks: 'The great theme that runs through Aquinas's epochal work is that 'grace does not abolish nature, but completes it.' There can be no conflict between religion and natural science, between the loving Creator and understanding his creation, so long as one correctly defines and demarcates both realms of thought.' However, the synthesis proved unsustainable and within a century of St Thomas's Aristotelian labours, two Franciscan Aristotelians, Duns Scotus and William of Ockham, argued powerfully that science and theology were twain and could *not* finally be reconciled.

3   Steven Pinker, T*he Blank Slate: The Modern Denial of Human Nature*, Allen Lane, Penguin Press, 2002, p. xi. Pinker's principal target was not old-time religion so much as modern romanticism and post-modernist nonsense.

4   Simone Weil, *Letter to a Priest*, Routledge Classics, London, 2002, p. 42.

5   For a fascinating exploration of this, see Israel Finkelstein and Neil Asher Silberman, *The Bible Unearthed: Archaeology's New Vision of Ancient Israel and Its Sacred Texts*, Touchstone, New York, 2002. Daniel Lazare wrote an excellent review of this book and its context, 'False Testament: Archaeology Refutes the Bible's Claim to History', *Harper's* Magazine, March 2002, pp. 39-47. As Lazare remarks, 'Beginning in the 1950s, doubts concerning the Book of Exodus multiplied, just as they had about Genesis. The most obvious concerned the complete silence in contemporary Egyptian records concerning the mass escape of what the Bible says were no fewer than 603,550 Hebrew slaves ... Not only was there a dearth of physical evidence concerning the escape itself, as archaeologists pointed out, but the slate was blank concerning the nearly five centuries that the Israelites had supposedly lived in Egypt prior to the Exodus, as well as the forty years that they supposedly spent wandering in the Sinai. Not so much as a skeleton, camp site or cooking pot had turned up ...'

6   The single most famous invocation of the resurrection of Jesus is surely that by St Paul in the fifteenth chapter of his first epistle to the Corinthians, especially verse 14: 'And if Christ be not risen, then is our preaching vain, and your faith is also vain.' There is a quite illuminating discussion of all this by Dominic Crossan and Jonathan Reed in *Excavating Jesus: Beneath the Stones, Behind the Texts*, HarperCollins, 2001, pp. 254-270.

7   Ibn Warraq, *Why I Am Not A Muslim*, Prometheus Books, New York, 1995, pp. 86-103, provides an excellent introduction to this subject. For a highly sanitised account of Muhammed's life, see Karen Armstrong's *Islam: A Short History*, Phoenix Press, London, 2001, pp. 3-20. Her chief claim on behalf of the Prophet is that he brought peace (albeit by means of the sword) to the barbarism and anarchy of pagan Arabia. This claim is endorsed by Ira Lapidus, in *A History of Islamic Societies*, Cambridge University Press, 1988, Chapter 2, 'The Life of the Prophet', pp. 21-36.

Neither of them so much as addresses the long-established claims about his corruption and opportunism that lie behind Ibn Warraq's remarks.

8   For a fine survey of the latest scientific views regarding the emergence of hominids over the past six to seven million years, see the special edition of *Scientific American*, August 2003, *New Look At Human Evolution*. Those who remain confused about the very idea of evolution itself, or who believe that the Biblical creation myth should somehow be accorded equal credibility, should read Ernst Mayr's magisterial *What Evolution Is*, Weidenfeld and Nicolson, London, 2002, not least his two appendixes, A, 'What Criticisms have Been Made of Evolutionary Theory' and B, 'Short Answers to Frequently Asked Questions About Evolution'.

9   Derek Bickerton, *Language and Species*, University of Chicago Press, 1990, is a classic study of the rise and significance of language. If you just dip into his book, I recommend you go straight to pp. 200-201 and read his remarks under the heading 'Mind and Machine'. On the cognitive networking of human beings as language animals, see Merlin Donald's *Origins of the Modern Mind: Three Stages in the Evolution of Culture and Cognition*, Harvard University Press, 1991.

10   For a good overview of this history, see Christopher De Hamel, *The Book: A History of the Bible*, Phaidon, 2001. De Hamel's book is beautifully illustrated and scrupulous in its scholarship, but it is not, as he remarks, 'a theological book'. For something closer to the latter, which is nonetheless accessible to believer and unbeliever alike, see Dennis Nineham, *The Use and Abuse of the Bible: A Study of the Bible in an Age of Rapid Cultural Change*, SPCK, London, 1978.

11   Jean-Jacques Rousseau, *The Social Contract*, Book I, Chapter 1 (ed. and trans. Maurice Cranston), Penguin, 1974, p. 49.

12   On the early history of human civilisation, in the fifteen millennia between the end of the last ice age and the beginnings of urban settlement in the great river valleys, see Steven Mithen's *After the Ice: A Global Human History 20,000 – 5,000 BC*, Weidenfeld and Nicholson, London, 2003.

13   G. W. Butterworth (trans.), *Clement of Alexandria: Exhortation to the Greeks, The Rich Man's Salvation, To the Newly Baptised*, Loeb Classical Library, Harvard University Press, 1999, p. 79.

14   Friedrich Nietzsche, *The Gay Science; With a Prelude in Rhymes and an Appendix of Songs*, trans. Walter Kaufmann, Vintage Books, New York, 1974, p. 181.

15   This has been an issue in the case of Biblical monotheism since at least the eighteenth century, when Voltaire made the famous remark 'If God did not exist, we would have had to invent Him.' Long before the rise of Christianity, Polybius, an educated Greek working as an historian in the Roman world, made a similar observation about Greek scepticism and the old religion of the people. 'The sphere in which the Roman commonwealth seems to me to show its superiority most decisively is that of religious belief. Here we find that the very phenomenon which among [the Greeks] is regarded as a subject of reproach, namely superstition, is actually the element which holds the Roman state together. These matters are treated with such solemnity and introduced so frequently both into public and into private life that nothing could exceed them in importance. Many people find this astonishing, but my own view is that the Romans have adopted these practices for the sake of the common people. This might not have been necessary had it ever been possible to form a state composed entirely of wise men. But as the masses are always fickle, filled with lawless desires, unreasoning anger and violent passions, they can only be restrained by mysterious terrors or other dramatizations of the subject. For this reason, I believe that the ancients were by no means acting foolishly or haphazardly when they introduced to the people various notions concerning the gods and belief in the punishments of Hades, but rather that the moderns are foolish and take great risks in rejecting them.' *The Rise of the Roman Empire*, Penguin, 1979, book VI #56, p. 349.

16   Friedrich Nietzsche, *Beyond Good and Evil: Prelude to a Philosophy of the Future*, Penguin, 1981, p. 183. He went on to remark: 'Such books of profundity and ultimate significance require for their protection an external tyranny of authority, in order that they may achieve those millennia of *continued existence* which are needed if they are to be exhausted and unriddled. Much has been gained when the feeling has at last been instilled into the masses (into the shallow-pates and greedy guts of every sort) that there are things they must not touch; that there are holy experiences

before which they have to take off their shoes and keep their unclean hands away – it is almost their highest advance towards humanity.'

17   Ibid., p. 133.

18   Ibid., p. 181.

19   Jean Seznec, *The Survival of the Pagan Gods: The Mythological Tradition and Its Place in Renaissance Humanism and Art*, Bollingen Series XXXVIII, Princeton University Press, 1972, is one outstanding reflection on this theme.

20   Alister McGrath, *The Future of Christianity*, Blackwell Manifestos, Oxford, 2002, pp. vii-viii.

21   James Carroll, *Constantine's Sword: The Church and the Jews*, Mariner Books, Houghton Mifflin, 2002.

22   The first such occasion was on p. 23, where Carroll asks, 'What kind of God shows favour to a beloved Son by requiring him to be nailed to a cross in the first place?' I wrote in the margin 'Indeed! So why are you a Catholic at all? Why not a fellow whose heroes are, say, Pythagoras, Spinoza and Leonard Cohen?'

23   For an outstanding scholarly introduction, see Ibn Warraq (ed.), *What the Koran Really Says: Language, Text and Translation*, Prometheus Books, New York, 2002.

24   On the earliest Christian communities and their relationship to the classical Judaic and Greco-Roman world, see Wayne A. Meeks, *The First Urban Christians: The Social World of the Apostle Paul*, Yale University Press, 1983.

25   John Henry Newman's classic *An Essay on the Development of Christian Doctrine*, Penguin, 1974, is a wonderfully lucid statement of the sense of coherence that has had orthodox Catholics adhere to Rome over many centuries. It is still worth reading and reflecting on. On Newman himself, see Ian Ker's outstanding biography *John Henry Newman*, Oxford University Press, 1988.

26   Belief in the plurality of worlds was one of the heresies for which Giordano Bruno was burned alive in Rome by the Catholic Church in 1600. For the best discussion of Bruno's ideas, see Hilary Gatti's, *Giordano Bruno and Renaissance Science*, Cornell Iniversity Press, Ithaca and London, 1999, especially Chapters 6 and 7, 'The infinite Universe' and 'The Infinite Worlds'.

27   On the biological side, see Melvin Konner, *The Tangled Wing: Biological Constraints on the Human Spirit* (rev. edn.), Henry Holt & Co., New York, 2002. David Sloan Wilson's *Darwin's Cathedral: Evolution, Religion and the Nature of Society*, University of Chicago Press, 2002 is an interesting attempt to explain the history of religion in 'Darwinian' terms.

28   This is not, of course, an entirely novel suggestion. A good deal of thought was given to it during the eighteenth century European Enlightenment. See J. G. A. Pocock, *Barbarism and Religion*, Cambridge University Press, 1999, volume 1, *The Enlightenments of Edward Gibbon 1737-1764*; volume 2, *Narratives of Civil Government*. The matter was also given close attention by G. W. F. Hegel, whose reflections on the significance of religion, in his master work *The Phenomenology of Mind*, are still worth reading.

29   I am not, of course, advocating the *suppression* of the existing monotheistic religions. There is a history to this which is unfortunate. See Simon Schama on the French Revolution, *Citizens*, Viking, 1989, especially pp. 830-36; and Daniel Peris on the Russian Revolution, *Storming the Heavens: The Soviet League of The Militant Godless*, Cornell University Press, 1998.

30   Johann von Goethe, *Italian Journey*, Penguin, 1982, p. 129.

It is only by becoming sensible of our natural disadvantages that we shall be roused to exertion, and prompted to seek out opportunities of discovering the operations now in progress, such as do not present themselves readily to view.
-Charles Lyell (1830)[1]

We are in the midst of a seismic shift in thinking about the nature of ourselves and the world we live in. It is no hyperbole to describe the magnitude of the shift as an intellectual revolution.
-Richard Leakey (1995)[2]

A window seat in the stratosphere certainly provides a better place from which to contemplate the world than most philosophers ever had.
-William H. Calvin (2002)[3]

3. On the Climate of Human Flourishing

*E*verything we normally think of as 'world history' – the beginnings of agriculture, the building of cities, the invention of writing and mathematics, the rise and fall of civilisations, the scientific and industrial revolutions – has occurred since the end of the last ice age. The next ice age should, on the law of averages, be very close and would have an apocalyptic impact on the human world. But there is another, hitherto unsuspected, danger of a deep freeze. If you think global warming is a serious danger, you don't know the half of it.

Let's start with conventional ice ages. The last ice age ended around 15,000 years ago. It had lasted for just over 100,000 years. It was preceded by a warm period, known by geologists and climatologists as the Eemian epoch, which had lasted 13,000 years. Before that was the next to last ice age, which had lasted around 100,000 years. Over the past three million years, since the ice ages began, the warm periods have averaged about 10,000 years and the ice ages ten times that span.[4] We have long since overshot the point at which the next ice age might have been expected to begin.

In short, while we now perceive that global warming is threatening us, we should consider that we are, in all likelihood, close to the end of the warm period that has been the climatological precondition for the whole of 'world history' – the climate of human flourishing. This has nothing whatsoever to do with human agency, but with massive cosmological and geological cycles. It would overwhelm our species with sublime indifference to our religious beliefs, metaphysical speculations and secular ideologies of progress. It may well be more than our sciences or politics can handle.

However, the slow onset of a new ice age is not the most alarming possibility that geologists tell us we should be considering. For the long cycles of ice and warmth, it transpires, have been accompanied by a remarkable phenomenon that William Calvin described in 2002 as 'one of the most shocking scientific realizations of all time ... [that] the earth's climate does great flip-flops every few thousand years and with breathtaking speed.'[5] This is a hitherto altogether unsuspected danger.

The long cycles of ice and warmth appear to be causally related to perturbations in the earth's orbit around the sun and the tilt of its rotational axis. The climate flip-flops are another matter. They are known among climate scientists as Dansgaard-Oeschger events or D-Os.[6] They are best illustrated by the most 'recent' example: the Younger Dryas, a D-O which interrupted the current warm period a few millennia before we invented agriculture.

In Calvin's words, '12,900 years ago, Europe cooled down to Siberian temperatures within a decade … the rainfall likely dropped by half, and fierce winter storms whipped a lot of dust into the atmosphere. Such conditions lasted for over 1,300 years, whereupon things warmed back up, even more suddenly. The dust settled and the warm rains returned, again within a decade.'[7] Then 'world history' began.

This, it should be emphasised, was not an 'ice age'. It was just a D-O. Such non-linear and drastic climate fluctuations, it turns out, have occurred hundreds of times since the ice ages began three million years ago. They had, Calvin argues, a rigorous and manifold 'sculpting' effect on our species – shaping hominids into canny generalists with a suite of physical attributes and cognitive skills unmatched in the rest of the animal kingdom.[8]

This was a decidedly Darwinian process. Through these aeons, the clay of humanity was assuredly shaped by Richard Dawkins's 'Blind Watchmaker' and not by the hand of a loving God.[9] Over the last three million years, our hominid ancestors were again and again compelled to retreat into narrow refugia as their numbers were severely culled by the harsh impact of D-Os.[10]

Most of this culling and adaptation afflicted and shaped the many species of hominid that preceded our own, but it eventually produced the hardiest and canniest of hominids – *Homo sapiens*, as we call ourselves – by about 100,000 or perhaps 150,000 years ago. We arose during the ice age before last (before the Eemian), in a world long since populated by other hominids, then colonised the biosphere during the last ice age, when the seas were low.[11]

When it ended, we emerged as the only surviving species of hominid. *Homo erectus* (a species that had endured for two million years and spread throughout Africa and Eurasia) and *Homo Neanderthalensis* (a hardy species of hunter that invented skin clothes and long throwing spears) did not survive the last ice age. A group of scientists at the Godwin Institute of Quaternary Research, Cambridge University, is currently trying to reconstruct the climate fluctuations between 60,000 and 25,000 years ago in order to ascertain their effect on the demise of the Neanderthals.[12]

The climate anomaly that has cradled our rise from nomads and slayers of megafauna to city-builders and inventors of weapons of mass destruction has been an unprecedented period of climate stability since the Younger Dryas.[13] Why have things been so stable for about 11,500 years? 'No one knows yet', Calvin tells us. 'But we know it's unusual, and see no reason why it should persist.'[14]

If you're temperamentally or theologically inclined to attribute this run of good luck to a benign human 'manifest destiny' or to the hand of a loving God, think again. Short of the next ice age, there could be a D-O in the near future (anywhere from the next few decades to the next few centuries), which would cut mean global temperatures by 5 to 10 degrees Celsius, with drastic implications for temperate climate agriculture and all that flows from it. This alone would be likely to cause a population crash of unprecedented proportions.[15] James Lovelock might call this 'Gaia's revenge', but it would merely be planet earth behaving as it has for millions of years before we woke up to the reality of the case.

What causes D-O events? They seem causally linked to irregularities in the flushing of salt from the North Atlantic. This, at least, is the hypothesis of American geochemist Wallace Broecker, who has been exploring the matter since the mid-1980s.[16] There is a conveyor operating in the Atlantic, in which salt sinks to the cold depths of the ocean and is flushed south, with warm water from the tropics (what we call the Gulf Stream) flowing north.[17] This is what keeps north-western Europe so much warmer than any other region at the same latitude.

If you dilute the salinity of surface water in the North Atlantic, you reduce the precipitation of salt to the lower, colder depths of the ocean. This retards flushing, which, in turn, retards the reverse flow of warm water north, dropping temperatures, which retards evaporation, reinforcing the cycle and sending cool, dry winds crossing the Atlantic to Europe. Such dilution can be caused by a massive discharge of fresh water from the Canadian or Greenland ice caps, or excessive precipitation.

A persistent failure of late winter sinking of the ocean surface near Greenland and Iceland is a likely cause of most of the abrupt cooling episodes which have punctuated the long cycles of ice and thaw. In other words, they've been the result of adding surplus fresh water to the ocean surface. Broecker's hypothesis is that this, in turn, triggers a worldwide rearrangement of ocean currents, with a consequent drastic reduction, by around 30%, in evaporation in the tropics. That, in itself, would cool the planet by 5 degrees Celsius.

For the longest time, such D-O events have occurred due to natural contingencies. The irony of global warming is that, for the first time, human agency could bring one about. By melting the Canadian and Greenland ice caps and increasing precipitation coming off North America, global warming could trigger – could already be causing – excessive discharges of fresh water into the North Atlantic. If a D-O event follows, it will not entail a mere climate correction, but a colossal shift that will wreak unprecedented havoc on human civilisation.

Can anything be done about this? We might begin by taking more seriously both global warming and the scientific research needed to understand it better.[18] But there is more to the matter than global warming. If the Broecker hypothesis is correct, the underlying issue is geophysical. It has to do with the ocean currents in the Atlantic. They, geologists have deduced, have a history altogether independent of our species.

The pivotal event in that history was the closure of the ocean gap between North and South America – what William Calvin calls the 'Old Panama Canal' – between 4.6 and 3.2 million years ago, as a result of continental drift.[19] This epochal geological event blocked the easy route for disposing of excess salt from the Atlantic, creating an instability via the salt build-up. The planet's climate had been slowly cooling for several million years, and then the closure of the Old Panama Canal created the Gulf Stream and the salt conveyor.

Evaporative cooling of the Gulf Stream in northern latitudes created excess moisture, which led to the build-up of ice mountains in the North Atlantic region. Thus, quite apart from perturbations in the Earth's orbital or rotational motion, continental drift may have had a major role in the onset of the ice ages from around three million years ago. It also created the mechanism that triggers D-Os.[20]

Given this geo-historical understanding, is it possible for our species to interfere with what Pope Paul VI, in 1968, called 'God's felicitous design', by fundamentally altering the conditions that cause D-Os?[21] It would involve terraforming: building dams to restrict the flow of fresh water into the Atlantic from Canada and Greenland and dramatically widening the Panama Canal to divert the salt flow, altering the ocean currents of three million years.

The problem is that, even if we were capable of these things, we don't yet know enough to hazard them, despite the huge risks in leaving things as they are. Far more research is required, along with sophisticated and redundant computer simulations of climate changes and the environmental implications of the engineering projects in question. We know enough, in other words, to apprehend that our ecological 'childhood' could be about to end, but not enough to have any confidence about taking our destiny into our own hands.

Scientific findings of this nature are stunning in their implications, at a number of levels. They confront us with the stark possibility of a global catastrophe on a scale we associate with the extinction of the dinosaurs or a nuclear winter. They challenge profoundly our hubris as presumptive masters of the biosphere. They make nonsense of pieties about our fate being in the hands of a divine providence. They cast the darkest of shadows over our recent faith in the material 'progress' ahead of our civilisation.

At the most fundamental level, however, they are a reality check. They challenge us to clarify our priorities at every level from bedtime stories for children to international politics. They underscore what a few seers have been proclaiming since the eighteenth century: that we live in a natural world, not a supernatural one, and that the education we impart to rising generations of our species must be based on a sure grasp of the natural sciences and natural history.

Consider that the immense time-frame William Calvin draws us into, in talking of the era of the ice ages, is but the very recent history of the biosphere we have colonised so voraciously. Our common – universal – frame of reference should be the vast geophysical history and cosmology to which these 'recent' events are but a coda. Our common and universal sense of ourselves as a species should, in this century, become firmly rooted in an understanding of what we are and how we have become what we are.

Such a common sense would begin with the most basic chronological grasp of the geophysical history of the earth, as something altogether truer and more fundamental than any religious creation myth.[22] It would consist of three time parameters: first, the nature and history of life since it emerged in the oceans and around hot vents in the earth 3.8 billion years ago, including the five great mass extinctions between 440 million and 65 million years ago.[23] Second, the long history of primate evolution, since the extinction of the dinosaurs, 65 million years ago. Third, the awesome history of hominid evolution, from seven million years ago through to the super nova explosion that we think of as 'world history'.[24] There is no myth on earth as dramatic and powerful as this natural history of the earth we inhabit. Nor is there anything in religious scriptures as sobering or instructive, as awesome and inspiring as the aeons of life on earth.

What we need is an integration of natural history with our common moral and cultural sense of what it is that we are. That is possible, but it is a social and cognitive challenge every bit as daunting in its complexity as the terraforming projects William Calvin pondered as prophylactics against a new D-O. It requires a degree of what Harvard biologist Edward O. Wilson calls 'fluency across the boundaries' between the humanities and the natural sciences.[25]

Only such fluency, based on an assured curriculum, Wilson writes, 'will provide a clear view of the world as it really is, not as it appears through the lens of ideology and religious dogma, or as a myopic response solely to immediate need.'[26] This fluency, grounded in a unity of all knowledge, Wilson calls 'consilience'. It has its origins in the first natural scientific speculations, those of Thales of Miletus, for which reason the physicist Gerald Holton has dubbed it 'the Ionian Enchantment'.[27]

The vision of consilience is what links the ancient Greeks, the Renaissance and the Enlightenment. It is the highest aspiration of *Homo sapiens* as a natural being and has been the glory of the species in the few millennia since we created cities and initiated systematic inquiry. It is, Wilson, urges, 'the way to renew the crumbling structure of the liberal arts', which has been undermined in recent decades by abandonment of the ideal of unity, dissolving into 'a slurry of minor disciplines and specialized courses.'[28]

But far more is at stake than the coherence of the liberal arts. What is at stake is the collective capacity of *Homo sapiens* to exercise sapience as the climatic golden age that has enabled our exuberant and violent civilisations to flourish quite possibly comes to its end. It could be, of course, as Thomas Homer-Dixon has speculated, that the challenge will be beyond us, because its complexity now exceeds our collective ingenuity.[29]

If we are to rise to the challenge, however, before a D-O or a new ice age – or a century of global warming – overwhelms us, we shall need a global culture both consilient and resilient. The strenuous effort to achieve that would place us in the grand tradition of Leonardo da Vinci and Giordano Bruno, Francis Bacon and Giambattista Vico. It would constitute a global, twenty-first century Renaissance, after the totalitarian and relativist 'plague years' of the twentieth century.

Yet even to evoke such visions is to hope against hope, given the overwhelming natural forces that may confound us. Religious visions and the perverse psychology of invoking 'Gaia's revenge' quite apart, it could be that our species faces a looming apocalypse about which we can do nothing. A century ago physicists realised that our sun, our superabundant star, would die in the remote future, setting a term to our natural existence. But that was billions of years off. The apocalypse of ice may be imminent, by comparison, merely centuries or even decades away.

That is a stunning thought. There is no cheerful way to confront it. It gives profound resonance to an early reflection by Friedrich Nietzsche, dated to 1873, in which he contemplated the future demise of the species: 'In some remote corner of the universe, poured out and glittering in innumerable solar systems, there once was a star on which clever animals invented knowledge. That was the haughtiest and most mendacious minute of "world history" – yet only a minute. After nature had drawn a few breaths the star grew cold, and the clever animal had to die.'[30]

Nietzsche was writing just as the physical sciences were beginning to bite deeply into traditional ways of thinking about man and the universe. He was thinking of entropy and the sun and, even in astrophysical terms, he was incorrect to describe the life of the sun as a mere few breaths in nature. It will not die any time soon and, when it does, it will fry our planet, not freeze it. But it could, all the same, be cold that brings down civilisation and humbles the clever animal who invented knowledge.

And here's the thing to get. Knowing this is neither haughty nor mendacious. What is haughty is thinking that our climatic golden age was intended for us and that our future is assured. What is mendacious is telling ourselves and our children that all is well or all is in the hands of God, or that business as usual will suffice to see us through the next D-O, or that scientists like Wallace Broecker are just making up all these climate stories.[31]

The natural world is real, it is the only one there is, and it is not designed by Providence for our use and benefit. There have been five mass extinctions before. We are already causing the sixth and there our grounds for considering that we could ourselves become its most spectacular victims. How you respond to that proposition will tell you a great deal about what manner of human being you are. Confused or consilient? Resigned or resilient? Think about it, clever animal.

3. *Endnotes*

1   Charles Lyell, *Principles of Geology*, vol. 1 (1830), Penguin, 1997, p. 33.

2   Richard Leakey and Roger Lewin, *The Sixth Extinction: Biodiversity and Its Survival*, Weidenfeld & Nicolson, London, 1995, p. 223.

3   William H. Calvin, *A Brain for all Seasons: Human Evolution and Abrupt Climate Change*, University of Chicago Press, 2002, p. 228. Calvin is Affiliate Professor of Psychiatry and Behavioural Sciences at the University of Washington School of Medicine in Seattle. The book was *Scientific American* book of the month and won the *Phi Beta Kappa* prize for scientific writing.

4   Richard B. Alley, *The Two-Mile Time Machine: Ice Cores, Abrupt Climate Change and the Human Future*, Princeton University Press, 2000.

5   Calvin, *A Brain for all Seasons*, p. 3.

6   Named by Wallace Broecker after the Danish climatologist Willi Dansgaard and the Swiss geophysicist Hans Oeschger, who pioneered the research into the phenomenon in the early 1980s. See Willi Dansgaard et al., 'Evidence for General Instability of Past Climate From a 250 kyr Ice Core', *Nature*, vol. 364, 1993, pp. 218-19. See also Thomas Levenson, *Ice Times*, Harper & Row, 1989 and Thomas Stocker's obituary for Hans Oeschger at http://www.climate.unibe.ch-/oeschger/obituary.html.

7   Calvin, *A Brain for all Seasons*, p. 232.

8   The literature on human origins has taken giant strides in recent years. See *inter alia*, Ian Tattersall and Jeffrey Schwartz, *Extinct Humans*, Westview Press, 2000; Frans de Waal (ed.), *Tree of Origin: What Primate Behaviour Can Tell Us About Human Evolution*, Harvard University, 2001; William H. Calvin and Derek Bickerton, *Lingua ex Machina: Reconciling Darwin and Chomsky with the Human Brain*, MIT Press, 2000. For a beautifully presented digest of some of the best recent research, see the special edition of *Scientific American*, 'New Look at Human Evolution' (2003).

9   Richard Dawkins does not, of course, actually believe that a blind deity is responsible for overseeing evolution. The term, which is the title of his book *The Blind Watchmaker: Why the Evidence of Evolution Reveals a Universe Without Design* (Norton, New York, 1986), simply means that the classical argument from design, famously articulated by William Paley in the seventeenth century, which saw in the natural world the sort of evidence of a designing deity as a discovery of a watch would imply the existence of a watchmaker, is a misinference. The actual nature of the universe is such as to suggest no designer, because if one existed we would have to infer from his handiwork that he was blind. David Hume anticipated this critique of Paley's argument in his eighteenth century *Dialogues Concerning Natural Religion*, in which it was pointed out that, even if the basic premise was granted (that an artifact points to there being an artificer), the actual evidence around us would lead us to infer not the omniscient, omnipotent and benevolent (loving) God of the Christians, but either an infant deity who had flubbed the job; a superannuated deity whose powers were failing him, or a team of deities who had failed to coordinate their plans and handiwork sufficiently.

10   Refugia are 'regions that, in the midst of major population downsizings, still provide the essential elements of the species' niche for small sub-populations. Shorelines and mountain tops are often refugia locales, with a little bit more climate change resulting in the extinction of the sub-population. Europe has many fewer plant species than one might expect because so many were, in effect, pushed into the Mediterranean during an ice age.' Calvin, *A Brain for all Seasons*, p. 310.

11   This is, actually, the chief theme of Calvin's book. It is the great virtue of his argument that he seeks a climatological explanation for the emergence by natural selection and not by design of the specific and distinctive attributes of our kind.

12   The project is explained on the Institute's website: http://www.esc.cam.ac.uk/oistage3

13   One of the most dramatic findings of recent years has been the apparent link between the arrival of *Homo sapiens* in Oceania and the Americas and the extinction of the Pleistocene megafauna – the giant mammals and marsupials of the last ice age. The impact was most severe in Australia, where the great beasts all died out at the end of the last ice age, some 15,000 years ago. Since our species did not arrive in Madagascar or New Zealand until about 1,000 years ago, the impact there was delayed. But once we landed on those islands, the megafauna populations collapsed precipitously. See Leakey and Lewin, *The Sixth Extinction: Biodiversity and Its Survival*, pp. 171-94, especially the table on p. 187. The key to these extinctions is that human beings had developed their hunting skills for tens of thousands of years in Africa and Eurasia, before they colonised Oceania and the Americas. The megafauna of those regions, however, had had no time at all to become wary of the canny and ruthless hominid predators and suffered disproportionately in consequence.

14   Calvin, *A Brain for all Seasons*, p. 224.

15   There are certain historical catastrophes that come to mind, chief among them the devastation and de-urbanisation wrought throughout the Levantine coastlands at the end of the Bronze Age, circa 1250 BCE; the devastation inflicted on both Roman Europe and Han China by Hunnish barbarians between the third and sixth centuries CE; the death toll inflicted by the bubonic plague in the fourteenth century CE; and the wholesale depopulation of the Americas by disease and violence in the sixteenth and seventeenth centuries CE. Concerning the last of these, Fernand Braudel remarked that they consisted of a 'colossal biological slump ... quite incommensurate with the Black Death', which killed about a third of the European population. *The Structures of Everyday Life: The Limits of the Possible*, Collins, 1985, pp. 35-8. None of these, however, was on a global scale and none altered the underlying physical conditions that make human flourishing possible.

16   See, in particular, Wallace S. Broecker and George H. Denton, 'What Drives Glacial Cycles?', *Scientific American*, vol. 262, no.1, January 1990, pp. 48-56; Wallace S. Broecker, 'The Great Ocean Conveyor', *Oceanography*, vol. 4, 1991, pp. 79-89; Wallace S. Broecker, 'Massive Iceberg Discharges as Triggers for Global Climate Change', *Nature*, no. 372, 1994, pp. 421-4; and Wallace S. Broecker, 'Abrupt Climate Change: Causal Constraints Provided by the Palaeoclimate Record', *Earth-Science Reviews*, vol. 51, August 2000, pp. 137-54. 'The climate record kept in ice and in sediment', wrote Broecker in 1997, 'reveals that since the invention of agriculture some 8,000 years ago, climate has remained remarkably stable. By contrast, during the preceding 100,000 years, climate underwent frequent, very large, and often extremely abrupt shifts. Furthermore, these shifts occurred in lockstep across the globe. They seem to be telling us that earth's climate system has several distinct and quite different modes of operation and that it can jump from one of these modes to another in a matter of a decade or two. So far, we know of only one element of the climate system which has multiple modes of operation: the oceans' thermohaline circulation. Numerous model simulations reveal that this ['conveyor'] circulation is quite sensitive to the fresh water budget in the high-latitude regions where deep waters form.'

17   This process is known by oceanographers as thermohaline circulation, 'the circulation path determined by temperature and salt – downwellings due to surface water density created by low temperature and high salinity. Because dense water tends to eventually sink through less dense underlying layers, it contributes a vertical aspect to ocean currents. The sinking waters do not always mix with the underlying layers. Sometimes they slide down a continental slope to the ocean bottom (the most dense bottom waters are formed this way near Antarctica). Or they

may become so dense from evaporative cooling (and evaporative augmentation of salt near the surface) that they plunge through the underlying layers. More organized thermohaline circulation occurs in giant whirlpools at some places in the Greenland Sea, 10-15 km across, slowly conveying surface waters to the ocean floor in a hard-to-see column.' Calvin, *A Brain for all Seasons*, p. 311.

18   I am compressing and paraphrasing Calvin here. The original passage reads: 'We are near the end of a warm period in any event; ice ages return even without human influences on climate. The last warm period abruptly terminated 13,000 years after the abrupt warming that initiated it 130,000 years ago, and we've already gone 15,000 years from a similar warm-up starting point. But we may be able to do something to delay an abrupt cooling. Do something? This tends to stagger the imagination, immediately conjuring up visions of terraforming on a science fiction scale ... Surprisingly, it may prove possible to prevent flip-flops in the climate [by building dams to restrict the flow of fresh water into the Atlantic from Canada and Greenland and widening the Panama Canal to divert the salt flow] ... But relying on such simple fixes presumes that you know what you're doing ... It would be especially nice to see another dozen major groups of scientists doing climate simulations, discovering the intervention mistakes as quickly as possible and learning from them.' Calvin, *A Brain for all Seasons*, pp. 275-7.

19   C. H. Haug and R. Tiedemann, 'Effect of the Formation of the Isthmus of Panama on Atlantic Thermohaline Circulation', *Nature*, no. 393, 18 June 1998, pp. 673-6; Neil D. Opdike, 'Mammalian Migration and Climate Over the last Seven Million Years', in Elisabeth S. Vrba, George H. Denton, Timothy C, Partridge and Lloyd H. Burckle, *Palaeoclimate and Evolution, with Emphasis on Human Origins*, Yale University Press, 1995, pp. 109-114. See also Mark A. Cane and Peter Molnar, 'Closing of the Indonesian Seaway as a Precursor to East African Aridification Around 3-4 million Years Ago', *Nature*, no. 411, 10 May 2001, pp. 157-62, for an argument that the northward movement of the island of New Guinea had a similar effect, by constricting circulation between the Indian and Pacific Oceans.

20   Calvin, *A Brain for all Seasons*, pp. 249-50.

21   Pope Paul VI was referring to the reproductive cycle in his famous encyclical *Humanae Vitae*.

22   There is an interesting debate, of course, about how common intelligent life is in the universe. See Amir Aczel, *Probability 1: Why There Must Be Intelligent Life in the Universe*, Little Brown & Co, Boston 1998, and Peter D. Ward and Donald Brownlee, *Rare Earth: Why Complex Life Is Uncommon in the Universe*, Copernicus, Springer-Verlag, New York, 2000.

23   The five great mass extinctions were: the end-Ordovician (440 million years ago), the Late Devonian (365 million years ago), the end Permian (225 million years ago), the end Triassic (210 million years ago) and the end Cretaceous (65 million years ago). Of these, the end-Ordovician was the most sweeping because 70% of all phyla of living beings vanished, never to be revived. The end Permian was the most lethal, however, because an estimated 90% of all extant species became extinct at that time, for reasons we have still not been able to reconstruct. For this reason, the Third Extinction has been dubbed 'the Great Dying'. See Vincent Courtillot, *Evolutionary Catastrophes: The Science of Mass Extinction*, Cambridge University Press, 2002. In the ancient world, saurian and other extinct megafauna remains were occasionally found, but the ancients lacked a scientific theory as to what they were and interpreted them in vague terms as the remains of monsters which, in a sense, of course, they were. See Adrienne Mayor, *The First Fossil Hunters: Paleontology in Greek and Roman Times*, Princeton University Press, 2000.

24   For a careful examination of the most explosive part of 'world history' – the twentieth century – see John McNeill, *Something New Under the Sun: An Environmental History of the Twentieth Century*, Allen Lane, Penguin, 2000.

25   Remarkably, in the 332 pages of his book, published in 1998, Wilson makes no mention of D-Os or of Wallace Broecker and his investigations. In his final chapter, he dwells at some length on the danger of global warming, but seems entirely unaware of the danger of its causing a D-O, or of the recursion of the ice ages. It is sobering to consider that even so immensely well informed a natural scientist and general scholar, as recently as the late 1990s, should have been unaware of this whole line of inquiry and its implications. It is, therefore, doubly ironical that he should quote, of all literary characters, Hotspur, from Shakespeare's *Henry IV*, in expressing his optimism that humanity will cope successfully with the environmental challenges it now faces: 'I tell you, my lord fool, out of this nettle, danger, we pluck this flower, safety.' Edward O. Wilson, *Consilience: Unity of Knowledge* (1998), p. 289.

26   Ibid., p. 13.

27   Ibid., p. 4, Ionian, of course, because Miletus was one of the coastal cities of Ionia, on the Aegean coast of modern Turkey, which was very much part of the Greek world, although incorporated into the Persian Empire in the sixth century BCE.

28   Ibid., pp. 12-13.

29   Thomas Homer-Dixon, *The Ingenuity Gap: How Can We Solve The Problems of the Future?*, Vintage, Random House, 2001.

30   Walter Kaufmann (ed. and trans.), *The Portable Nietzsche*, Penguin, 1982, p. 42.

31   'At the moment, we are an ignorant species, flummoxed by the puzzles of who we are, where we came from and what we are for,' wrote Lewis Thomas in 1979. 'It is a gamble to bet on science for moving ahead, but it is, in my view, the only game in town.' *Late Night Thoughts in Listening to Mahler's Ninth Symphony*, Viking, 1983, p. 15.

War is both king of all and father of all, and it has revealed some as gods, some as men; some it has made slaves, others free.
- Heraclitus (c. 500 BCE)

... while practically all the arts have made a great advance and we are living today in a very different world from the old one, I consider that nothing has been more revolutionised and improved than the art of war.
- Demosthenes (341 BCE)

... the Allies thought that the Germans might attempt a surprise attack, but knowing that the Germans knew that the Allies expected it, the Allies assumed that the Germans probably wouldn't do it!
But they did.
- David P. Barash (2003)[1]

4. On the Origins of Warfare

*E*ven scientists, when engaging in debates, often disregard much of the evidence of those with whose opinions they disagree. This is symptomatic of the universal human tendency to take a position and staunchly never budge. Rather than carefully consider opposing arguments, we are markedly prone to ignore them, ridicule them or attack those who present them.

It is highly refreshing, therefore, to find two authors who have serious disagreements, but don't lapse into such behaviour. Lawrence Keeley and Raymond Kelly are specialists on the origins of human warfare. The subject itself is clearly important. The way they have engaged in inquiry and debate, however, is truly impressive and because they have been so meticulous in their reasoning, we can learn a good deal from them about both the subject they address and the best methods for inquiring into *any* subject.

'The subject of war among ancient and modern tribal peoples remains prone to glib speculation, the caprices of intellectual fashion, and the deeper currents of secular mythology', wrote Keeley in 1996.[2] Disillusioned with modern war, twentieth-century scholars had 'pacified the past', arguing that pre-literate and pre-agricultural human beings were peace-loving and warfare rare, ritualised and not very deadly. In reality, he contended, pre-civilised warfare was more endemic, more brutal and took a higher toll of the populations involved than have the historical wars of Western civilisation.

In 2000, Raymond Kelly published a re-examination of the evidence and argued that war had originated very late in the evolution of the species and the human world had been warless for countless millennia before that.[3] In other words, the pre-civilised past was, indeed, warless. Though it seemingly represented a challenge to his own argument, Keeley greeted the book as 'important, interesting, plausible' and 'fascinating reading'. He did not dismiss its claims, even though they might be seen as antithetical to his own.

Apart from his remarks about Kelly's book, Keeley merits serious respect as a scholar. The preface to his own book is remarkably honest in explaining how he'd been driven by the archaeological evidence to *change* his whole way of thinking about warfare. That preface is worth reading even on its own because it shows a truly scientific mind at work, examining its own thinking, testing it, learning and revising its beliefs. As his response to Kelly shows, he remained alert after his work was published, looking to learn, rather than just entrenching himself behind his prejudices and his erudition, determined to drive off all challenges.

'This book had it genesis in two personal failures,' his preface begins. 'One of a practical academic sort, the other intellectual ... My practical failure involved two unsuccessful research proposals requesting funds to investigate the functions of recently discovered fortifications surrounding some Early Neolithic (c. 5000 BCE) villages in north-eastern Belgium.' Archaeologists reviewing the proposal for the US National Science Foundation refused to accept that the nine-foot-deep ditches and palisades around the villages constituted 'fortification' and therefore declined to recommend funding for the research. A third proposal was accepted 'only after I rewrote it to be neutral about the function' of the ditch/palisade structures, referring to it as an 'enclosure' rather than a 'fortification.' 'In other words, only when the proposal was cleansed of references to that archaeological anathema, warfare, was it acceptable to my colleagues.'

Nonetheless, Keeley and his Belgian colleague Daniel Cahen were 'shocked' when their new research confirmed that these early villages and others they discovered during their research were, indeed, fortified. 'Our mutual amazement was based on the prejudices we shared with the very colleagues who had given my early unsuccessful proposals a sceptical review. Subconsciously, we had not really believed our own arguments ... Later, reflecting on my own education and career, I realised that I was as guilty as anyone of pacifying the past by *ignoring or dismissing evidence* of prehistoric warfare – even evidence I had seen with my own eyes.' The archaeological evidence had, in fact, long been readily available, but 'I [had] dismissed this data as either unrepresentative, ambiguous or insignificant.' The rejection of his research proposals, he relates, had made him aware of the prejudices of his colleagues, but then he realised how he had, for many years, 'worn the same blinders'.

By the early 1990s, the view among archaeologists had shifted to accepting that the Early Neolithic had not, after all, been a warless golden age, but an epoch in which warfare had been endemic. Keeley was impressed by the fact that resistance had been overcome by evidence and argument. His faith in the robustness of archaeology as a science was strengthened, because he had seen that physical evidence had exhibited 'an extraordinary ability to overcome even the most ingrained ideas.' Like all disciplines, he conceded, archaeology 'has unacknowledged blind spots, unconscious prejudices and declared theoretical biases', but because it is based on hard, physical evidence, archaeology was less able than some disciplines to dismiss uncongenial facts 'by selective *ad hominem* scepticism, clever sophistry or the currently fashionable denial that there is any "real past".'

The real past, revealed by archaeology far more than by any mythology, is of immense antiquity. Hominids have been around for some millions of years, our own species, *Homo sapiens*, for at least 100,000 years.[4] Yet, as of 500 years ago, Keeley points out, only a third of the world was civilised.

'Australasia and Oceania, most of the Americas and much of Africa and north Asia remained preliterate and tribal.' Such 'peoples without history' are the province of anthropologists.

Yet what had anthropologists revealed about warfare among these prehistoric and tribal peoples? Almost nothing. 'Less by sustained argument than by studied silence or fashionable reinterpretation, pre-historians have increasingly pacified the human past. The most widely used archaeological textbooks contain no references to warfare until the subject of urban civilisations is taken up. The implication is clear: war was unknown or insignificant before the rise of civilisation.'

Keeley quotes a 1991 book by two military historians as exemplifying this view of human affairs. 'In less than 2,000 years, man went from a condition in which warfare was relatively rare and mostly ritualistic to one in which death and destruction were achieved on a modern scale ... The Iron Age also saw the practice of war firmly rooted in man's societies and experience and, perhaps more importantly, in his psychology. War, warriors and weapons were now a normal part of human existence.'[5] He quotes a professor of sociology in a letter to the editor of the *Chronicle of Higher Education* in 1991. The sociologist wrote of 'the emotional richness and cultural diversity of traditional African tribal life' compared with 'the enhanced capacity for destructiveness that the emergence of all civilisational structures brought forth, such as organised mass warfare.'

This idea that warfare as such and its deadly nature in particular arose with civilisation, whereas prehistoric societies had lived lives of emotional richness, peace and mutual goodwill, is the myth Keeley set out to confute. He adduced three kinds of evidence: first, archaeological remains of mass killings dating back to prehistoric times in Africa, Europe and the Americas; second, the brutality exhibited in prehistoric warfare; third, the remarkable statistical evidence showing that, on a *per capita* basis, tribal and prehistoric warfare was both more endemic and *far more deadly* than the famous wars of the historical and civilised world.

At a Late Palaeolithic site called Gebel Sahaba in the Sudan which dates back as far as 12,000 BCE, 40 per cent of the skeletons recovered from a burial ground used over several generations showed signs of having been killed in combat or having been executed by blows to the head and neck.[6] At Talheim, Germany, a Neolithic mass grave has been found with the remains of eighteen adults and sixteen children who had been killed and thrown into a large pit.[7] At Crow Creek, South Dakota, 'archaeologists found a mass grave containing the remains of more than 500 men, women and children, who had been slaughtered, scalped and mutilated during an attack on their village' in around 1325 CE, i.e. more than 150 years before Columbus crossed the Atlantic. In short, as Keeley observes, well 'before any possible contact with civilisations, the tribesmen of Neolithic Europe, like those of the prehistoric United States, were wiping out whole settlements.'[8]

Regarding the brutality of prehistoric warfare, Keeley points out that adult males of an enemy tribe were very rarely taken prisoner, allowed to surrender or spared from execution. Usually they were summarily killed. If prisoners were taken, it was usually in order to use them as sacrificial victims or to torture them to death over several days and possibly to eat them. Such behaviours have been documented, he relates, 'for the Maoris and Marquesans of Polynesia, Fijians, a few North American tribes, several South American groups and various New Guinea groups.'[9]

He then adds, 'Of course, many tribal societies took no prisoners and retained no captives of any sex or age ... Perhaps the harshest treatment of captives was meted out in Polynesia. The Tahitians are described as leaving enemy children pinned to their mothers with spears or 'pierced through the head and strung on cords'. The Maoris sometimes disabled captive women so that they could not escape, permitting the warriors to rape, kill and eat them when it was more convenient to do so.'[10]

The most interesting category of evidence, however, is the statistical. Keeley makes the startling observation that the absolute numbers of people killed in civilised warfare (70,000 in one day at Cannae in 216 BCE, for example, or 50 million in the six years of the Second World War) *mislead* us into thinking that tribal or prehistoric warfare was a trivial and relatively harmless affair by comparison. Calculated as a percentage of the populations at risk, he argues, this is so far from being true that we need to rethink our whole way of looking at warfare. 'A typical tribal society lost about .5 per cent of its population in combat each year.' This may sound trivial, but that is a statistical illusion. Consider that the total number of people killed through all causes in the wars of the twentieth century was between 100 and 150 million. Then consider that, if you applied the *tribal* death rate from war to twentieth century populations, there would have been 'more than 2 *billion* war deaths since 1900.'[11]

Keeley adds a thought-provoking footnote for the benefit of those sceptical of this approach to the evidence. 'Some readers may be unconvinced by percentage comparisons between populations of hundreds or thousands of people and populations of millions or tens of millions – that is, they are more impressed by absolute numbers than ratios. However, consistent with such views, such sceptical readers must also disdain any calculations of death *rates* per patient or passenger mile and therefore always choose to undergo critical surgery at small, rural Third World clinics and fly on small airlines.

At such medical facilities and on such airlines, the *total* number of patient or passenger deaths is always far fewer than those occurring on major airlines or at large university and urban hospitals. These innumerate readers should also prefer residence on one of the United States' small Indian reservations to life in any of its metropolitan areas, since the annual *absolute* number of deaths from homicide, drug abuse, alcoholism, cancer, heart disease and automobile accidents will always be far fewer on the reservations than in major cities and their suburbs.'[12]

We are unlikely to rationally or effectively address the scourge of war 'while we are in the thrall of nostalgic delusions', Keeley concludes. The prehistoric and tribal world was not one of peace, plenty and emotional richness and it does not offer the solutions we need to the problems of the modern world. In sober fact, 'the only practical prospect for universal peace must be more civilisation, not less. Adherence to the doctrines of the pacified past absolve us from considering the difficult question of what a truly global civilisation should consist of and, more importantly, what its political structure should be.'[13]

Raymond Kelly was not prepared, however, to give up the doctrines of the pacified past. And, despite all the evidence he had mustered against those doctrines, Keeley took Kelly's argument seriously. Why? Because Kelly argues carefully and dispassionately. He *accepts* almost all of what Keeley points out about tribal peoples *since the Neolithic*. He also accepts that tribal peoples everywhere have been violent, not peaceful and gentle.

Yet, sifting the evidence finely, he still believes that warfare originated *very late* in human evolution and that he can pinpoint what led to its emergence and proliferation. He observes that 'excepting a single Upper Palaeolithic site, archaeological evidence points to a commencement of warfare that postdates the development of agriculture. This strongly implies that earlier hunter-gatherer societies were warless and that the Palaeolithic was a time of universal peace.'[14]

Kelly, like Keeley, is acutely interested in argument and evidence, the testing of hypotheses and the rejection of ill-considered prejudices or theoretical biases. 'The issue is too important', he remarks, 'to limit ourselves only to knowledge that makes us feel good, and to consequently fail to consider all the relevant data.'[15] It is principally for this reason, I think, that Keeley warmed to his writing. He does not reject any of Keeley's evidence, yet he finds Keeley's argument inconclusive.

He is aware of Keeley's 'survey of a substantial body of relevant ethnographic and archaeological data and ... his denunciation of what he sees as "the pacification of the past."'[16] He specifically argues, however, that the 'earliest conclusive archaeological evidence of warfare' is that described by Keeley at Gebel Sahaba.[17] Prior to that, however, going back over hundreds of thousands of years of human evolution, he finds evidence of *violence*, not evidence of *warfare*. This is a distinction very important to his case and, as it happens, one that Keeley had not paid very much attention to. It led Kelly to a fascinating hypothesis.

'If war is *not* a primordial feature of human society', he reflects, 'then it must have originated at some point in the human past'. The question is, was that far back in the Upper Palaeolithic, as long as 35,000 years ago, or perhaps even before the emergence of modern humans 100,000 to 150,000 years ago, or was it much more recently? Here, he urges, we must define our terms with care. 'It is not the case', he argues, 'that one definition of war is as good as another. Rather, there are explicit logical criteria for establishing a superior definition.' Thus, we need to differentiate between homicide, capital punishment, raiding, feuding and warfare. To make this point he observes that 'pongicide (apes killing one another) is an analogue of homicide and both are undoubtedly ancient. However, chimpanzees lack both capital punishment and war.'[18]

'War', he argues, 'is grounded in the application of a calculus of social substitution to situations of conflict such that these are understood in group terms.' Warless societies are *not* non-violent. 'On the contrary, physical violence is ... a principal vehicle of conflict resolution, as manifested in regulated, contest-like fighting and in the removal of a killer or sorcerer by execution. However, what warless societies do uniformly manifest are intrinsic limitations on the extent to which one act of lethal violence leads to another.' The emergence of war in the Neolithic, he argues, must be seen, therefore, as 'a transition from one form of collective violence to another, rather than a transition from peaceful non-violence to lethal armed conflict.'[19]

Kelly corroborated Keeley's contention that primitive societies are very violent ones. 'Homicide rates in simple foraging societies', he accepts, 'are considerably *higher* than those reported for agricultural societies with more developed forms of socio-political organisation.'[20] But 'the calculus of social substitution that is the hallmark of war is clearly absent', and 'delineating this boundary makes it possible to rigorously discriminate between the warless societies and those in which warfare is present.'[21] He argues that it can be empirically shown that this boundary was probably not crossed by the overwhelming majority of human societies before the beginnings of agriculture.

Then a transformation set in. The engine of that transformation, he argues, consisted of an adaptive modification towards war among societies competing for *reliable and abundant*, not scarce, resources. This last point is as counter-intuitive and important as Keeley's observation about absolute numbers and ratios, so it is worth pondering. Kelly called it a paradox, but he drew attention to the fact that it is *only* under such conditions (of relative abundance) 'that a society can afford to have enemies for neighbours'. The demographics of warfare, as described by Keeley, dictate that warlike societies would have been *selected against* right down through the Upper Palaeolithic, because they would have been '*unable to colonise* environments characterised by *low* resource density, diversity and predictability.'[22] The ethnographic case material Kelly used to make this point is analysed with a scrupulous care that won Keeley's admiration.

This is a beautiful argument, proceeding by the careful weighing of evidence, the testing of hypotheses and the refinement of definitions. Therefore, it throws light back into the past, rather than generating consoling myths in the present. This is social science at its finest and most illuminating. Keeley remarks in a footnote that successive waves of existentialism, structuralism, post-structuralism and postmodernism 'have left American universities a "burned-over district" like those areas of nineteenth century New England exhausted by a succession of religious evangelisms.'[23] His engagement with Kelly, however, was fresh, lucid and scientific. Wherever scholarship of this calibre is possible, universities will remain laboratories of learning and not mere cloisters of ideological evangelism.

4. *Endnotes*

1   David P. Barash, *The Survival Game: How Game Theory Explains the Biology of Cooperation and Competition*, Times Books, Henry Holt & Co., New York, 2003, p. 60.

2   Lawrence H. Keeley, *War Before Civilisation: The Myth of the Peaceful Savage*, Oxford, 1996.

3   Raymond C. Kelly, *Warless Societies and the Origins of War*, University of Michigan Press, Ann Arbor, 2000.

4   Donald Johanson and Blake Edgar, *From Lucy to Language*, Cassell Paperbacks, 2001.

5   Ibid., p. 22, quoting R. Gabriel and A. Metz, *From Sumer to Rome: The Military Capabilities of Ancient Armies*, Greenwood Press, New York, 1991, pp. 3, 19.

6   Keeley, *War Before Civilisation: The Myth of the Peaceful Savage*, p. 37.

7   Ibid., p. 38.

8   Ibid., p. 68.

9   Ibid., p. 85.

10 Ibid., p. 87.

11 Ibid., p. 93.

12 Ibid., p. 214, n. 21.

13 Ibid., p. 179.

14 Kelly, *Warless Societies and the Origin of War*, p. 2.

15 Ibid., p. 18.

16 Ibid., p. 125.

17 Ibid., p. 148.

18 Ibid., p. 123.

19 Ibid., pp. 41-3.

20 Ibid., p. 20.

21 Ibid., p. 10.

22 Ibid., p. 135.

23 Keeley, *War Before Civilisation: The Myth of the Peaceful Savage*, p. 221, n. 1.

It is absurd to hold that a man ought to be ashamed of being unable to defend himself with his limbs, but not of being unable to defend himself with speech and reason, when the use of rational speech is more distinctive of a human being than the use of his limbs.

- Aristotle[1]

The interaction between the orality that all human beings are born into and the technology of writing, which no one is born into, touches the depths of the psyche ... Writing is consciousness raising.

- Walter J. Ong[2]

Writing is regarded as the threshold of history, because it ended the reliance upon oral tradition, with all the inaccuracies this entailed ... [A]mong the innumerable benefits created by a script, writing allows us to capture our ideas when they arise and, in time, to sort and scrutinize, revise, add, subtract and rectify them to arrive at a rigor of logic and a depth of thought that would otherwise be impossible ...

- Denise Schmandt-Besserat[3]

5. On Language and Poetry

*I*f we are language animals, is it not vital that we master language and use it with all the richness and flexibility of which it is capable? Don Watson's *Death Sentence: The Decay of Public Language*, published in 2003, is a passionate polemic on this subject and a useful point of entry for reflecting on it. *Death Sentence* is a cry from the heart for a richer and also more disciplined use of the English language in Australian public life. The book is a stimulus to thinking about our use of language for this reason, but also because, as it happens, it is itself characterised by a poor use of the language.

There are some witty lines in the book and lots of quotes from interesting European and American writers. There is, also, a basic attitude towards the value of the Western canon from which it is hard to dissent. What is lacking is anything resembling a systematic argument. For this reason, Watson failed by his own implicit standards. His book is a mildly cantankerous ramble, punctuated by outbursts of political and social spleen. It lacks any but the most rudimentary structure, seems to be based on no particular theory or hypothesis, slides from topic to topic and allusion to allusion barking at all manner of bogies and, while deploring a general state of affairs, is exceedingly vague and inconsistent as to its causes.

Watson's theme is anything but new. It is that the gracious and eloquent ways in which language used to be spoken and written in the glory days of civilisation have been giving way to a vulgarised, demotic inarticulateness that is unpleasant, depressing and even dangerous. Such complaints can be found already in Plato and Confucius and at many points since. The contemporary complainant needs to be a little cautious, therefore, in assessing just what the problem is before getting too splenetic or, to use Watson's preferred term 'indignant'.

Much more recently than Plato or Confucius, the likes of Jacques Barzun and Allan Bloom deplored the decline of education and the vulgarising of Western (especially American) language and culture.[4] In 1994, the rambunctious literary critic Harold Bloom began one of his books with a prelude titled 'An elegy for the Canon'.[5] He declared, 'We are destroying all intellectual and aesthetic standards in the humanities and social sciences, in the name of social justice ... What has been devalued is learning as such, as though erudition were irrelevant in the realms of judgment and misjudgment.'[6]

Watson clearly sought to place himself in this company. What, then, did he add to the reflections of the great masters? He was more concerned with Australian public life than with that of the Western world at large. He remarks that the British and Americans are better off linguistically than we are because they are more self-confident, more naturally given to elevated self-expression and loquacity, respectively. Alas, he sighs, Australians suffer from 'a stubborn refusal to be articulate.'[7]

Yet if this was his principal contention, it would surely have made sense to begin by pointing to the larger arguments of the Barzuns and Blooms and then to have indicated an intention to reflect on the specifically Australian case of cultural decline and linguistic barbarisation. He might, from such a starting point, have offered us a survey of nineteenth century Australian speech and writing, educational standards and public rhetoric, before proceeding to examine how we have arrived at what he regards as a sorry state of affairs, more than two centuries after British settlement.

He did nothing of the kind. He opened with an introduction that consists not of an outline of the coming argument, but of a rap on managerial jargon, during which he invoked George Orwell, Primo Levi, Norman Mailer and Simone Weil, but not a single Australian writer. He then gave us four pieces of writing which have neither number nor title, nor sub-titles nor symmetry. More fundamentally, they lack any evident rationale for their separateness. In this respect they remind one of the stanzas of much contemporary poetry, which seem separated more or less at random.

The first 'chapter' runs for fifty-four pages. It is followed by what looks like a sort of intermezzo, of just thirteen pages, in which Australia is dwelt on more specifically than elsewhere. Then the ramble recommences, with a 'chapter' sixty pages long, followed not by a conclusion but by a final 'chapter', thirty-eight pages in length, which does not round out any argument – for, indeed, there has been none – but merely continues the ramble until, it would appear, the author has run out of breath.

It is plain throughout that Watson was irritated and upset by various ways in which, as he expresses it, 'every day we vandalize the language'. He states, 'We can only be indignant and we should resist.'[8] Yet he does not make at all plain why this vandalism is occurring, how 'we' should express our indignation or what he would propose by way of resistance. One pictures Henry Higgins, played by the inimitable Rex Harrison of course, singing indignantly 'Why can't the English [Australians] teach their children how to speak?'

Should we create a sort of *Academie Anglaise* to defend the language against its vandalisers? Or is he even more democratic than that? Would he see a unionisation of writers as the way to go? Public demonstrations orchestrated by the linguistic equivalent of Trotskyite activists, earnest and outraged at the inroads of managerialism and globalisation on the language of ordinary people?

Doubtless, these whimsical rhetorical questions sound unfair to Watson. He was surely not calling for any of these things. The problem is, he didn't make it at all clear what he *was* calling for. He failed even to make clear to whom he was referring when he asserted that 'we' vandalise the language every day and 'we' should be indignant and resist. If he believed, as well he might have, that resistance begins at home, he might reasonably have started with himself and have resisted the temptation to see into print so undisciplined a piece of writing as he gave us.

I stated that *Death Sentence* seems to be based on no particular theory or hypothesis, slides from topic to topic and allusion to allusion barking at all manner of bogies and, while deploring a general state of affairs, is exceedingly vague and even inconsistent as to its causes. Let me illustrate what I mean by these remarks. Since he nowhere set it out explicitly, we are left to puzzle out from Watson's prose what he sees as having brought us to the condition of unsanctioned vandalism of the language. His allusions to issues and assertions about causes provide clues. As for any clearly articulated theory or hypothesis, there is neither.

It is possible to pick out from Watson's persiflage an astonishing variety of putative causes for the condition he deplores. The first and most evident is the advance of business jargon into politics and academia – which it entered, he trumpets, as the German army entered Poland (in 1939). He returns to this idea again and again. It's just that he never develops it into the form of a rational inquiry. Indeed, quite early in his book, he confesses that he lacked sufficient interest in the subject to attempt such an inquiry. This is a little odd, considering that he felt so indignant, but let that pass. It was after all, by his own account, only an 'essay' he was writing.

Lacking sufficient interest in chasing down his principal quarry did not, however, discourage Watson from darting off in at least a dozen different directions in pursuit of other possible causes for the alleged linguistic malaise of our time. These pop up in the book more or less at random. The rise of Communications Theory (along with Media, Cultural and Women's Studies) is one. Grubby practicality seems to be a second. 'Modern language', Watson remarks airily at one point, 'handcuffs words to action, ideas to matter, the pure thought to the dirty deed.'[9]

A third is a certain 'pomposity' which 'afflicts most people when they write formally or write formal speeches.' A fourth is consumerism. 'It seems', Watson expatiates, 'that consumer choice expands in inverse proportion to our vocabulary. We use fewer words and words of less variety. We arrange them with less imagination and dexterity. We tangle and abuse them. We take the richest soil the culture has and turn it into a few clods.'[10]

His 'we' again seems more than a little problematic here. All the more so because he has, earlier in the book, declared cheerfully that English has survived 'everything that's been thrown at it: political and social revolutions, industrial and technological revolutions, colonialism and post-colonialism, mass education, mass media, mass society. More than just surviving these upheavals, it adapts and grows, is strengthened and enriched by them. *And never has it grown more than now*: by one estimate at the rate of more than 20,000 words a year, and for every new word several old ones change their meanings or sprout additional ones. *It is wondrous on this level* [emphases added].'[11]

The contradiction here is twofold. On the one hand, Watson celebrates the capacity of the language to survive several centuries of enormous social change and upheaval, yet fears that it is succumbing in the face of that dread social phenomenon 'consumerism'. On the other hand, he specifically states that the language is growing wondrously, by tens of thousands of new words per annum, only to then lament that 'we' 'use fewer words and words of less variety'. At the very least, he needs to disaggregate his 'we' here in order to clarify whatever point he is trying to make.

A fifth cause of linguistic decay that pops out of Watson's prose is the 'din, chaos and method acting' of popular culture, or 'the language of twitching narcissism' – whatever that is.[12] A sixth, 'the diabolical environment of politics' in which 'unreasoning forces throw up unreasoning things like red herrings and dead cats and fling them in the path of journalists ... Reason goes up in smoke. The truth is less significant than the political contest.'[13] Pardon me? This is new? This is peculiar to Australia, or the globalised world?

A seventh cause he points to is the recent ascendancy of film and pictorial images. Here Watson gets himself into another unnecessary tangle. 'Pictures rule: but words define, explain, express, direct, hold together our thoughts and what we know. They lead us to new ideas and back to older ones. *In the beginning was the Word* [emphases added].'[14] He seems to have been unable to resist here the allusion to the opening verse of the Gospel of St John. But what was his precise intention in using it?

This is no mere quibble. For, as it happens, the Word was *not* in the beginning, before pictures, at all. Biologically and cognitively, the eye and the picture-forming brain are incomparably older than the word.[15] The word was not even in the beginning of the life history of hominids. It was something that emerged as an added capacity within a brain primarily configured for visuo-spatial orientation and standard animal responses.

Pictures – the seeing eye and visual brain – have ruled for about half a billion years. Words have only existed for, at the very most, half a million and segmented speech for as little as 100,000 – and 'the word' has never altogether ruled. Even when used, words build on the visual underpinnings of cognition, in so far as they spin meaning out of metaphors – re-picturings of reality.[16] Their use in systematic reasoning has always been partial and problematic.

If Watson had understood this, he might have been less distressed by some of what he saw (or heard) around him and also more interested in inquiring into it rather than merely sounding off. He might also have been more discriminating in his praise of words, since by his own account they do *not* always define, explain, express, direct or hold together our thoughts. To the contrary, they all too readily become jumbled and confused – even in the hands, as it were, of a tolerably well educated person such as himself.

An eighth cause for our linguistic problems, according to Watson, is that Australia is a colonial backwater that, alas, never had a violent revolution or a civil war to put some definition or fire into our public language. 'Self-government came without the necessity to fight for it,' he complains.[17] Why precisely this should be a cause of linguistic impoverishment he fails to explain. In any case, he implies that the problem goes back to the very beginnings of settlement, in that 'When the British reached Australia's shores they seemed to lose their inspiration.'[18]

This last point, apart from its apparent inconsistency with his lament about the lack of a violent revolution, is itself rather incoherently advanced. He quotes a speech against slavery by William Pitt before the House of Commons, commending his 'words of distinction; words marrying intellect with moral passion', immediately before observing that the British lost their inspiration on arriving at Botany Bay. It doesn't seem to occur to him that it was not, after all, William Pitt who arrived at Botany Bay. Nor does he adduce any evidence that those who did arrive were any more inspired before they departed Britannia's shores than after they settled at 'the arse end of the Earth' – his beloved Paul Keating's notorious description of Australia.

Other causes, nine through twelve, are science, because it has radically reduced 'the element of chance in life,'[19] the 'fading of the King James Bible' (funny, but my copy hasn't faded at all),[20] the fact that the baby boomers 'lost God and took to sociology and marijuana,'[21] and the rise of sport, since nothing breeds cliché more than sport, except perhaps 'film and television and celebrity and news and business.'[22] Quite a list! Yet none is followed through, none is shown to have had any specific effect and none even of the presumed effects is weighed against the proliferation of new words – non-clichés – to which Watson himself testifies.

His *piece de resistance* (so to speak) is surely Watson's resigned and plaintive remark 'I have written this essay in the hope that awareness might increase in some small degree and with it, indignation: a small degree and a not very sanguine hope because I know that powerful forces, including possibly the whole tide of history, are against us.'[23] The whole tide of history? What, Hegel in reverse? Spirit turning into a linguistic black hole? What a grimly fascinating notion. But of course he neither sets out a case for this being so, nor defines his terms, nor tells us who 'us' is. And to paraphrase Bill Clinton, much would seem to depend on what the meaning of 'us' is.

As a piece of argument, *Death Sentence* is a dismal showing. Nor is it improved by Watson's tendency to lash out at the usual suspects; usual, that is, if you are a paid-up member of what exasperated conservatives have long since dubbed 'the chattering classes'. Then prime minister John Howard, in particular, is pilloried again and again for using plain language, rather than something that would stir Watson's heart. The Americans generically are, in passing, casually accused of 'lynching people' in the late nineteenth century. The claim is entirely unsupported by actual historical evidence and is offered as if that practice had been as endemic in the United States of the Gilded Age as headhunting in Borneo before colonial interference.[24]

One might go on to catalogue quite a long list of other solecisms and minor irritants in *Death Sentence*, but there's no need to flog a dead horse, or sentence. Indeed, the above remarks notwithstanding, one feels a certain kinship with Watson at times. This is especially so where he waxes lyrical about the beauties of Shakespeare and the vital role of language in elementary and secondary education. He aspires to high standards and his book is an anguished insistence that we do too little to cultivate them in our schools and universities.

Looking back to his own schooling in rural Australia, he reflects, 'If we picked up a feeling for the language, it was in English literature ... but especially Shakespeare. Shakespeare was the best thing they gave us. *Julius Caesar, Macbeth, King Lear* and a couple of the sonnets burrowed their way in and took up residence in our inhospitable souls ... It was the one hint we had that there were mysterious powers in language: that beautifully arranged words could liberate, possess, bewilder and intoxicate. They contained revelations. They could extend a person. There was pleasure in just reciting them.'[25]

Amen to all that; although my own pleasure in Shakespeare was not stirred by either school or university. It was discovered on my own time. Few of my contemporaries, so far as I could or can judge, experienced the intoxication to which Watson refers. The only time I can recall such a mood capturing a whole class, or most of it, was in 1967, when a young school teacher read to our fifth grade class J. R. R. Tolkien's *The Lord of the Rings*.

Yet for the most part, in his little book, Watson communicates not the pleasure of literature or the possibilities of language; but a cocktail of cultural deracination, personal frustration and vaguely populist politics, mixed with confused snobbery. One imagines him at the city gates, so to speak, in the character of Marullus, from Shakespeare's *Julius Caesar*, berating bureaucrats, academic jobbers and working folk with the outraged cry, 'You blocks, you stones, you worse than senseless things! O you hard hearts, cruel, mean and low. Know you not Poetry?'[26]

At a time when the language has become the world's dominant tongue and when an extraordinary range of scholarship is being produced in it, while sub-cultural and scientific neologisms proliferate, Watson's gloom is surely misplaced. Certainly there is a dearth of inspired speech in our parliaments (but, as Gilbert and Sullivan attested, this was true of the British Parliament well over a century ago); certainly there are, as there have always been, a multitude of citizens who are less than eloquent; certainly our schools and universities could do with revitalisation.

Death Sentence was never likely, however, to contribute much to such a revitalisation. And even were things far grimmer, the best response would surely have been that of Edgar to Gloucester, in Shakespeare's *King Lear*: 'What, in ill thoughts again? Men must endure their going hence, even as their coming hither. Ripeness is all. Come on!'[27] There is no way things *ought* to be and there is no prize for mere indignation. The language is as open to creative and rigorous use as it ever was. What's required is leadership from creators rather than sullen 'resistance' by those who believe that fine language has suffered a death sentence.

5.   *Endnotes*

1   Richard McKeon (ed.), *The Basic Works of Aristotle*, Random House, New York, 1941, p. 1328.

2   Walter J. Ong, *Orality and Literacy*, Routledge, 1988, p. 175.

3   Denise Schmandt-Besserat, *How Writing Came About*, University of Texas Press, Austin, 1996, p. 1.

4   Jacques Barzun's numerous works have this as a consistent theme. See in particular *The House of Intellect* (1959), *The American University: How It Runs, Where It Is Going* (1970), *Simple and Direct: A Rhetoric For Writers* (1975), *The Culture We Deserve* (1989), *Begin Here: The Forgotten Conditions of Teaching and Learning* (1991) and *From Dawn to Decadence: 500 Years of Western Cultural Life* (2000). Allan Bloom's *The Closing of the American Mind: How Higher Education Has Failed Democracy and Impoverished the Souls of Today's Students* (1987) is a classic of the genre.

5   Harold Bloom, *The Western Canon: The Books and School of the Ages*, Harcourt, Brace and Co., New York, 1994

6   Ibid., p. 35.

7   Don Watson, *Death Sentence: The Decay of Public Language*, Random House Australia, Sydney, 2003, p. 70.

8   Ibid., p.8.

9   Ibid., p. 27.

10   Ibid., pp. 42-3.

11   Ibid., p. 12.

12   Ibid., p. 46.

13   Ibid., p. 60.

14   Ibid., p. 65.

15   For an illuminating history of the evolution of the eye, see Andrew Parker, *In The Blink of an Eye: The Cause of the Most Dramatic Event in the History of Life*, Free Press, London, 2003. Parker is an Australian who received his PhD from Macquarie University, Sydney, while working in marine biology for the Australian Museum. In 1999, he became a Royal Society University Research Fellow at Oxford University's Department of Zoology. His book is a model of the clarity and elevated, unpretentious expression that Watson called for but did not himself provide.

16   For a systematic development of this idea, see George Lakoff and Mark Johnson, *Philosophy in the Flesh: The Embodied Mind and Its Challenge to Western Thought*, Basic Books, New York, 1999.

17   Watson, *Death Sentence*, p. 67.

18   Ibid., p. 68.

19   Ibid., p. 98.

20   Ibid., p. 150.

21   Ibid., p. 158.

22   Ibid., p. 158.

23   Ibid., p. 181.

24   Ibid., p. 74.

25   Ibid., pp. 164-5.

26   William Shakespeare, *Julius Caesar*, act 1, scene 1, lines 40-42, where, of course, the last part reads '... cruel men of Rome. Knew you not Pompey?'

27   William Shakespeare, *King Lear* act 5, scene 2, lines 9-11.

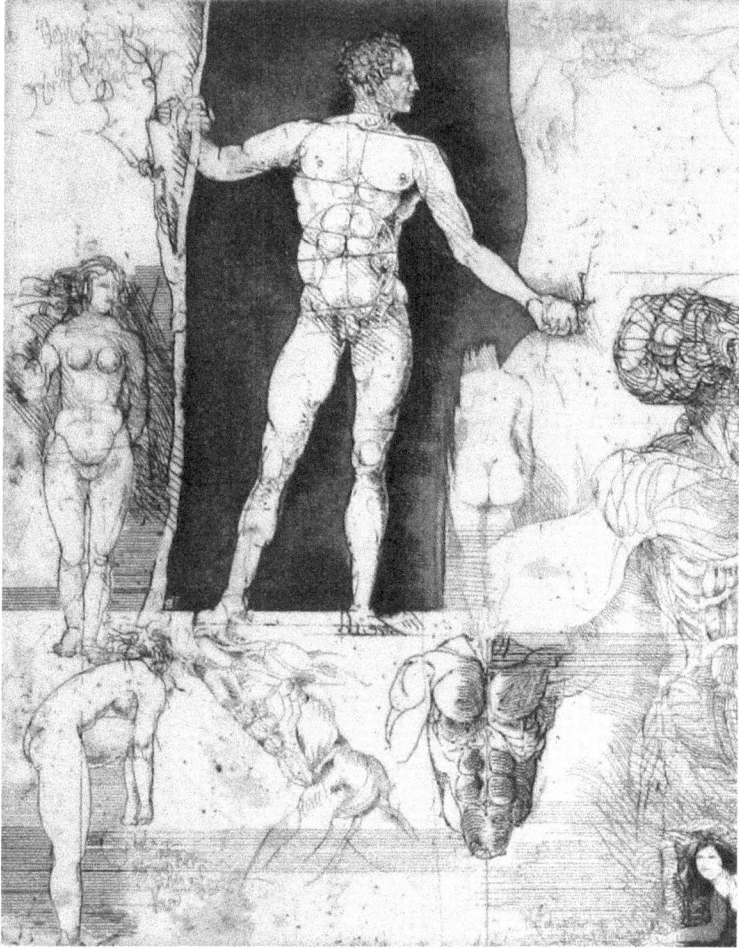

JS 110, *Sketches Also After Dürer* (detail)
Jörg Schmeisser, etching, 1974.

Everything hinged on this. If Erasmus won, Christ was done for. The history of the next five hundred years would prove Luther right ... Once reason is given an inch, the questioning begins, and there is no way to stop it. This is why Luther calls 'Mistress Reason' the 'Devil's whore'. She is seductive, deceiving, offering a moment of pleasure in order to seize the whole soul.
- John Carroll (2004)[1]

Not infrequently, Luther himself was bewildered by the new world he encountered, and his instincts in such moments were conservative.
- John Dillenberger (1961)[2]

Objection, evasion, happy distrust, pleasure in mockery are signs of health: everything unconditional belongs in pathology.
- Friedrich Nietzsche (1886)[3]

6. On Reason and Revelation

*J*ohn Carroll is a sociologist with an idiosyncratic take on Lutheran Christianity who believes he has 'diagnosed what has gone wrong with Western culture'. The West, he believes, has replaced faith with reason and has, in consequence, wrecked its culture.[4] Things started to go wrong about 500 years ago and now 'we live amidst the ruins of the great, five-hundred-year epoch of humanism'. To recover, he asserts, we must restore the authority of Christianity, since without Christ crucified there is nothing but death in store for us.

To see the past 500 years as an epoch of failure and decline within Western culture surely requires a heroic inversion of reality. We can grant Carroll that some things are wrong, perhaps very seriously so, within Western culture, without embracing his startling assertion. Many observers, from very different vantage points, have claimed that much is wrong with the West in our time. One thinks, for instance, of Jacques Barzun,[5] or Harold Bloom,[6] both teachers, as John Carroll is, and also prolific authors. Carroll's diagnosis and prescription, however, are not those of Barzun, to say nothing of Bloom.

The assertion that something is fundamentally wrong with Western culture, in fact, is commonplace. The ideological Left, whether Marxist, anarchist, anti-globalist or environmentalist, has asserted it for many decades. The ideological and religious Right has been prone to make it at least since the French Revolution. The great early nineteenth century Austrian reactionary, Count Metternich, saw the decline as dating from the sixteenth century because the invention of the printing press undermined ecclesiastical authority, while the invention of gunpowder and the conquest of the Americas undermined the old aristocratic order.[7]

Carroll belongs somewhere in this Metternichian tradition. With his complaint that Western culture has lost its sources of authority and his appeal to some kind of ill-defined Christianity, he is certainly not a man of the Left. His renunciation of virtually the entire Western canon, from Erasmus and Shakespeare to Voltaire and Kant, Marx and Darwin makes him sound at times like a pre-Vatican II Catholic Cardinal Secretary of the Holy Office providing grounds for the prohibition of books. He calls Marx and Darwin 'wreckers', but comments that 'There were, of course, hundreds of others in all cultural areas.'[8] One wonders what would be done with libraries full of such humanist writings, or the works (or even the persons) of 'wreckers', in Carroll's ideal Christian state.[9]

Would he ban such works, as Plato, in the fourth century BCE, wanted to ban the old poets from his Republic? Carroll's theological enthusiasm brings to mind the marvellous legend, judged by Edward Gibbon to be apocryphal, concerning the destruction of the great library at Alexandria. Asked by his general, Amrou, who had captured the city, what to do with the vast collection of books, the Muslim Caliph Omar is said to have responded, 'If these writings of the Greeks agree with the book of God, they are useless and need not be preserved; if they disagree, they are pernicious and ought to be destroyed.'[10]

Carroll does not advocate book-burning, but his anti-humanism is so irrational that something like Omar's legendary judgement is implicit in it. He asserts that things have been going fundamentally wrong for centuries because we have strayed from collective belief in the God of the Bible. He asserts that we have so strayed because we relied on reason instead of on faith.[11] He asserts that all humanism is vitiated by the nihilism that follows from reliance on reason, so that humanism 'was doomed from the start' and that it is now in ruins. He even sees the attack on New York by Muslim fanatics on 11 September 2001 as having demonstrated the hollowness of humanism, rather than the ugliness of fundamentalist religious fanaticism. He is, surely, in error on every count.

The simplest response to his sweeping polemic is plainly to turn his own rejection of reason back on him. Reason, he asserts, is the root of the problem – give it an inch and there is no stopping it: Christian faith is done for. There is, of course, a rich Catholic tradition which takes a very different view of the matter. But consider that, having repudiated reason itself, he tries to *reason* us into agreeing with his interpretation of the last 500 years. That he does a bad job of this is the least of his problems. By his own account, he has no business even attempting the task. After all, either he (with Luther) is right, and reason is the Devil's whore, or else they are wrong, in which case the entire polemic against reason and humanism is misconceived.

Our choice as a civilisation is between Christ and reason, the 'I Am' of the God of the Bible or the 'I am' of the human individual, Carroll asserts.[12] 'The early men of the Renaissance were not aware that they would have to choose,' he comments. 'They were Christians. The most instructive example, Erasmus himself, tried in his moderate Christian humanism to adapt his religion to the methods of the new secularism. It took Luther to smell a rat ... When Luther said to Erasmus ... "You are not devout!", he had, philosophically speaking, hit the nail on the head. He had prophesied the inevitable path of humanism once it had chained itself, as it must, to a belief in free will. This simple and direct, uncouth German peasant had told the most refined, best educated, wittiest and most eloquent man of his time, a man he admired, "You stand on nothing." '[13]

This is Carroll's most constant refrain: humanism stands on nothing – and we have to have a place to stand. He asserts, with Luther, that the place to stand is unreasoning faith in Christ, which he freely describes as 'the darkness of faith'. Yet he appears to expect either that we shall agree with him because of the reasons he provides, or attempt to reason him out of his stand. He provides us with a superficially impressive display of knowledge, only to declare that 'The path of knowledge leads in circles, spiraling down – into the heart of darkness.'[14] If, however, knowledge leads us inexorably on such a downward spiral, why does he take us on that path? If, conversely, as he seems to imply by his use of it, knowledge can lead to enlightening insights, why does he assert that it leads only to the heart of darkness?

In short, we could dismiss Carroll out of hand because of his radical intellectual incoherence. But consider the case he makes that the past 500 years have seen a rebellion against Christianity, with the consequence that our whole culture is now in ruins. For if such a case could plausibly be made, even if Carroll himself makes it badly, and even though he undermines his own case by repudiating reason and knowledge, it would be useful to respond to it. Otherwise, we run the risk of a Christian counterpart to Islamism arising in our midst, based on the proposition that ours is an apostate culture and must be overturned in the name of revealed truth.[15]

Throughout his book, Carroll seems to imagine himself to be presiding over the funeral of 'Caesar', by which he fairly clearly means modern humanism. 'A requiem must be sung,' he writes in his prologue, 'one that gets the story right, in all its magnificence and its meanness. We come less to honour Caesar than to bury him, that there be no mistaking that he is dead, that we understand him so as not to choose him again.'[16] This is, surely, not merely an appropriation of Shakespeare's famous setting of the funeral of Caesar, but also a deliberate inversion of celebrated passages in Hegel, Heine and Nietzsche which proclaimed the death of God. But what does he mean by 'getting the story right'?[17]

He means, if I have understood him at all, telling the truth about the history of ideas from Leonardo da Vinci and William Shakespeare to Charles Darwin, Sigmund Freud and Albert Einstein. He also means telling the truth about the consequences of those ideas in the 'real world' – the world of material life and social institutions. He thinks he knows the truth in both cases and can rationally compel us to agree with him. He has, of course, no place to stand in this regard. But that aside, the story he tells about our culture is unconvincing.

The truth in the first case, he argues, is that humanism's idea of the self-sufficiency of the human individual is negated by the reality of death. 'Without God, without a transcendental law, there is only death,' he declares.[18] This is quite a characteristic statement by Carroll, but we have to work hard to ascertain what precisely he means by it. His interpretation of Holbein's *The Ambassadors* and Shakespeare's *Hamlet* suggests that he means without God, life is futile, because it ends in death.[19] He places particular emphasis on *Hamlet* in this regard. Shakespeare, 'the greatest of all humanists,'[20] he writes, was the principal teacher of the futility of humanism and *Hamlet* 'the hub of Shakespeare's whole work.'[21]

Hamlet is death obsessed, Carroll suggests, and is 'obliteratingly alone.'[22] His dilemmas are those of the humanist individual and modern humanism has been obsessed with him. '*Hamlet, Hamlet and Hamlet* again – the West would come to know it by heart,' he writes.[23] Harold Bloom has a similar view of the importance of Shakespeare and of *Hamlet*. He even agrees with Carroll that Hamlet's fierce intelligence leads him to nihilism.[24] But Bloom has a far wider and less theologically frantic perspective on what we can learn from the great play and its central character. His view of our fate as mortal beings is elegiac. Carroll's might be called *Elijiac* since he seems, like the Old Testament prophet Elijah, to be bent on slaying the priests of Baal – the 'wreckers' of culture.[25] More generally, Carroll completely fails to address the question of why Christianity is any better a response to the reality of death than are any number of other sets of beliefs.

It is with regard to material life and social institutions, however, that Carroll most exposes how tendentious his interpretation of both past and present really is. In his preface, he states that *The Wreck of Western Culture: Humanism Revisited* 'has given stronger acknowledgement to the achievements of liberal democracy' than did his 1993 book, *Humanism: The Wreck of Western Culture*. He states:

> Humanism succeeded in building its city of light ... The wreck of humanist culture is in stark contrast to the physical edifice that its drive to know, channelled into science and technology, and applied in factories, has produced. Humanism's lasting achievement has been industrial civilisation and its brilliant triumph over most of the trials inflicted by age-old necessity: poverty, starvation, disease and brute labor ... Who in their right mind would give up clean water, sanitation and sewers, antibiotics, reliable supplies of varied foodstuffs, civic police, the jumbo jet, computers and skyscrapers in exchange for what came before – the filth, contagion and stench of medieval Europe?[26]

Given his repudiation of reason and knowledge as futile and as leading to the heart of darkness, this acknowledgement of the benefits of Western culture borders on the hilarious.

It brings to mind the celebrated scene in Monty Python's *The Life of Brian* in which Reg (John Cleese), leader of the Judean People's Front, asks his fellow malcontents 'What have the Romans ever done for us?' One after another, the others mention things the Romans have (allegedly) introduced to Judea: the aqueduct, sanitation, roads, irrigation, medicine, education, wine, public baths, public order and peace. The retort of an irritated Reg is classic, 'Yeah, alright, fair enough ... Alright, but *apart* from sanitation, the public baths, education, wine, public order, the roads, the fresh water supply and public health, *what* have the Romans *ever* done for us?'[27] In the case of Monty Python, the list of benefits brought by Rome is inaccurate in various respects, but Carroll's list of the benefits of modern humanism is perfectly accurate and could easily be extended. This makes it all the more bemusing that he takes the stand he does.

One could argue with Carroll endlessly about his various observations, but there is little point. He is lost in the labyrinth of the past and we need not chase him there.[28] For his assertion that humanism stands on nothing is empty. It is a mere metaphysical conceit. What humanism has, in fact, done is to open up the world to our understanding in ways never even remotely approximated by religious 'revelation'. Has it wrecked our culture in the process? Not compared with anything *before* 1500.[29] On the contrary, it has made available to countless millions of people the riches of human culture across all the ages.[30] This has put a great deal, including Carroll's kind of Christian fundamentalism, into the melting pot, but it has no more wrecked Western culture than the cosmopolitanism of Athens and Rome 'wrecked' archaic Mediterranean culture.

All that said, there are more interesting speculations on the condition and possible fate of Western culture, and even human civilisation as a whole, than Carroll's. One of them is a little book written more than thirty years ago by Carroll's own mentor, George Steiner – *In Bluebeard's Castle*. Steiner ruminated on the possibility that scientific humanism would open up too much knowledge and bring self-destruction on humanity in the process. What he did *not* do was to quail at this thought and try to take refuge in some unreasoning and atavistic form of religion. He allowed at least two possible responses: stoic acquiescence or a 'Nietzschean gaiety in the face of the inhuman, the tensed, ironic perception that we are, that we always have been, precarious guests in an indifferent, frequently murderous, but always fascinating world.'[31]

Carroll fails to rise even to the level of his mentor, in allowing that there is more than one possible way to respond to how the world looks, at what Steiner dubbed 'this cruel, late stage in Western affairs'.[32] Yet even Steiner could have allowed other possibilities. There is no self-evident reason to believe that the stage we are at in Western or human affairs is either especially cruel or 'late'. St Augustine believed, 1,600 years ago, that he lived in the senescence of the world.[33] We have far more reason to believe that, God or nature permitting, we stand on the cusp of extraordinary possibilities.[34] Death – which, like sex, has been a pre-condition for the mutability and variety of life on our planet – does not make any of these futile. On the contrary, it gives them their savour and their magnificence.

6.   *Endnotes*

1   John Carroll, *The Wreck of Western Culture: Humanism Revisited*, Scribe, Melbourne, 2004, pp. 53-5.

2   John Dillenberger, *Martin Luther: Selections From His Writings*, Anchor Books, Doubleday & Co., New York, 1961, Introduction, p. xii.

3   Friedrich Nietzsche, *Beyond Good and Evil*, 'Maxims and Interludes' #154, trans. R. J. Hollingdale, Penguin, 1973, p. 85. Consider, also, Nietzsche's remarks in his famous polemic against Christianity, *The Antichrist* (1888): 'Truth has had to be fought for every step of the way, almost everything else dear to our hearts, on which our love and our trust in life depend, has had to be sacrificed to it. Greatness of soul is needed for it: the service of truth is the hardest service. For what does it mean to be *honest* in intellectual things? That one is stern towards one's heart, that one despises 'fine feelings', that one makes every Yes and No a question of conscience! – Belief makes blessed: *consequently* it lies ...' #50. Or again: 'One should not let oneself be misled: great intellects are skeptics ... Convictions are prisons ... The believer is not free to have a conscience at all over the question "true" and "false": to be honest on *this* point would mean his immediate destruction. The pathological conditionality of his perspective makes of the convinced man a fanatic – Savonarola, Luther, Rousseau, Robespierre, Saint-Simon – the antithetical type of the strong, emancipated spirit. But the larger than life attitudes of these sick spirits, these conceptual epileptics, impresses the great masses – fanatics are picturesque, mankind would rather see gestures than listen to *reasons* ...' #54.

4   Carroll, *The Wreck of Western Culture*, Prologue, p. 1.

5   Jacques Barzun, *From Dawn to Decadence: 1500 to the Present*, HarperCollins, New York, 2000.

6   Harold Bloom, *The Western Canon: The Books and School of the Ages*, Harcourt, Brace and Co., New York, San Diego and London, 1994; and *Where Shall Wisdom Be Found?*, Riverhead Books, New York, 2004. In an interview following publication of *The Western Canon*, Bloom remarked, 'I am aware that I am fighting a rear-guard action and that the war is over and we have lost.' *Newsweek*, 7 November 1994, p. 62.

7   Henry Kissinger, in his study of the Congress of Vienna, rather curiously, extolled what he called Metternich's 'lucid and powerful' world view. 'Up to the sixteenth century, Metternich maintained, the forces of conservation and of destruction had been in an increasingly spontaneous balance. But then there occurred three events which, in time, caused civilisation to be supplanted by violence and order by chaos; the invention of printing and of gunpowder and the discovery of America. Printing facilitated the exchange of ideas, which thereby became vulgarized; the invention of gunpowder changed the balance between offensive and defensive weapons; and the discovery of America transformed the situation, both materially and psychologically. The influx of precious metals produced a sudden

change in the value of landed property, which is the foundation of a conservative order, and the prospect of rapid fortunes brought about a spirit of adventure and dissatisfaction with existing conditions. And then the Reformation completed the process by overturning the moral world and exalting man above the forces of history.' *A World Restored: The Politics of Conservatism in a Revolutionary Age*, Universal Library, Grosset and Dunlap, New York, 1964, p. 201.

8   Carroll, *The Wreck of Western Culture*, p. 162.

9   One thing which springs to mind in this context is the notorious *Syllabus of Errors* promulgated by Pope Pius IX (1792-1878, Pope 1846-1878), in December 1864 – 140 years ago. In it, the Pope denounced eighty common ideas of the Enlightenment as 'errors'. It would be of interest to ask Carroll which of these ideas he, too, would denounce as errors and which of them he would endorse against the teaching of the Vatican. They included the following: '14. Philosophy is to be treated without taking any account of supernatural revelation; 15. Every man is free to embrace and profess that religion which, guided by the light of reason, he shall consider true; 18. Protestantism is nothing more than another form of the same true Christian religion in which form it is given to please God equally as in the Catholic Church; 21. The Catholic Church has not the power of defining dogmatically that the religion of the Catholic Church is the only true religion; 23. Roman pontiffs and ecumenical councils have wandered outside the limits of their powers, have usurped the rights of princes, and have even erred in defining matters of faith and morals; 55. The Church ought to be separated from the state and the state from the Church; 76. The abolition of the temporal power of which the Holy See is possessed would contribute in the greatest degree to the liberty and prosperity of the Church; 77. In the present day it is no longer expedient that the Catholic religion should be held as the only religion of the state, to the exclusion of all other forms of worship; 80. The Roman Pontiff can and ought to reconcile himself and come to terms with progress, liberalism and modern civilisation.' Walter Kaufmann, *Religion From Tolstoy to Camus*, Harper Torchbooks, New York, 1961, 1964, pp. 163-170.

10   Edward Gibbon, *The Decline and Fall of the Roman Empire*, Everyman Library, J. M. Dent and Sons, London, 1969, vol 5, pp. 345 6. Gibbon rejected the story on several grounds: that it was not related by the Christian sources one would expect to have testified to it; that such an action would have been contrary to Muslim practice which was that 'works of profane science, historians or poets, physicians or philosophers, may be lawfully applied to the use of the faithful'; and that there is evidence of gradual destruction of the library by Roman or Christian actions centuries before Amrou's armies stormed the city in 638 CE. He went on to observe: 'I sincerely regret the more valuable libraries which have been involved in the ruin of the Roman empire; but when I seriously compute the lapse of ages, the waste of ignorance, and the calamities of war, our treasures, rather than our losses, are the object of my surprise [p. 347]'. For a brief history of the library of Alexandria, see Luciano Canfora, *The Vanished Library*, Vintage, London, 1991.

11   For a careful reflection on the struggle between the darkness of faith and the uses of reason in the millennium and a half before Erasmus and Luther, see Charles Freeman, *The Closing of the Western Mind: The Rise of Faith and the Fall of Reason*, William Heinemann, London, 2002. It is also well worth reading Richard E. Rubinstein's *Aristotle's Children*, Harcourt Inc, 2003.

12   For a very different understanding of who 'Christ' actually was – one which relies on painstaking scholarship and reasoning and which puts in serious question all Christian claims about the man from Galilee, see Geza Vermes, *The Authentic Gospel of Jesus*, Allen Lane, Penguin, 2003. Vermes is of Hungarian origin and is Professor Emeritus of Jewish Studies at Oxford University, as well as Director of the Forum for Qumran Research of the Oxford Centre for Hebrew and Jewish Studies. This is his fifth book on Jesus.

13   Carroll, *The Wreck of Western Culture*, p. 4.

14   Ibid., p. 199.

15   David Daniell dedicates his superb work of humane scholarship, *The Bible in English* (Yale University Press, 2003) 'To the memory of William Tyndale, 1494-1536, translator of genius, martyred for giving English readers the Bible from the original languages.' The freedom to inquire and not merely accept ecclesiastical authority or some kind of literalist 'revelation' was hard won and Western culture would, indeed, be wrecked if this freedom was to be undermined.

16   Ibid., p. 2.

17   In his *History of Religion and Philosophy in Germany*, published in 1834, Heine wrote: 'We will speak of this catastrophe, the downfall of deism, in the next instalment. A strange misgiving, a mysterious reverence, forbids us to write further today. We are rent with the most terrible pity. It is Jehovah himself who is preparing for his death. We have known him so well, from his cradle onwards in Egypt, where he was brought up amidst divine calves, and crocodiles, sacred onions, ibises and cats. We saw him say farewell to the play-fellows of his childhood, and to the obelisks and sphinxes of his native Nile valley to become a little god-king in Palestine to a tribe of poor shepherd folk where he lived in a palace of his own. We saw him later coming into contact with the Assyrian and Babylonian civilisations and putting off his all-too human passions [a title later used by Nietzsche]; no longer fulminating wrath and revenge at every turn, or at least not thundering about every paltry trifle. We saw him emigrate to Rome, the capital, where he renounced all his national prejudices and proclaimed the divine equality of all peoples, and organized an opposition against Jupiter with such fine phrases and went on intriguing so long that at last he came into power and reigned over the city and the world-urbem et orbem-from the Capitol. We saw him becoming more and more spiritual; whimpering tenderly, a loving father, a friend of humanity, a universal benefactor, a philanthropist at the last. And nothing could save him. Do you hear the passing bell? Kneel down. They are bringing the sacraments to a dying god.'

18   Carroll, *The Wreck of Western Culture*, p. 32.

19   Ibid., p. 33: 'The pursuit of knowledge is futile. What is the point if it provides no defence against the skull? Plotting the motion of the stars will not help to provide direction in life. Playing the lute will not soothe raw nerves once the whiff of a corpse has penetrated the nostrils, not unless the music intimates a greater frame, one within which the human individual can stand. The greatest of all humanist institutions, the university, is a mausoleum of dead ideas, a rattling of dry bones. Its teaching is incapable of reaching out to hold the hand through the darkness. Holbein has put it with brutal simplicity: there is no humanist solution. The most learned men have no answer to death. Once faith is gone, fate is reduced to necessity – and the ultimate necessity is death.'

20   Ibid., p. 45.

21   Ibid., p. 38.

22   Ibid., p. 43.

23   Ibid., p. 50.

24   Harold Bloom, *Hamlet: Poem Unlimited*, Riverhead Books, New York, 2003.

25   1 Kings 18:40, 'And Elijah said unto them, "Take the prophets of Baal; let not one of them escape." And they took them: and Elijah brought them down to the brook Kishon, and slew them there.' *Holy Bible: King James Version*, Salt Lake City, Utah, 1979, p. 499.

26   Carroll, *The Wreck of Western Culture*, p. 8.

27   *Monty Python's Life of Brian*, Handmade Films, 1979, directed by Terry Jones. DVD Criterion Collection 1999.

28   'Few are made for independence – it is a privilege of the strong. And he who attempts it, having the completest
       right to it, but without being compelled to it, thereby proves that he is probably not only strong but also daring
       to the point of recklessness. He *ventures into a labyrinth*, he multiplies by a thousand the dangers which life as such
       already brings with it, not the smallest of which is that no-one can behold how and where he goes astray, is cut off
       from others and is torn to pieces limb from limb by some cave minotaur of conscience. If such a one is destroyed,
       it takes place so far from the understanding of men that they neither feel it nor sympathize – and he can no longer
       go back! He can no longer go back even to the pity of men!' Nietzsche, *Beyond Good and Evil*, #29, p. 42.

29   The work of Norman Cohn offers rich insights into the religious culture and psychology of pre-1500 Europe. He
       draws particular attention to the roots of apocalyptic thinking and fear of demons and witches in both Biblical
       and popular religion. See his *The Pursuit of the Millennium* (Paladin, 1970), *Europe's Inner Demons* (Paladin, 1975)
       and *Cosmos, Chaos and the World to Come* (Yale Nota Bene, Yale University Press, 2001).

30   Walter Kaufmann, *Religions in Four Dimensions: Existential, Aesthetic, Historical, Comparative* (Readers Digest Press,
       New York, 1976), is far more enlightening and inspiring than any dogmatic scholarship. His *Critique of Philosophy
       and Religion* (Princeton University Press, 1978) is a classic. But something like Roy Rappaport's *Ritual and Religion
       in the Making of Humanity* (Cambridge University Press, 1999) is only possible because of humanistic scholarship and
       demonstrably transcends the dogmatic claims of any particular religion.

31   George Steiner, *In Bluebeard's Castle*, Faber and Faber, London, 1971, p. 106.

32   Ibid., loc. cit.

33   'Augustine thought of himself as living in the Sixth, the last, the old Age of the World. He thought of this not as a
       man living under the shadow of an imminent event; but rather, with the sadness of one for whom nothing new could
       happen. All that needed to be said had been said: a man is old at sixty, Augustine thought; even if he drags on, as some
       had done, to one hundred and twenty. It is futile to calculate the end of the world: for even the shortest spell of time
       would seem too long for those who yearned for it.' Peter Brown, *Augustine of Hippo*, Faber and Faber, London,
       1967, p. 296.

34   'With the strength of his spiritual sight and insight the distance and as it were the space around man continually
       expands: his world grows deeper, ever new stars, ever new images and enigmas come into view. Perhaps everything
       on which the spirit's eye has exercised its profundity and acuteness has been really but an opportunity for its exercise,
       a game, something for children and the childish. Perhaps the most solemn concepts which have occasioned
       the most strife and suffering, the concepts "God" and "sin", will one day seem to us of no more importance than
       a child's toy and a child's troubles seem to an old man – and perhaps "old man" will then have need of another
       toy and other troubles – still enough of a child, an eternal child!' Nietzsche, *Beyond Good and Evil*, #57, p. 64.

Bible legend states that the trouble started after Eve ate the Golden Apple of Discord. This was the forbidding fruit. An angry God sent his wraith. Man fell from the space of grace. It was mostly downhill skiing from there.
- Anon.

Plato invented reality. He was teacher to Harris Tottle, author of The Republicans. *Lust was a must for the Epicureans. Others were the Vegetarians and the Synthetics, who said, 'If you can't play with it, why bother?'*
- Anon.

Diderot became a famous encyclicalist. Voltare wrote a book called Candy *that got him into trouble with Frederick the Great, who is credited personally with increasing the population of Prussia by almost a third during his lifetime. Rousseau wished only to unchain the normal savage.*
- Anon.

7. On the Common Ignorance of History

It is often lamented that we never learn from history. But what history do we know in any collective sense? Consider the following little gems as indications of the common ignorance of history: 'History, a record of things left behind by past generations, started in 1815. Thus we should try to view historical times as the behind of the present. This gives incite into the anals of the past.' 'There was Upper Egypt and Lower Egypt. Lower Egypt was actually farther up than upper Egypt, which was, of course, lower down than the upper part.' 'Calvinists were the only ones who believed in predetonation. It is not surprising that their preaching consisted mainly of dogmatic explosions.'

If you've ever read W. C. Sellar and R. J. Yeatman's classic, *1066 And All That*, you'll recognise the basic style. If you've been a Goons fan all your life, or an *aficionado* of the Monty Python crew's fearless jousting with the absurd, you'll delight in the sheer inanity of such lines. They are not, however, lifted from the world of parody or the collected buffooneries of John Cleese and his comic colleagues. They come directly from term papers and blue book exams written by college and university history students in North America.

Did you know that 'Zorroastrologism was founded by Zorro' and that 'This was a duelist religion'? Well, you can be excused in that case, because the subject matter is admittedly obscure. Consider, however, the origins of the Biblical faiths. 'The history of the Jewish people begins with Abraham, Issac and their twelve children. Judyism was the first monolithic religion. It had one big God named "Yahoo". Old Testament profits include Moses, Amy and Confucius, who believed in Fidel Piety. (One of the only reasons Confucius was born was because of a Chinese tradition.)'

Now, ask yourself honestly, whether you ever really knew these things. If you didn't, you've obviously been missing out on the absolutely standard North American high school education in recent decades. All the above pearls of wisdom come from a wonderful little book called *Non Campus Mentis: World History According to Uni Students*, which is simply the real thing edited into an hilarious anthology by Professor Anders Henriksson, an historian at a place called Shepherd College, a four-year West Virginia state college, not far from Washington D.C.[1]

Henriksson has been collecting and publishing material of this nature since 1983. This, however, is what must be regarded as his *piss de resistance*. It's hysterically funny and deserves instant classic status alongside Sellar and Yeatman. As editor, he has, he tells us, 'taken the liberty of weaving their sentences into a more or less coherent fabric, but the words belong to them. The spelling may be avant garde and the logic experimental, but no one can fault these young scholars for lack of creativity.' As witness to this, consider the observation of one of his contributors: 'From the secondary sources we are given hindsight into the future. Hindsight, after all, is caused by a lack of foresight.' Experimental indeed.

The book begins from first principles and then takes us on a guided tour of world history, assisted by excellent illustrations and three remarkable maps. We are instructed first of all concerning prehistory and the 'Stoned Age'. 'Prehistory, a subject mainly studied by anthroapologists,' we are told, 'was prior to the year 1500. When animals were not available, the people ate nuts and barrys. Social division of labour began when a tribe would split into hunters and togetherers. Crow Magnum man had a special infinity for this. Advances were most common during the inter-galactic periods.'

Thereafter, the narrative continues, 'Civilization woozed out of the Nile about 300,000 years ago. The Nile was a river that had some water in it. Every year it would flood and irritate the land. This tended to make the people nervous.' And of what did this civilisation consist, having woozed out of the Nile? Well, naturally, of the building of sacred structures and concern with salvation. Thus, 'The pyramids were large square triangles built in the desert. O'Cyrus, a god who lived in a piramid, would give you the afterlife if your sole was on straight.'

It was not only in Egypt, however, that civilisation woozed out before 1500. We learn that 'Babylon was similar to Egypt because of the differences they had apart from each other. Egypt, for example, had only Egyptians, but Babylon had Summarians, Acadians and Canadians, to name just a few.' Isn't that interesting? It gets better, though. It seems that the Sumerian culture 'began about 3,500 years before Christmas' and that at its height, it featured such illustrious characters as 'King Nebodresser,' who 'lived in a hanging garden to please his Hutterite wife' and 'Hammurabi' who 'was a lawyer who lived from 1600 BC to 1200 BC.'

One of the beauties of this little book is that it shows just how hugely entertaining history can be in the right hands. Even the most ancient history and Biblical history. Did you know, for example, that 'David was a fictional character in the Bible who fought with Gilgamesh while wearing a sling' and saved his people 'from attacks by the Philippines'? Have you ever heard of 'King Xerox of Persia,' or of Jason's 'hunt for the Golden Fleas'? No? Then you're missing out. Read this little book and catch up on what you failed to learn during a misspent youth.

I would be remiss, however, if I led you to believe that *Non Campus Mentis* was confined to history only of the kings and captains kind. It includes many other important subjects, each in its due place. For example, the origins of Western philosophy are admirably and concisely recounted. The Atomists, the Sophists and the beginnings of scientific method – i.e. never taking anything 'for granite when solving a problem' – are all summarised. Then we are sagely informed that the Pre-Socratics 'lived long before Plato and were not decisively influenced by his work.' A pity, perhaps, since 'Plato invented reality' and 'was teacher to Harris Tottle, author of *The Republicans*.' Socrates, meanwhile, 'was accused of sophmorism and sentenced to die', poor man, 'of hemroyds'.

Nor is literature neglected. 'Scrophicles', we are told, wrote a play called *Antipode* 'about a young girl, her boyfriend and her misfunctional family.' What is actually quite fascinating here, as with so much of the material in *Non Campus Mentis* is that there is a faint echo of the real in these unintentional parodies. Whoever it was that recalled Sophocles as Scrophicles and *Antigone* as *Antipode* had plainly had the real thing pass through his or her brain. It had simply left an indistinct impression, of a Harris Tottelian nature. The result, a burlesque of the classical, is itself an absolute classic.

The stunners just keep coming. 'Christianity', we discover, 'was just another mystery cult until Jesus was born. The mother of Jesus was Mary, who was different from other women because of her immaculate contraption.' If such lines had been uttered by Monty Python they would have been taken – as *The Life of Brian* was – for wilful blasphemy. Henriksson's innocents, on the other hand, offer them without the slightest malicious intent. This is actually their understanding of the way things were.

It does get difficult at times to sustain the belief that this is so. Consider the statement that 'Eventually Christian started the new religion with sayings like, "The mice shall inherit the earth".' Later Christians fortunately abandoned this idea.' Yet Henriksson collected the material himself and assures us that it is completely authentic. Sellar and Yeatman surely invented much of their delightful take on British history, but Henriksson assures us he is simply quoting the youth of America in relating that 'When they finally got to Italy, the Australian Goths were tired of plungering (sic) and needed to rest. Italy was ruled by the Visible Goths, while France and Spain were ruled by the Invisible Goths' and 'During the Dark Ages it was mostly dark.'

This sort of priceless little blooper litters the pages of *Non Campus Mentis*. There are so many of them in fact that I found myself nearly choking with laughter while trying to read the book over lunch the day I bought the book and laughing all but uncontrollably for page after page once lunch was safely ingested. Most of the excerpts I've quoted so far are from the first thirty-five pages, but it runs to more than 140 pages and the hilarity is unrelenting.

Did you know that 'In the 1400 hundreds most Englishmen were perpendicular'? Or that John Huss was 'burned as a steak' because he 'refused to decant his ideas'? Or that 'Victims of the Black Death grew boobs on their necks'? Did you know that, during the Renaissance, 'It became sheik to be educated'? That 'Henry Bourbon married Edict of Nantes and became King of France with the promise to reconstipate the country to Catholicism'? I kid you not. Or that 'The German Emperor's lower passage was blocked by the French' – presumably reconstipating him as effectively as ever France was by Catholicism.

The scatolological aspect of history is further developed with the revelation that the Thirty Years War 'began with the Defecation of Prague'. But the tone improves when it comes to the capacity for movement brought about by the industrial and scientific revolutions. The nineteenth century 'was called the Roamantic Age, because everybody moved around', in case you hadn't known that before. And social mobility was such that 'successful businessmen could be raised to the porridge'. No doubt this was especially the case with those who had successfully sown their wild oats.

Such mobility contributed to political reform. The Chartists, for instance, demanded 'universal suferage and an anal parliament.' Most people would agree, I think, that they achieved both of these aims in the long run. They also wanted voting to be 'done by ballad', a reform of the political process which has yet to be brought into being. Meanwhile, similar sentiments must have been at work in the Vatican more than a century ago, since Pope Leo XIII reportedly issued 'a book of conservative ideas' called not *Rerum Novarum*, but *Rectum Novarum*. Truly.

All this anal progress, year by year, was not without its difficulties. It transpires, for example, that the American Civil War 'began in 1830' when 'slavery spread its ugly testicles across the West'. Such outrages were common in that era, an age in which 'Most Englishmen believed in the missionary position' and when European nations in general 'tried to take over other lands' in order to 'gain so-called "cleavage"'. This was the era of 'Rudward Kissinger's 'The White Man's Burden''.

When we enter the twentieth century, the malapropisms and mental meandering become steadily more assonant in their asininity with those of George W. Bush. We discover, for example, that 'World War I broke out around 1912-1914. The deception of countries to have war and those who didn't want one led the countries of Europe and the world to an unthinkable war which became thinkable.' The British 'used mostly Aztec troops to fight at Gallipoli' and 'Florence of Arabia fought over the dessert'. Most unforgettably of all, however, 'When peace broke out the men excitedly relieved themselves wherever they were.' Have you ever read history as good as this?

There's ever so much more, but one of my favourite snippets is the account of how a depressed Hitler, in the last days of the Second World War, 'crawled under Berlin. Here he had his wife Evita put to sleep, and then shot himself in the bonker.' The where? The what? The bonker. Don't ever do it to yourself, because if you do you'll never again be able to spread your ugly testicles across the West. Hitler never has, anyway. On the other hand, we do still – some of us – listen to *Sleeping Beauty* and think of Evita, don't we? So, the tragedy of the Berlin bonker was not entirely without its Roamantic side.

Anders Henriksson is justly proud of his compilation. 'At the very least', he writes in his suitably brief postscript, 'this excursion into the past should disabuse anyone of the silly notion that history is stuffy and boring. Does it send any deeper message? Is it disturbing evidence of a generation raised in ignorance by incompetent schools and disengaged parents? Should we sound the alarms and, in imitation of the late Joseph Stalin, grab some handy "escape goats"? Maybe.' Deeper reflection, however, suggests a less punitive and more thoughtful response, he suggests.

Hasty, sweeping courses in world history, with their daunting swirl of strange names, dates, places and events, he argues, followed by cramming and disconnected from any sense of social meaning combine to make most learning for most students haphazard. It is these pressures, he thinks, that generate such 'mind-numbing absurdities' and 'bizarre free associations'. When you've stopped holding your sides and have wiped the tears of mirth from your cheeks, it is these basic realities of the mass production system of 'higher education' that you are left with. And that should give you pause. For here, as in America and Canada, those whose heads are full of these wonderful inanities are our fellow citizens and the voters who give us our anal parliaments. Read and be mortified.

7. *Endnote*

1   Anders Henriksson, *Non Campus Mentis: World History According to Uni Students,* Workman Publishing, New York, Hardie Grant Books, 2002.

Dar'st thou ayd mutinous Dutch,
and dar'st thou lay
Thee in woodden Sepulchers, a prey
To leaders rage, to stormes, to shot, to dearth?
Dar'st thou dive seas, and dungeons of the earth?
Hast thou couragious fire to thaw the ice
Of frozen North discoveries?
- John Donne[1]

When I contemplate the ardor with which the Anglo-Americans prosecute commerce, the advantages which aid them, and the success of their undertakings, I cannot help believing that they will one day become the foremost maritime power of the globe. They are born to rule the seas, as the Romans were to conquer the world.
- Alexis de Tocqueville[2]

The most powerful cause of the breakdown of the closed society was the development of sea-communications and commerce.
- Karl Popper[3]

8. On the Mastery of the Oceans

If history is important, the history of human maritime endeavour is among its more notable chapters because it was the means by which the world of ancient times became the modern world. 'Maritime supremacy is the key which unlocks most, if not all, large questions of modern history, certainly the puzzle of how and why we – the Western democracies – are as we are', as Peter Padfield writes.[4] From the sixteenth century, command of the seas gave the Atlantic European powers the capacity to break open the whole world for plunder, for trade, for science and for geopolitical mastery.

From the end of the seventeenth century through to the beginning of the twenty-first, that maritime supremacy has been the near monopoly of the English-speaking peoples. It has underwritten their defeat of other empires in all quarters of the globe, their immense commercial prosperity, their pre-eminence in the advancement of the natural sciences and their development of systems of constitutional government which remain the envy of all oppressed peoples.

The critiques of Western imperialism since the nineteenth century have concentrated on the plunder of the globe by the Europeans, beginning with the appalling ransacking of the Americas by the Spanish and Portuguese in the sixteenth century and climaxing with the colonising of 'darkest Africa' in the late nineteenth century. Yet all the fruits of what we now summarise as 'globalisation' also have grown out of the world-opening epic of the modern era.

The emergence of a global order of commerce, science and law, not the plunder and rapine that have too often accompanied it, was the extraordinary achievement of the last few centuries. Maritime supremacy was the foundation of that achievement. English has been the language of its institutional articulation. Critical reason has gradually refined and civilised it, reducing plunder and inducing order. At the beginning of the twenty-first century, such maritime supremacy is exercised by the United States of America. Critical reason, however, has much work still to do in refining and 'democratising' the world order over which it presides.

The argument that there is a crucial link between the opening-up of the world and the exercise of naval power is not a new one. Karl Popper argued in the 1940s that 'the most powerful cause of the breakdown of the closed society was the development of sea-communications and commerce.'[5]

He identified this development with fifth-century BCE Athenian imperialism. Certainly it is the naval victory by Athens over the Persian Empire at Salamis in 480 BCE that has always been seen as the dawn of Western civilisation vis à vis the East. That triumph of Athenian arms was a bloodbath, in which tens of thousands of Persians and Persian slave-rowers perished, but it has ever since been 'synonymous with abstract ideals of freedom and "the rise of the West".'[6]

The most famous paean to those ideals is still the funeral speech delivered at Athens in the winter of 431 BCE by Pericles. 'Our fathers stemmed the tide of foreign aggression', he told his fellow citizens. 'Our greatness has since grown on the basis of a constitution which affords equal justice to all, a career open to merit and processes of vigorous and free discussion in public policy making. We throw open our city to the world and never by alien acts exclude foreigners from any opportunity of learning or observing (as was done regularly by Sparta), although the eyes of an enemy may occasionally profit by our liberality.'[7]

This 'open society' Pericles dubbed 'the school of Hellas'. He boasted that it would enjoy the admiration 'of the present and succeeding ages', but acknowledged that it had built an empire. 'We have forced every sea and land to be the highway of our daring', he declared. Athenian arms have enjoyed striking success, he told his listeners, because the free citizens of the trading state, unlike the slaves of unfree states, 'act boldly and trust in themselves', die resisting rather than live submitting, meet danger face to face and leave behind them, when they perish, 'not their fear, but their glory.' 'These take as your model', he exhorted his listeners, 'and judging happiness to be the fruit of freedom and freedom of valour, never decline the dangers of war.'[8]

It was also Greeks who provided the model and the methods of scientific inquiry which the modern world was to take so much further. 'It was the Ionian cities of coastal Asia Minor – all of them thriving ports – that became the hothouses of the new debate about the nature of the world' in classical times.[9] It was in the port city of Miletus, in the sixth century BCE, that Thales and Anaximander took the first deliberate steps towards the formulation of 'a new world view based not on myth but on natural science.'[10]

One heir to this early natural science was a maritime explorer called Pytheas, a fourth century BCE citizen of the Greek city of Massilia (Marseilles). Driven not by greed for gain or lust for power, but by scientific curiosity, he undertook a stunning journey into the Atlantic, around the coasts of ancient Britain, as far north as Iceland and the Arctic Circle and back down the 'Amber Coast' of north-western Germany, in the 320s BCE. Pytheas is now regarded as having completed a journey of scientific discovery 'more fruitful than any preceding the age of Henry of Portugal' 1,750 years later. He has been described as a precursor of Columbus and Darwin, Cook and Galileo.[11]

Those four names bring us back to the opening-up of the globe and the human mind in the modern era which began with the revival of classical ideals. Columbus, Darwin, Cook, Galileo are names to conjure: heroes of modernity. Reading Cook's journals is still an astonishing journey of discovery.[12] His was not a voyage of plunder or trade, but of humanistic science. It was Cook who commented, in late August 1770, of the Australian Aboriginals that while 'they may appear to some to be the most wretched people upon Earth ... in reality they are far more happier than we Europeans ... They live in a Tranquillity which is not disturb'd by the Inequality of Condition.'[13]

It is still breathtaking to read his account of the ice floes in the deep southerly oceans near Antarctica. His mandate on his second voyage was, in fact, to venture as he saw fit with the intent of making 'further discoveries towards the South Pole.'[14] This he did, in his little wooden barque, *Resolution*. And so we find him, in the summer of 1772-3, confiding in his journal that the ice islands 'can only be discribed by the pencle of an able painter'. They were such as to fill the mind 'with admiration and horror. The first is occasioned by the beautifullniss of the Picture and the latter by the danger attending it, for was a ship to fall aboard one of these large pieces of ice she would be dashed to pieces in a moment.'[15]

Cook did more than any other explorer to chart the actual geography of the Pacific and Southern Oceans. His voyages are emblematic of science opening up a world otherwise confected from myth and fancy. Only half a century earlier, it had still been possible for Jonathan Swift to locate his fanciful islands of Lilliput and Blefuscu in the Indian Ocean south-west of Sumatra and in the vague vicinity of 'Dimens Land'; Brobdingnag on a peninsular extrusion of the continent of North America; Laputa, Balnibarbi, Luggnagg and Glubbdubdrib in the vicinity of a geographically contorted Japan.[16] Cook's voyages substituted actuality for fancy.

Cook's journals were prepared for publication in their present form by one J. C. Beaglehole of the Hakluyt Society. Nothing could be more appropriate, for it was Richard Hakluyt, a contemporary of Bacon and Shakespeare who, two hundred years before Cook, had first systematically compiled a scientific account of the voyages and discoveries of English adventurers. His compilation was enormous, extending to some 1,500,000 words; a veritable encyclopedia of the newly appreciated extent and strangeness of the sixteenth-century world. It was with such records in his keeping that Bacon himself remarked, in *The Advancement of Learning*, that the 'proficience in navigation and discoveries may plant also an expectation of the further proficience and augmentation of all sciences.'

All the characteristics of the newly invented capacity to break open the whole world were represented in Hakluyt's various informants. Plunder, trade, science and the quest for mastery were all active in sixteenth-century Englishmen. Nowhere is this more evident than in the fact that the first edition of Hakluyt's huge work was dedicated to Sir Francis Walsingham, Queen Elizabeth I's chief of secret intelligence.[17]

Yet it was the *science* which made the enduring achievements of the epoch both extraordinary and fruitful. Hakluyt and his informants were the inheritors of Dutch ingenuity – that of Mercator and Ortelius – in generating 'an augmented picture of the world in the light of the new discoveries.'[18] Well after Hakluyt's time, in the seventeenth century, Holland, not England, was the centre of both commerce and intellectual liberty in Europe.[19] The augmented Dutch map of the world was an early fruit of this commercial and intellectual freedom. It is alluded to in Shakespeare's *Twelfth Night*, where Maria tells Sir Toby that Malvolio 'does smile his face into more lines than is in the new map, with the augmentation of the Indies.'[20] And what was 'the augmentation of the Indies'? Why, the Americas of course.

This 'augmented map of the world' was the chart of the beginnings of 'globalisation'. The scientific wonder of it stands out amid all the blood and greed and thunder of those far-off days. The world was becoming more *thinkable*, more *navigable*, not just for bodies but for minds. One of the earliest theorists of how all this could be turned to common advantage was Hugo Grotius, the great seventeenth-century Dutch humanist and pioneer of international law.

His arguments still provide the foundation for efforts to build a more rational, peaceful and prosperous world order. In the Americas, in African waters, in the East Indies, the Dutch, Padfield reminds us 'protected their monopoly spheres as forcibly as the Portuguese or Spanish ever had.'[21] Yet Grotius held out the possibility of a more open and correspondingly more prosperous world – a world of 'freedom of the seas.'

The Dutch did not get to enjoy 'freedom of the seas,' because England took over the seas in the eighteenth century. After that, until the mid-twentieth century, 'freedom of the seas' meant the *Rule Britannia*. Yet England's dominance was not due to guns alone. Its principles of enterprise, governance and finance time and again left the ambitions of its great continental rival, France, in tatters. English *liberty* grew with English commerce and inspired both the American and French revolutions.[22]

It is, in fact, the distinguishing characteristic of the British Empire that it spread liberty around the globe as no other empire had done. To be sure, this was by no means a uniform or always even intentional process, but it occurred all the same and without it there is no good reason to believe that the ascendant form of government in the present day world would be a liberal democratic one. Where British settlement and institutions took root, without too much admixture of illiberal continental cultures, as in the North American colonies, Canada, Australia and New Zealand, liberal democratic institutions also took root.

It is against this background that the twentieth-century rise and present overwhelming supremacy of *American* sea power needs to be considered. For much more self-consciously than Britain in its heyday, the United States of America declares itself to be the torchbearer of liberty and the champion of democracy around the world. The maritime power of the United States rose in the shadow of British naval supremacy from the 1880s to the 1930s. Then, as a direct consequence of saving the British Empire from the challenges of Germany and Japan in two world wars, it took Britain's place as the supreme naval and global power after 1945.

Just as Voltaire and Montesquieu had remarked on the virtues of England in the 1720s, so another brilliant Frenchman, Alexis de Tocqueville, remarked on those of the Anglo-Americans, a century later. Observing the United States in its Jacksonian prime, Tocqueville commented:

> When I contemplate the ardour with which the Anglo-Americans prosecute commerce, the advantages which aid them, and the success of their under-takings, I cannot help believing that they will one day become the foremost maritime power of the globe. They are born to rule the seas, as the Romans were to conquer the world.[23]

It was, in the light of Tocqueville's prescient remarks, fitting that the most famous theoretician of sea power in the twentieth century should have been an American, Alfred Thayer Mahan. His classic study, *The Influence of Sea Power Upon History, 1660-1783*, was published in Boston in 1890, the year that Congress approved the construction of America's first coal-powered battleships.

In the same year that he published his famous study of sea power, Mahan published an article headed 'The United States Looking Outward'. In authorising construction of the first battleships that year, Congress had deliberately limited the supply of fuel they would be able to hold, so as to limit their range to coastline defence. It wanted nothing to do with European-style imperialism. Mahan had much more robust views. In 1892, in an address to officers at the US Naval War College, he asked, 'All the world knows, gentlemen, that we are building a new navy ... Well, when we get our navy, what are we going to do with it?'[24]

His study of sea power had led Mahan to one irreducible conclusion: that everything depended on being able to concentrate enough naval fire power to destroy the enemy fleet. Herein lay the origins of the great white fleet which was to become the principal instrument for President Theodore Roosevelt's policy: speak softly but carry a big stick. Mahan's question, 'What are we going to do with our navy when we get it?' found its answer in the twentieth-century rise of America to precisely that maritime supremacy Tocqueville had foreseen in the 1830s.

As American naval power grew, however, the occasion almost never presented itself for practising Mahan's doctrine. In the First World War, the need was for convoy escorts and massive logistics capability, not for the battleships beloved of the Mahanians. In the Second World War, the need was for aircraft carriers and amphibious assault capabilities in the Pacific and, once again, in the Atlantic, for convoy and logistics capacity. In the Korean War, it was amphibious capabilities. In the Vietnam War carrier-based aircraft and gigantic logistics operations were the naval contribution. In the Cold War, the navy was again and again challenged to define its role, but it could not invent a plausible Mahanian one, because the Soviet navy was not itself developed on Mahanian lines. In the Gulf War, its role was again anything but Mahanian.

Mahanian or not, however, the American navy has become the most tangible embodiment of the Anglophone Western ascendancy, carried over from the twentieth century. Headquartered still at Pearl Harbor, the Pacific Fleet is the keystone in the Roman arch of the Pax Americana in the Asian Pacific world. This was demonstrated very deliberately in 1996, when President Clinton sent two aircraft carrier battle groups to the vicinity of Taiwan. This was in response to China's attempt to intimidate Taiwanese voters in the lead-up to what were, significantly, the first democratic Presidential elections in the history of the Chinese world.

All the themes of the long history of Western maritime supremacy were in play in this case. For China is a classic continental power. Like Persia it seeks to subdue 'Athenian' Taiwan. Like Spain it seeks to subdue 'Dutch' Taiwan. Ironically, Taiwan (Formosa) was briefly colonised by the Dutch in the seventeenth century. Like absolutist France, China seeks to challenge the maritime-based hegemony of 'English' America. And Taiwan is where this challenge is concentrated, as light is concentrated by a magnifying glass held up to the sun.

China is systematically building up its naval power – submarines, destroyers, strike aircraft, surveillance radars, anti-ship missiles – with a view to being able to deter or destroy American aircraft carriers and rapidly subdue Taiwan by force.[25] It will not attempt for the foreseeable future to develop the capacity to challenge the US Navy in Mahanian fashion – as the Japanese did in the Second World War. Rather it seems clearly to be working on the assumption that it can deter the US from intervening on a decisive scale, by threatening or even sinking an aircraft carrier very early in a sudden attack on Taiwan.[26]

Will China actually initiate such an attack? Will it do so once the 2008 Olympic Games are over? It seems very hard to believe. Yet the build-up is real and the strategic calculations of Chinese hardliners are, fairly clearly, to develop a genuine blue water naval capability. The so-called 'Blue Team' in the US, the American hardliners, are pushing hard for China's build-up to be met by US force deployments in East Asia and support for Taiwan. These are the dynamics of which conflicts are born: mutual suspicion, escalation in armaments, increased fear, misperception and miscarried 'pre-emption' – that is to say, acting 'while there is still time'.

No power in the present world has more to brood on concerning the rise of the West and its maritime supremacy than does China. A thousand years ago, it was the wealthiest and most technologically sophisticated state in the world. Six hundred years ago, its navies were larger and better equipped than any in Europe and sailed from the East China Sea to the Persian Gulf and the eastern coasts of Africa. But then China turned inward, just before Hakluyt's voyagers and Donne's hardy explorers turned outward to break open all passages to 'the Indies' – China being the ultimate goal. The Europeans conquered 'the Indies', then came to China and overpowered it.

Now China is coming out into the world again. Its more passionate citizens dream of its becoming once more the greatest state on earth.[27] In the way stands the United States of America and, right under China's nose, its supreme naval power – the maritime supremacy of 'the augmented Indies'. That is why China's naval build-up is no small matter. That is why Taiwan is no small matter. That is why we Australian children of Cook and grandchildren of Hakluyt, we great-great-grandchildren of Pericles and South Sea cousins of the Americans, have reason to think long and hard about where we stand and why, in the brewing Sino-American naval confrontation in the East China Sea. It is a good time to study both big history and Grotius.

8.   *Endnotes*

1   John Donne, *Satyre: of Religion*, in Helen Gardner (ed.), *The Metaphysical Poets*, Penguin, 1972, p. 48.

2   Alexis de Tocqueville, *Democracy in America*, vol. 1, Vintage Books, New York, 1980, p. 447.

3   Karl R. Popper, *The Open Society and Its Enemies*, vol. 1, Plato, 5th edn., Routledge, London, 1974, p. 177.

4   Peter Padfield, *Maritime Supremacy and the Opening of the Western Mind: Naval Campaigns That Shaped the Modern World*, 1588-1782, John Murray, London, 1999, p. 1.

5   Popper, *The Open Society and Its Enemies*, p. 177.

6   Victor Davis Hanson, *Carnage and Culture: Landmark Battles in the Rise of Western Power*, Doubleday, New York, 2001, p. 29. It was after this victory that Aeschylus wrote his tragic drama *The Persians*, in which he has the Persian chorus declare 'Long ago the heavenly Powers laid upon the Persian name terms: to seek on land her fame; din of horsemen, crash of towers, sack of cities – these were ours.' Philip Vellacott (ed.), *Aeschylus: Prometheus Bound, The Suppliants, Seven Against Thebes, The Persians*, Penguin 1961, p. 125.

7   Robert B. Strassler (ed.), *The Landmark Thucydides: A Comprehensive Guide to the Peloponnesian War*, Touchstone Books, Simon and Schuster, New York, 1998, pp. 112-113.

8   Ibid., pp. 114-115.

9   Barry Cunliffe, *The Extraordinary Voyage of Pytheas the Greek*, Allen Lane, Penguin, 2001, p. 27.

10   Ibid., p. 30.

11   Ibid., p. 169.

12   Philip Edwards (ed.), *The Journals of Captain Cook*, Penguin, 1999.

13   Ibid., p. 174.

14   Ibid., p. 221.

15   Ibid., p. 257.

16   Jonathan Swift, *Gulliver's Travels*, Oxford University Press, 1998.

17   In the 1580s, Hakluyt even approached Sir Francis Drake, plunderer of the Spanish treasure ships and circumnavigator of the globe, with the idea of his subsidising a chair at Oxford University in navigation.

18 Richard Hakluyt, *Voyages and Discoveries: The Principal Navigations, Voyages, Traffiques and Discoveries of the English Nation*, Penguin,1985. Introduction by Jack Beeching, pp. 12-15.

19 Padfield, *Maritime Supremacy and the Opening of the Western Mind*, Chapter 4, 'The Dutch Golden Age'.

20 *Twelfth Night*, act 3, scene 2, lines 79-81. Indeed, Shakespeare's oeuvre more generally reflects the rich age of opening and adventure in which he lived. Writing even of the documents Hakluyt compiled, Beeching comments: '... there is a perceptible enrichment of language. Cadence and vocabulary begin to reflect the intoxicating taste of success in complex endeavour. The English language is, visibly, becoming equivalent to Shakespeare's mind.' Loc. cit., p. 13.

21 Padfield, *Maritime Supremacy and the Opening of the Western Mind*, p. 83.

22 As Padfield points out (ibid., p. 175), both Voltaire (1726) and Montesquieu (1729) were amazed at the contrast between arbitrary monarchical authority and religious censorship in France and the freedom of expression and *habeas corpus* which prevailed in England.

23 de Tocqueville, *Democracy in America*, p. 447.

24 George W. Baer, *One Hundred Years of Sea Power: The U. S. Navy 1890-1990*, Stanford University Press, 1993, p. 22.

25 *Annual Report on the Military Power of the People's Republic of China*, June 2002, pp. 46-55.

26 Richard Fisher, 'To Take Taiwan, First Kill a Carrier', *China Brief*, vol. 2, issue 14, 8 July 2002.

27 Paul Monk, *Thunder from the Silent Zone,: Rethinking China*, Scribe, Melbourne, 2005, esp. Part One: Grand Strategic Perspective, pp. 11 – 64.

Supercelestial thoughts and sub-terrestrial conduct are two things, let me tell you, that I have always found to agree very well together.
- Montaigne[1]

I was driven from the Church by the strangeness of its dogmas and the approval and the support which it gave to persecutions, to the death penalty, to wars and by the intolerance common to all sects ...
- Tolstoy[2]

It is not their love for men, but the impotence of their love for men which hinders the Christians of today from – burning us.
- Nietzsche[3]

9. On Religion and Coercion

We may not realise it yet, but the great and now rather dated experiment in radical secularism is ending', wrote Sydney's Catholic Archbishop, Cardinal George Pell, in a 2002 essay.[4] To cure the ills of secular society, he argued, we 'need to get religion.'[5] This last statement warrants serious reflection. For, while secular society certainly has its ills, as we conceded in reflecting on the anti-humanism of John Carroll, the prescription of religion as a cure for them should be approached with great circumspection.

When Cardinal Pell declared that we need to 'get religion', we might have been forgiven for thinking that he meant Roman Catholicism, in particular. It could be that he believed Judaism, Protestant Christianity, Islam or Mormonism, Buddhism or Hinduism would serve almost as well. What claims he would make for *other* religions, however, he would certainly have wanted to make even more strongly for Roman Catholicism. It may, therefore, be considered the test case for his central claim. Do we, then, need to 'get religion'? And what would we be getting, if we did?

These very questions were addressed not long before Pell's essay was published, by Pell's co-religionist, George Weigel, a prolific and intensely serious American theologian. Weigel is the author of such books as *Catholicism and the Renewal of American Democracy, Freedom and Its Discontents: Catholicism Confronts Modernity* and *Soul of the World: Notes on the Future of Public Catholicism*. He is also the author of a massive and impressive biography of Pope John Paul II.[6]

Weigel is also the author of a book with the bold title *The Truth of Catholicism*.[7] It is dedicated 'to the faculty, staff and students of the Pontifical North American College in Rome, 1995-2000'. He headed his foreword to the book, 'An Invitation to Come Inside'. He doesn't indicate this, but any theologically literate person is likely to recognise in his title the famous – and disturbing – passages from the Gospels of Matthew and Luke about invitations to attend a great wedding feast.[8] Of these, Matthew is the more unsettling.

Here Jesus tells his disciples that the Kingdom of Heaven is like a wedding feast prepared by a king for his son, to which many people were invited, but declined to come. Some even set upon the king's messengers and beat and killed them. 'The king was very angry', the evangelist relates, 'so he sent his soldiers, who killed those murderers and burned down their city. Then he called his servants and said to them, 'My wedding feast is ready, but the people I invited did not deserve it. Now go to the main streets and invite to the feast as many people as you find'. So the servants went out into the streets and gathered all the people they could find, good and bad alike; and the wedding hall was filled with people.

'The king went in to look at the guests and saw a man who was not wearing wedding clothes. "Friend, how did you get in here without wedding clothes?" the king asked him. Then the king told the servants, "Tie him up hand and foot and throw him outside in the dark. There he will cry and gnash his teeth." And Jesus concluded, "Many are invited, but few are chosen."'

This scriptural passage, along with that from Luke, later formed the basis for debates about whether Jews and pagans should be *compelled* to accept the 'invitation' to become Christians or Catholics.[9] Specifically, those guests who, having initially been invited, declined to come to the feast and killed the king's messengers, all too readily equate to the Jews, whose refusal of the Christian 'invitation' festered in the minds of Popes and lay Catholics for two millennia until it erupted in the era of the Holocaust.[10]

There is, in short, quite a history behind George Weigel's innocent-sounding 'invitation to come inside'. It is a history which prompted Nietzsche to remark, in the 1880s, 'It is not their love for men, but the impotence of their love for men which hinders the Christians of today from – burning us.'[11] To 'get religion' in Cardinal Pell's sense means, whether he avows this or not, to 'get' this history. If as a Catholic or one in dialogue with Catholics one does not know that history, one simply does not know Catholicism; one lives, as Walter Kaufmann once phrased it, 'in a fool's paradise'.

Modern liberalism and secularism had their origins in prolonged attempts to break *out* of the 'wedding feast' and to get the Catholic Church to abandon all efforts at coercion in favour of a *genuine* invitation – which is to say one that, being open-ended, might be declined without causing rancour. For such rancour had produced both the Inquisition, from the thirteenth century, and the bloody wars between Catholics and Protestants of the sixteenth and seventeenth centuries. All too many people at that time 'got' religion. It was the great sceptic Michel de Montaigne who remarked, in the sixteenth century, however, 'Supercelestial thoughts and sub-terrestrial conduct are two things, let me tell you, that I have always found to agree very well together.'[12]

Propagating 'the truth of Catholicism', George Weigel asks: 'Is Jesus the Only Saviour?' To all those who aren't already loitering outside his Church's door, this must surely seem either an offensive question or simply a silly one: offensive if you happen to have strong religious beliefs of a non-Christian kind; silly if you are bemused by the whole idea that there is a need for a 'saviour'. Yet he proceeds to answer his own question in the manner of classic Jesuitical apologetics: 'God gives everyone the grace necessary to be saved, including those who have never heard of Jesus Christ. Yet everyone who is saved is saved because of what God did for the world and for humanity in Jesus Christ.'[13]

This is not a promising beginning to a book for the twenty-first century. It reads as if the past two hundred years of both liberal scholarship and physical science had never taken place. It reads as though the author is addressing barely literate American undergraduates who, in their innocence and immaturity, are trying to choose between Jesus, Marx and Buddha as masters of their souls. That twenty-first century world civilisation will require transcending all these old icons and actually creating new structures for responsibility and freedom is not something Weigel asks his readers to so much as consider. He simply wants them to 'get [his] religion'.

What George Weigel plainly believes is that a self-confident, dogmatic Catholicism has much to offer a world beset, as he sees it, by radical secularism. His book was spurred by a sense that Catholic spirituality offers a bright and bracing challenge to the 'brave new world' of 'rationally organised self-indulgence'. 'The brave new world', he asserts, 'is flat, painless, essentially carefree. The world of the saints is always craggy and sometimes painful; it includes dark nights of the soul as well as moments of ecstatic love.'[14] But this is a false dichotomy, if ever there was one and the 'brave new world' trope is rather a tired one.

It is clearly the case that the figure of Jesus still inspires many people. It is also very likely true that the Roman Catholic Church may have a remarkable role to play in the creation of a truly humane, global civilisation. Yet where that inspiration and that role are most to be admired is not where religion has been 'got'. It is, rather, where ethical integrity and compassionate imagination have triumphed over mundane motivation and narrowness of mind. And if, at Easter or Christmas, there is something to celebrate about the Christian tradition, it is not the 'truth' of the religion, but its too-often suppressed ontological commitment to *overcoming* 'religion' and setting the minds and hearts of humankind free.

George Weigel, like George Pell, loves the Roman Catholic Church. That's why they want the rest of us to 'get' their religion. That's why they invite us to come inside. And if the harsher aspects of its history and its doctrines are put aside, it can be admitted that the Church is an imposing edifice. Those who invite us into it still, in their own way, share the dream of those great Catholic figures of late Roman times who witnessed the Church's triumph over classical paganism and believed it, in Peter Brown's words, 'capable of bringing to the masses of the known civilised world the esoteric truths of the philosophy of Plato, a church set no longer to defy society, but to master it.'[15]

It is time, however, that they gave up this old dream. For the human world has changed immeasurably in recent centuries and the 'esoteric truths of Plato's philosophy' are no longer – if they ever were – what the 'masses of the known civilised world' stand in need of – to say nothing of the arcane dogmas of the Nicene Creed. As William James remarked nearly a hundred years ago, 'The philosophers are dealing in shades, while those who live and feel know truth ... They are [now] judging the universe as they have hitherto permitted the hierophants of religion and learning to judge them.'[16]

What has supplanted 'the esoteric truths of Plato's philosophy' since the path-breaking work of Lyell and Darwin, is the realisation that we are biologically evolved creatures, not lost souls. We have not been sent to this speck of earth as a place of trial and suffering before passing to a better (or much worse) place. A sophisticated understanding of cognitive archaeology and religious history can explain to us why there have been and are religions, but the 'truth' of religion has long since gone out the window.[17]

What we have left are *structures* and *practices* which represent much of our cultural patrimony and still contain profound challenges to us as beings, but whose *truth* claims disappear when we gaze into the depths of our Pleistocene origins. Consider that, to insist now on the 'truth' of one's religion is like insisting that one's native *language* is the 'true' language. If George Weigel had written a book called *The Truth of English* (never mind *The Truth of Latin*), we would be very puzzled by his title and would wonder what on earth he was on about.

Linguistics has enabled us, since the late eighteenth century, to gradually explore what language is and how many languages there are, but there is no such thing as a 'true (natural) language'. The term is a nonsense. One might consider English or Chinese to be a richer linguistic house to dwell in than, say, Hopi or Swahili, but that does not make the latter pair false.[18] Similarly, philosophical ontology and anthropology make it possible to explore what religion is and what William James called 'the varieties of religious experience', but there can no more be a true religion than there is a true language.

It goes without saying that there are hundreds of millions of human beings in the world today who would disagree with the proposition that there can no more be a true religion than there can be a true language. Cardinal Pell and George Weigel are, presumably, among them. However, if these hundreds of millions were to find a way to resolve their differences and agree on what the 'true religion' is, a philosophical method would plainly be required, which would transcend any of their particular dogmatic claims. Find such a method and you might save a great deal of mutual recrimination and frustration. You would also have shown that the way to truth is not via any one of the religions in and of itself. QED.[19]

In the meantime, what will be required is a set of rules to restrain the true believers of various faiths from vilifying, persecuting and even slaughtering one another. No better means has been discovered thus far than liberalism and the secular state. It was the rudiments of such 'liberal' or secular insight that prompted the Roman official Gallio to tell the Jews who wanted to punish St Paul for 'trying to persuade people to worship God in a way that is against the law' that theirs was 'an argument about words and names and your own law' and that he declined to judge such things.[20] It was also the great wisdom of the Jewish sage Gamaliel not to persecute the followers of Jesus, but to let them stand or fall by the merits of their own actions.[21]

When, therefore, Cardinal Pell declares that we need to 'get religion', or George Weigel invites us into the communion of the Roman Catholic faith as a cure for the presumed ills of our liberal secular society, we might respond by saying: if it is more ethical integrity you seek to inculcate in society, well and good. If it is greater compassionate imagination, that is admirable. If you seek, through the traditions of your religion, to enhance the general sense among our people that human life is rich in possibilities and that there is more to life than superficial hedonism suggests, then good for you.

If, on the other hand, all you seek to do is insist that your old creed is true and that ethics and imagination, possibility and human well-being are all to be dictated by it, then you are on an ontological treadmill and in no condition to 'save' anyone. The twenty-first century is bound to be an era of unprecedented challenges and extraordinary possibilities for humankind. You are welcome to contribute. At its best, your old traditions – your Bible, your music and art, your better austerities and solemn ceremonies – are a treasure for the ages and for the whole world, as Shakespeare is. Shakespeare's richness, however, does not make English the true language. Nor do your riches make yours the true religion.

Let's not pour new wine into old bottles; let's not write of dogmas and ceremonies as if human beings were made to subscribe to them, but rather the reverse. Who was it who said 'Man was not made for the Sabbath, but the Sabbath for man'? If being 'saved' means anything, let's consider that it means being open to possibility and full of the courage to let go of our dogmas. That way lies some form of resurrection from the deadliness of being whited sepulchres and fanatics about the 'law'.

9.   *Endnotes*

1   Michel de Montaigne, *Essays*, Penguin, 1958, 'On Experience', p. 405.

2   Leo Tolstoy, 'My Religion' (1884), in Walter Kaufmann (ed.), *Religion From Tolstoy to Camus*, Harper Torchbooks, 1964, p. 48.

3   Friedrich Nietzsche, *Beyond Good and Evil*, #104, Penguin, 1981, p. 78.

4   George Pell, 'The Failure of the Family', *Quadrant*, March 2002, pp. 16-22.

5   Ibid., p. 22.

6   George Weigel, *Witness To Hope: The Biography of Pope John Paul II*, Cliff Street Books, HarperCollins, 1999.

7   George Weigel, *The Truth of Catholicism: Ten Controversies Explored*, Cliff Street Books, HarperCollins, 2001.

8   Matthew 22:1-14; Luke 14:15-24.

9   St Augustine is often singled out for criticism in this regard. He does invoke the wedding feast from Matthew in talking about the winnowing out of the reprobates from within the Church in due course (*The City of God*, Book XVIII, Chapter 49, Penguin 1984, p. 831). For an intelligent discussion of his attitude to coercion, however, see Jean Bethke Elshtain, *Augustine and the Limits of Politics*, University of Notre Dame Press, 1995, pp. 94-101. See also Peter Brown, 'St Augustine's Attitude to Religious Coercion', *Journal of Roman Studies*, LIV, 1964, pp. 107-116.

10   David I. Kertzer, *The Popes Against the Jews: The Vatican's Role in the Rise of Modern Anti-Semitism*, Alfred A. Knopf, New York, 2001, makes this case compellingly.

11   Nietzsche, *Beyond Good and Evil*, p. 78.

12   Montaigne, *Essays*, p. 405

13   Weigel, *The Truth of Catholicism*, p. 6.

14   Ibid., p. 180.

15   Peter Brown, *Augustine of Hippo: A Biography*, Faber & Faber, London, 1969, p. 225.

16   William James, 'The Present Dilemma in Philosophy' (1906), *Pragmatism and Other Writings*, Penguin, 2000, p. 19.

17   Ian Tattersall, *Becoming Human: Evolution and Human Uniqueness*, Harcourt, Brace and Co., 1998, is a lucid introduction to the understanding of the nature of humankind achieved since Charles Darwin first raised the revolutionary idea that we had emerged from remote primate ancestors over millions of years.

18   In the 1930s Benjamin Lee Whorf famously believed that 'A change in language can transform our appreciation of the Cosmos' and this is true, also, of religion, but this only underscores that neither is, in itself, a vehicle for the truth about the Cosmos, only for a perspective on it. See John B. Carroll (ed.), *Language, Thought and Reality: Selected Writings of Benjamin Lee Whorf*, MIT Press, 1956. Languages also die. See Daniel Nettle & Suzanne Romaine, *Vanishing Voices: The Extinction of the World's Languages*, Oxford University Press, 2000.

19   QED: *quod erat demonstrandum*, 'that which was to be proved (or demonstrated)'. This Latin expression, particularly its abbreviated form, is a long-time favourite of logicians, including scholastic theologians, and was used here in a slightly ironic sense.

20   Acts of the Apostles 17: 12-17.

20   Acts of the Apostles 5: 33-39.

What we need now is good information and careful thinking, because in the years to come this issue [of climate change] will dwarf all the others combined. It will become the only issue. We need to re-examine it in a truly sceptical spirit – to see how big it is and how fast it's moving – so that we can prioritize our efforts and resources in ways that matter.

- Tim Flannery[1]

Each day decisions are made about global political priorities. We choose to support some worthy causes while others are disregarded. Unfortunately, political decisions seldom take into account a comprehensive view of the effects and costs of solving one problem in relation to another. Priorities are often set in an obfuscated environment involving the conflicting demands of the media, the people and the politicians. Despite all good intentions, the decision-making process is marred by arbitrary and haphazard methods.

- Bjorn Lomborg[2]

Consider a newspaper article in which it is reported that men with high cholesterol have a 50 percent higher risk of heart attack. The figure of 50 percent sounds frightening, but what does it mean? It means that out of 100 fifty year old men without high cholesterol, about 4 are expected to have a heart attack within 10 years, whereas among men with high cholesterol this number is 6. The increase from 4 to 6 is the relative risk increase, that is 50 percent ... The absolute risk increase is 2 out of 100, or 2 percent, no matter whether one counts those with or without heart attacks. Absolute risks do not leave room for playing with such numbers.

- Gerd Gigerenzer[3]

10. On Evidence and Apprehension

Once, during a workshop which had nothing specifically to do with religion, I was urged by a born again Christian to join his church because, he declared with radiant sincerity, the 'end times' are upon us and joining his church was the way to be saved. When I asked him why he believed that the 'end times' are upon us, he replied, 'The evidence is all around us.' The evidence? Well, yes: wars and rumours of wars, the looming danger of apocalyptic plagues and the omens of things portended long since in the Book of Revelation.

But these things have been all around us since before the Book of Revelation was written, I objected. What makes you interpret things *right now* as evidence that the 'end times' are upon us? He looked at me with the ardent eyes of the true believer, but offered no argument. What appeared to him a truth illuminated by signs and prophecies seemed to me to be a mere superstition, which is to say, an illusion based on perceiving a pattern that is not actually real.

Since our own culture has Christian roots, there are plenty of people who share or are willing to indulge this particular superstition and call it 'faith'. Fewer, I think, will be inclined to indulge, much less share, the belief of the Muslim enthusiasts around the current Iranian president that the Imam Mahdi is about to appear and overthrow the enemies of Allah. Yet that belief is based on just the same kind of thinking as the belief of born again Christians that the 'end times' are at hand.

What is especially interesting here, I suggest, is not any particular religious faith, but something more common in the way we human beings think. Whether with regard to religion or politics or everyday life, we look for patterns and apprehend outcomes in perceived sequences of events, we become emotional in regard to them and tell dramatic stories to ourselves and one another about them.

In a mnemonic formula, we are APES: Apprehensive, Pattern-seeking, Emotional, Story-telling creatures.[4] These characteristics make us very clever compared with other animals and enable us to share a great deal of information. Unfortunately, they do not reliably give us the truth. That requires painstaking unpicking and cross-examination of what we take to be patterns and probabilities. It requires discipline in governing our emotions and a willingness to radically revise our dramatic stories in the light of hard-won and provisional insights.

We are not, in general, very good at doing any of these things. There are good reasons for getting better at them right now, though; not simply because religious faiths are apparently on the rise again globally, but because religion entirely apart, it is all too easy to interpret the times we live in as 'apocalyptic' and we need to do a great deal of clear thinking about the patterns of events we are witnessing, or think we are, and what they portend

Whether with regard to climate change or the prospect of a global avian flu pandemic; terrorist strikes at critical infrastructure or at our cities with radiological bombs or biological toxins; the proliferation among rogue states of weapons of mass destruction, the 'failure' of the island states around our periphery, the global depletion of natural resources from fish to fossil fuels; or the sense that our society or our civilisation (perhaps all civilisation) is morally and politically decaying; we are surrounded by prophecies of gloom and doom which present themselves not as faith-based, but as both scientific and realistic. What are we to think? Or how are we to think about these things?

In 2006, I was asked by a research organisation called Australia 21 to think about these things with particular regard to Australia. The task was to reflect on how we think about actual or possible threats to the future security and prosperity of this country. Australia is, in a certain sense, a microcosm of the world at large. We are more fortunate than almost any other country in terms of our natural endowment, geo-strategic security and institutional soundness (economic, political and social). We are not, however, immune from the larger trends in the world and have begun to realise that our security and prosperity depend more and more on these larger trends.

What threats, then, can we at least imagine that we face or might face in a foreseeable time-frame? Apart from the large and dramatic ones mentioned above, there are, surely, quite a few others. Consider, for instance, the possible implications for Australia of a major war involving Iran and the disruption of global oil supplies, or a war in East Asia, whether on the Korean peninsula or across the Taiwan Strait.

Consider the possibility of a major economic or political crisis that seriously derails the rapid growth of China's economy and with it the resources boom that BHP Billiton and Rio Tinto understandably hope will go on, as Chip Goodyear and Ross Garnaut both forecast in 2006, for decades to come. Consider the possibility of a radical Islamic upsurge or even political ascendancy in Indonesia, or a significant economic recession in the United States, owing to a failure to correct long-standing patterns of dis-saving and deficit spending.

One could add to this list readily enough, but there are whole other categories of threat to consider. What kind of threat to our future security and prosperity is posed, for example, by the apparent generational decline in the quality of our educational institutions and the shrinking capacities of our universities to produce the scientists, engineers, linguists and even good teachers that we need in order to thrive? What kind of threat is presented by the sclerosis which seems to be affecting all the major political parties at the root and branch level? What of our health system, the endemic diseases of affluence and the aging of our population?

Once we begin to draw up a list of possible threats – and you may wish to draw up your own, as a kind of exercise – two tasks become quickly apparent. We need to group them and assign some kind of hierarchical order of importance or priority to them. One way to group them is to identify which ones are threats to the institutional fabric or underpinnings of Australian society; which are challenges arising from within Australia, that those institutions must manage; and which are possible shocks to Australia arising from wholly external sources.

The third category tends to seem the most dramatic, but the first is where attention needs, ultimately, to be focused, since it is the capacity of our institutions to cope with challenges and shocks that will determine how grave these actually turn out to be. This was very nicely illustrated by the manner in which the Australian economy rode out the Asian financial crisis in the late 1990s because timely and pro-active reforms had made our financial institutions and currency more resilient than they would otherwise have been.

Many countries demonstrably face extremely serious problems of the first and challenges of the second kind because they have corrupt, tyrannous or grossly incompetent governments or no effective government at all, or they suffer from widespread illiteracy, endemic disease, poorly developed infrastructure, rapidly increasing populations for whom there are too few jobs and so on In relative terms, Australia remains very much a 'lucky' country in these respects. But in reality, 'luck' has had very little to do with it. Australia has, on the whole, been very well managed and developed. Our collective well-being is merited, not merely fortuitous.[5]

Jared Diamond has laboured to draw attention to the failure of societies throughout history to cope with or adapt to threats to their viability in regard to resource usage and the maintenance of public infrastructure, or of climate shifts which they neither anticipated nor understood.[6] He singled out Australia as a 'First World' country that is already facing severe environmental problems – 'overgrazing, salinization, soil erosion, introduced species, water shortages and man-made droughts' – due to destructive use of its natural resource base.

Writing in 2004-05, Diamond was relatively optimistic about Australia's capacity to acknowledge and cope with these challenges because 'Australia has a well-educated populace, a high standard of living, and relatively honest political and economic institutions by world standards.'[7] If we can foresee threats to our future security and prosperity in this regard, they must, surely, arise from a sense that the standards of education are declining, or our standard of living is facing uncertainties or roadblocks, or that the relative honesty of our political and economic institutions is at risk.

There is another possibility, however: that none of these institutions has actually deteriorated in any absolute sense, but the challenges they face are becoming so stressful that they are beginning to struggle to cope with them. Or, again, that they are well enough suited to dealing with the challenges they face, but would be ill-equipped to cope with a shock of any one of the many kinds that may arise from beyond the country's shores. The impact of Hurricane Katrina on New Orleans and the failure of American emergency services to cope well with it provide something of a mental model in this regard.

While quite a long list of conceivable or latent 'threats' can be identified across the board, it is the 'big ones', the external shocks which seize the imagination. These are what economists refer to as 'exogenous shocks' – unexpected blows that overwhelm the institutions of societies otherwise coping more or less well with their standard challenges. Traditionally, the threat of foreign invasion has held the high ground in this category, but that has changed markedly over the past decade. It is the apparent *aggregation* of novel and ominous kinds of threats that can make the contemporary scene appear apocalyptic: wild terrorism, climate change, the danger of pandemic disease, alarm over the depletion of fossil fuel resources and the less comforting possible implications of the immense changes occurring in China.

Of all these novel threats, none seems more portentous than climate change. Tim Flannery is not alone in arguing that it confronts us all with an apocalyptic danger, though this proposition does not yet command universal, much less rational, assent. In a 2006 report for the Lowy Institute, Alan Dupont and Graeme Pearman urged that the Federal government and, indeed, Australian governments at all levels, set about 'thinking the unthinkable' as regards the future impact of climate change, but they were not very specific as to precisely how this could be done in a useful and economical manner.[8]

The threat of avian flu is, by some accounts, almost as apocalyptic. The difference is that its impact would be soon, sudden and relatively short term, rather than gradual, insidious and enduring. In another report for the Lowy Institute this year, Warwick McKibbin and Alexandra Sidorenko explored a number of scenarios for a possible global avian flu pandemic. They developed a model for estimating the economic costs and mortality rates of a pandemic at four different levels of severity. The four scenarios were based on previous epidemics and emerged from the model as follows:

Level	Flu Model	Global Mortality	Economic Cost
Mild	HK 1968-69	1.4 million	$330 billion
Moderate	Asian 1957	4.2 million	N/A
Severe	Spanish 1918-19	71.1 million	N/A
Ultra	Spanish 1918-19+	142.2 million	$4.4 trillion

They judged that the moderate scenario 'is probably the most likely outcome.'[9] Others are far more inclined to believe that an ultra scenario is all but upon us and that it could kill up to two or three times the maximum number estimated by the McKibbin/Sidorenko model.[10] McKibbin and Sidorenko allow that 'there are many unknowns in modelling pandemic influenza scenarios. There are very few observations in history to draw on and there is a great deal of uncertainty about how individuals and markets will respond.'

Even while estimating that a moderate scenario is more likely than a severe or ultra one, McKibbin and Sidorenko concluded their study with the remark, 'The extent of potential human and economic losses across the scenarios considered suggests that [a] large investment of resources should be dedicated to preventing an outbreak of pandemic influenza.' Yet they note that pandemics historically originate in eastern Eurasia, especially China; that the more severe the scenario the more skewed the costs are actually likely to be, with Africa and Asia hit far harder than North America, Western Europe or Japan. Whose resources, therefore, should be invested in preventing the outbreak? How large is a 'large investment in resources', given the enormous disparities between the scenarios and between the anticipated costs on a country by country basis? How would such an investment be made? These things they did not discuss.

The nightmare scenario, of course, is several of these threats becoming reality at the same time and compounding one another. Both the natural and historical worlds have known such things. Some 251 million years ago, the greatest mass extinction in the geological record nearly ended life on earth.[11] It has only been in the past ten years that a series of hypotheses has been advanced to account for it. There remain various uncertainties, but the best explanation seems to be that it was the consequence of three cataclysmic geophysical events occurring within a short space of time. First there was a gigantic flood basalt eruption which triggered a huge release of methane from the ocean floor as a result of global warming, thus increasing world temperatures cumulatively by 10 degrees Celsius. Then, altogether fortuitously, came the impact of an asteroid at least twice the size of that which wiped out the dinosaurs 65 million years ago.[12]

Suppose the calculations of the climate scientists to be broadly accurate and global warming to be occurring to an increasingly dangerous degree; suppose, further, that avian flu does break out in humanly transmissible form within the next few years and sweeps the world in an 'ultra' scenario, killing between 100 and 350 million people – starting with all AIDS sufferers, since they would be among the most vulnerable – and, at the same time, terrorists succeeded in obtaining nuclear weapons on the black market that Pakistan's A. Q. Khan did so much to establish and that they set off such bombs in New York, London and Singapore. What kind of world would Australia find itself in after the pandemic killed up to 1,000 times as many people as 9/11 and the nuclear bombs went off in three cities pivotal to the global economy?

Thoughts such as this prompt two sets of questions other than this last one, of course. The first set of questions has to do with how probable any one of these eventualities actually is and how probable the conjunction of them would be. The second set of questions has to do with how we could best either prevent any of them from occurring, or prepare against the consequences of them possibly occurring. What is vital is that they prompt us to think, rather than leap to conclusions about what must be done, or even what is likely to come to pass. It is, surely, arresting to consider that, having met in Copenhagen in 2004, and conferred at some length about where well-intentioned policy-makers around the world might most usefully commit resources for the common good, a group of distinguished economists did not include action on climate change, measures to head off avian flu or measures to secure all fissile material and nuclear weapons as among their top priorities.[13]

The heart of the problem here is psychologically integrating and mentally reconciling the relative probabilities, as well as the relative gravities, of various possible future courses of events. As a small illustration of how mentally interesting such challenges can be, consider the fact that, several years ago, 99 per cent success was declared in the recall and securing of the nuclear arsenal of the former Soviet Union. That sounds very impressive until one reflects that the arsenal in question had contained 20,000 strategic and 22,000 tactical nuclear weapons.

That 99 per cent of these had been secured meant that 'only' 420 of them had not been. This was one of the reasons both Western intelligence agencies and that eminent risk analyst Warren Buffett have declared that it is 'inevitable' that terrorists will obtain and use at least one nuclear weapon against a major Western city at some point in the next decade.[14] Yet American security specialist Graham Allison has described nuclear terrorism as a rigorously 'preventable' catastrophe. For a relatively modest expenditure of resources, he argues, all nuclear weapons and access to the key technologies could be secured against terrorists.[15]

This is not the place to discuss the merits of Allison's argument. Rather, it is important to register a few more fundamental cognitive challenges that confront all of us in trying to weigh up the range of threats or perceived threats we face and determine what it is most prudent or useful to do with regard to them. There are three core challenges here: assessing threats or risks in and of themselves; comparing the relative dangers that such threats entail; and correcting our own judgements, or misperceptions in the public mind or in policy circles with regard to threats.

Plainly, imagined threats of certain kinds can trigger responses which are either unwarranted or ill-advised, or both. Even where threats are reasonably well authenticated, however, weighing up their relative gravity and probability can encounter all but intractable problems, not simply because it is intrinsically challenging, but because so many parties to the debate will be affected by intuitive judgements that are not easily corrected by counter-intuitive reasoning. Just consider the case of those who believe that the 'end times' or the coming of the Imam Mahdi are upon us.

Such problems are accentuated by the reality that, even if the weighing is ever so well carried out by specialists, it will encounter the problem of intuitive fears and misapprehensions within the political elite and the general public. The propensity of human beings in general – and those not specifically trained in the disciplines of relative risk analysis in particular – to misconstrue or exaggerate the nature and severity of threats produces perverse incentives for political leaders and lobbyists to exploit public misapprehensions.

This is the gravamen of a recent argument that 'the creation and exploitation of irrational fears causes us to misdiagnose the problems we face, and to misdirect resources to programs and policies which have little or no benefit.'[16] However, the accusation that this is being done is too easy to level unfairly, given how genuinely difficult it is to assess threats accurately, to allocate resources proportionately in response, and to adjust policy rationally in 'real time'. APES, whether in political office or in the street, are naturally given to fear and misapprehension and are very susceptible to believing passionately in dramatic stories based on sometimes very dubious plausibility.

When it comes to weighing up and 'correcting' threat perceptions, truth and wisdom are seldom self-evident. We should not, therefore, be too hasty to assert that those who hold a contrary opinion to our own are either foolish or motivated by dubious agendas. We need, rather, to invent means to assess, compare and weigh threats more scrupulously and to communicate and correct our threat assessments and weightings more clearly than is almost universally the case. Do we face apocalyptic shocks to our security and prosperity in the near future or just over the horizon? Perhaps, but only the clearest and most careful thinking will give us a reasonable chance of first identifying and then, if need be, fending off the 'four horsemen' of the apocalypse and preserving our material civilisation against the various dangers posed by viruses and terrorists, an over-heating atmosphere or over-heated imaginations.

10.   *Endnotes*

1   Tim Flannery, *The Weather Makers: The History and Future Impact of Climate Change*, Text Publishing, Melbourne, 2005, p. 8.

2   Bjorn Lomborg (ed.), *How to Spend $50 Billion to Make the World a Better Place*, Cambridge University Press, 2006, p. xi.

3   Gerd Gigerenzer, *Reckoning With Risk: Learning to Live with Uncertainty*, Penguin, 2002, p. 205. Gigerenzer is Director of the Centre for Adaptive Behaviour and Cognition (ABC) at the Max Planck Institute for Human Development in Berlin.

4   I owe this rather charming acronym to the ingenuity of my friend and colleague Yanna Rider.

5   Thomas Barlow, *The Australian Miracle: An Innovative Nation Revisited*, Picador, Pan Macmillan, Sydney, 2006.

6   Jared Diamond, *Collapse: How Societies Choose to Fail or Succeed*, Penguin, 2005.

7   Ibid., p. 379.

8   Alan Dupont and Graeme Pearman, *Heating Up the Planet*, Lowy Institute, Sydney, 2006.

9   Warwick J. McKibbin and Alexandra Sidorenko, *Global Macroeconomic Consequences of Pandemic Influenza*, Lowy Institute, February 2006, p. 24.

10   Laurie Garrett, 'The Next Pandemic?', *Foreign Affairs*, July-August 2005, pp. 4-5 estimates that 'assuming a mortality rate of 20% and 80 million illnesses, the United States could be looking at 16 million deaths and unimaginable economic losses.' Michael T. Osterholm, 'Preparing for the Next Pandemic', *Foreign Affairs*, July-August 2005; Mike Davis, *The Monster at our Door: The Global Threat of Bird Flu*, New Press, New York and London, 2006.

11   Peter Ward, Professor of Biology at the University of Washington, Seattle, commented in early 2008 that there is a major caveat to this idea of 'life nearly ending': 'This cataclysmic event is often portrayed as the time when "life nearly died", but that is hardly fair. The oldest and most successful life forms on Earth – the bacteria and archaea – sailed through virtually unharmed. The Permian extinction period is better seen as a time when life almost went back to normal – when biological conditions that had prevailed on Earth for more than 3 billion years briefly re-established themselves. The microbes did not merely survive: it now appears that they played a leading role in the extinctions.' 'Precambrian strikes back', *New Scientist*, February 2008, pp 38 – 41.

12   Doug Erwin, *Extinction: How Life on Earth Nearly Ended 250 Million Years Ago*, Princeton University Press, 2006. See also Michael J. Benton, *When Life Nearly Died: The Greatest Mass Extinction of All Time*, Thames Hudson, London, 2003.

13   Bjorn Lomborg (ed.), *How to Spend $50 Billion to Make the World a Better Place*, Cambridge University Press, 2006.

14   Warren Buffett quoted in *Fortune* magazine, 11 November 2002, in an article by Andy Serwer titled 'The Oracle of Everything'.

15   Graham Allison, *Nuclear Terrorism: The Ultimate Preventable Catastrophe*, Times Books, Henry Holt and Co., New York, 2004.

16   Carmen Lawrence, *Fear and Politics*, Scribe, Melbourne, 2006, p. 5.

JS 455,*Chania*
Jörg Schmeisser, etching, 1991.

Part Two – The Classical Tradition

Why, it seems like only yesterday, or the day before, when our vast armada gathered, moored at Aulis, freighted with slaughter, bound for Priam's Troy.

\- Homer[1]

The end of the eastern Mediterranean Bronze Age, in the twelfth century BCE, was one of history's most frightful turning points ... Altogether, the end of the Bronze Age was arguably the worst disaster in ancient history, even more calamitous than the collapse of the western Roman Empire ... [T]he fall of [Troy] may have marked the beginning of the Catastrophe.

\- Robert Drews[2]

Troy was, perhaps, a client state of the Hittite empire, which was one of the chief Near Eastern powers at that time. This state was the cause of hostilities between Greeks and Hittites in the mid-thirteenth century BCE ... [T]he true background to the historical Trojan War ... can be adduced from first-hand primary sources, the diplomatic archives of the Hittite empire ... In short, the essential facts of Homer's story – the city, the location by the Dardanelles, the Greek expedition, the war – were all true.

\- Michael Wood[3]

1. On Myth and Pre-classical History

Wolfgang Petersen's 2004 Hollywood version of Homer, *Troy*, is the sword and sandal epic kind of introduction to ancient legend. It was not a serious effort to reconstruct for a twenty-first century world the tragic drama immortally penned by the poet 2,700 years ago. Rather, it seems to have been an attempt to use the distant and epic appeal of the legend of Troy to entertain an audience that flocked to films such as *Titanic, Pearl Harbor* and *The Lord of the Rings*. While this is commercially understandable, it made the film merely an ephemeral piece of entertainment. Given both the power of the legend and the realities of our time, Petersen could have made a much more powerful piece of theatre.[4]

Both Eric Bana (Hector) and Brad Pitt (Achilles) are reported to have said that, in preparing to play their heroic roles, they read *The Iliad* for the first time and were awed by Homer's poetry. One might hope that others who saw the movie were prompted to read *The Iliad* and were as stirred by it as were the two actors.[5] Having been so affected by the great poem of the storm of war, it would be interesting to learn what Bana and Pitt thought of David Benioff's screenplay or Wolfgang Petersen's direction of *Troy*.

Petersen and his team seem to have taken pains with some things, but they neglected or made a mess of others. Considerable efforts were made, for example, to have the sets for Sparta, Mycenae and Troy look authentic. But for some unaccountable reason, as the Greek armada sails east for Troy, the sun rises *behind it*. Indeed, it rises from the west throughout the film. Given that the film had begun with a map showing Greece to the west of Troy, you'd have thought something as elementary as this would have been taken into account.

Moreover, in legend, the siege of Troy lasted ten years. Petersen's film has the action over and done with in about three weeks, including a twelve-day pause for Hector's funeral. It is not at all apparent why. Less flagrant, but just as indicative of rather cavalier disregard for the realities behind the 'action', Petersen has Troy's main gate facing the sea. It didn't. The famous main gate of Troy was the Scaean Gate, which faced south – inland. There is no classical or archaeological evidence for a major gate facing the sea.[6]

Petersen and Benioff are just as capricious in deleting from their account of the Trojan War many of the dramatic details that have been the subject of drama and opera for two and a half millennia. The blood sacrifice by the Greek king Agamemnon of his own daughter, Iphigenia, at Aulis (so that the Greek fleet might have a fair wind for Troy),[7] the cutting of Polyxena's throat on the tomb of Achilles, the throwing of Hector's little son, Astyanax, from the city walls by the victorious Greeks – are all omitted. The audience is allowed to believe that Astyanax and his mother, Andromache, escape. Menelaus and Agamemnon, the 'bad guys' are both killed (by Hector and Briseis, respectively), contrary to the entire classical tradition.

Most melodramatically of all, Achilles, instead of being slain outside the city walls, runs through Priam's palace looking for Briseis amid the city's sack. This seems almost a direct reproduction of the scene in *Titanic*, in which Kate Winslett seeks desperately for Leonardo di Caprio as the ship fills with water. It is an extravagant concession to juvenile sentimentality and a betrayal of the grim spirit of Homer. What Petersen surrendered in crafting his film this way was the power of the great original to seize people by their throats and compel them to feel that surge of pity and fear which Aristotle believed was the purpose of tragic drama.[8]

Benioff's screenplay, unsurprisingly, has the same characteristics. It lacks gravity and caters too much to a superficial and mawkish taste. Perhaps its finest moment is where it draws most directly on Homer. The scene is that in which Priam comes to the Greek camp, kneels at the feet of Achilles, kisses his hands, then says gravely, 'I have endured what no-one on earth has ever done before – I put to my lips the hands of the man who killed my son.' Benioff here used the very words from Robert Fagles' acclaimed translation of *The Iliad*, Book XXIV, lines 590-91. In general, he does nothing of the kind.[9]

Benioff innovated with characteristic effect in having Hector and Achilles, in particular, consistently express scepticism about the gods. Whereas the gods are ever-present and active in *The Iliad*, in *Troy* they are absent. Their cults are depicted as harmless and colourful features of civic life, but they never intervene in response to invocations or blasphemies and are openly mocked by Achilles without the hero suffering for it. What is not clear, however, is precisely what the writer was trying to achieve by having the figures of legend exhibit this anachronistic scepticism.[10]

Hector, for example, is a model of good sense, but the underlying implications are only feebly followed through. After leading the Trojans to victory on the first day of the war, he counsels prudence on the second. 'The Greeks underestimated us yesterday,' he tells his father's war council. 'We should not return the favour.' But Priam's court seer declares he has seen an eagle soaring in the air with a snake clutched in its claws – a sign that Apollo will champion the Trojans in battle against the Greeks. 'Bird signs!' Hector exclaims in exasperation. 'You want to plan a strategy based on bird signs?!'

Was Benioff trying to make an anachronistic theological or philosophical point here? Or to poke fun at the WMD intelligence shambles in the war against Saddam Hussein? Or to highlight the tragic plight of Hector, caught up in a current he could not master and carried to his inevitable doom? The screenplay as a whole is too insubstantial for one to work out which, if any, of these possibilities was in Benioff's mind.

Had he wanted to challenge the delusions of ordinary people and their seers, he had a great classical tradition on which to draw. He could, for example, have studied Euripides' *The Women of Troy*, written and performed 2,420 years ago. Set after the fall of Troy, this play was a sombre reflection on the human catastrophe entailed in the sacking of cities. Its Athenian audience not only knew their Homer far better than a modern audience, but their soldiers had just that year (416 BCE) sacked Melos, killing its men and enslaving its women and children.

In Euripides, it is Hector's mother, Hecuba, who gives voice to Euripides' critique of the ancient gods – Hector being dead by the time the play begins. For some reason, Hecuba is entirely absent from *Troy*, as if Priam had been a widower. Yet she is a substantial figure in the classical legends, who exclaims, after the fall of the city: 'The man who finds his own wealth and security a cause for pleasure is a fool. Those forces which govern our lives are as unpredictable as capering idiots. Assured good fortune does not exist ... O dearest friends, I see the cold abyss of truth ... Gods, gods, where are you? Why should I cry to the gods? We cried out to them before and not one heeded us.'[11]

In such passages, Euripides, whom Aristotle called the most tragic of the great dramatists,[12] foreshadowed the darker passages in Shakespeare, notably in *Macbeth* and *King Lear*. Benioff, by contrast, throws in Hector's lines about bird signs like a mocking schoolboy, then lapses back into lines redolent more of *Days of Our Lives* than of Euripides or Shakespeare. Perhaps he simply was not capable of doing better. In any case, he and Petersen, in seeking to merely entertain an immediate public, sacrificed their chance to create something that would endure. After its run in the cinemas, the film almost immediately faded away into obscurity.

Some thirty years ago, Michael Cacoyannis filmed Euripides' play as *The Trojan Women*, starring Katherine Hepburn as Hecuba, Vanessa Redgrave as Andromache and the young Irene Pappas as Helen. That film was anything but 'entertaining'. It was intended as a reflection on the horrors of war and their tendency to lead to atrocities. Petersen, Benioff and their producers may well have decided that a film à la Cacoyannis simply would not sell, but they themselves sold out much of the moral force of the classical tradition.

Enough of Petersen and Benioff, though. As Troilus exclaims in Shakespeare's war-weary take on the legend of Troy, 'I cannot fight upon this argument. It is too starved a subject for my sword.'[13] Far more stirring and far more worthy of attention is the legend behind the film and, even more, the reality behind the legend. Shift from Petersen to Homer and your gain in the graphic grasp of the face of battle is immeasurable. Shift from Homer to history and you glimpse the very roots of Mediterranean civilisation, centuries before the founding of Rome.

It is a testament to the enduring power of Homer's writing that, even in the twentieth century, he should have remained the benchmark against which writing about war was measured, neither shrinking from its horror, nor denying its grim heroism. Ernst Junger's extraordinary memoir of the First World War, *Storm of Steel*, is a case in point. Both its warrior ethos and its unvarnished descriptions of violent death strike one again and again as 'Homeric'.

The very opening of the book takes the reader into a Homeric world – a world of clashing gods (overwhelming forces) and murderous furies: 'Full of awe and incredulity, we listened to the slow grinding pulse of the front, a rhythm we were to become mightily familiar with over the years. The white ball of a shrapnel shell melted far off, suffusing the grey December sky. The breath of battle blew across to us, and we shuddered. Did we sense that almost all of us – some sooner, some later – were to be consumed by it, on days when the dark rumbling yonder would crash over our heads like an incessant thunder?'[14]

Not for nothing do even more recent books on war in our time evoke Homer. Two very recent ones are Robert Kaplan's *Warrior Politics* and Philip Bobbitt's *The Shield of Achilles*. Both are concerned that the twenty-first century world faces the possibility of a catastrophe redolent of the end of the Bronze Age – civilisation under siege by marauding warriors intent on the sack of cities. 'The *ancientness* of future wars has three dimensions:' Kaplan writes, 'the character of the enemy, the methods used to contain and destroy him, and the identity of those beating the war drums.'[15]

Kaplan argues that the world has entered an age of increasing anarchy in which murderous gangs in West Africa, Russian or Albanian mafias, Latin American drug cartels, uprooted and religiously deluded Muslim jihadists all pose threats to the peace and prosperity of the 'walled city' of the liberal democracies. 'Like Achilles and the ancient Greeks harassing Troy,' he writes, 'the thrill of violence substitutes for the joys of domesticity and feasting.' He even quotes Achilles' words to Odysseus, from Book XIX of *The Iliad*: 'You talk of food? I have no taste for food – what I really crave is slaughter and blood and the choking groans of men.'[16]

Bobbitt took his very title from Book XVIII of *The Iliad* – 162 lines of which he used as a long epigraph to the book.[17] These lines describe the astonishing shield wrought by Hephaestus for the great warrior Achilles, a shield which depicted the entire Bronze Age world in miniature. Here were the heavens, with the sun, moon and stars; the earth and sea; two cities, one at peace, with all the arts of agriculture, law and civic life, and one under siege, Strife and Havoc spurring slaughter beneath the city's walls. All this was emblematic for Bobbitt in a book designed to help a contemporary readership understand the nature of war and peace.

'This is the main point I wish my readers to bear in mind:' Bobbitt writes, 'war is a product as well as a shaper of culture. Animals do not make war, even though they fight.[A] No less than a market and the law courts, with which it is inextricably intertwined, war is a creative act of civilized man with important consequences for the rest of human culture, which include the festivals of peace.'[18] It is those consequences Bobbitt sought to explicate in his book – written before 11 September 2001, though not published until the following year.[B]

In his foreword to *The Shield of Achilles*, historian of war Michael Howard writes, 'Bobbitt believes that mankind could be facing a tragedy without precedent in its history. It is not clear that he is wrong.'[19] 'We are entering a fearful time,' Bobbitt himself writes at the end of his book, 'a time that will call on all our resources, moral as well as intellectual and material.'[20] Both feared a cataclysm in which world order would disintegrate under the impact of anarchic and terrorist assaults, including the indiscriminate use of weapons of mass destruction.

This is not the place to explore Bobbitt's thesis.[C] I mention it because, in a major study of war and peace at the beginning of the twenty-first century, he saw fit to put the shield of Achilles right into his title.[21] My disappointment with *Troy* is that it was a missed opportunity to provide a mass audience with a serious 'thesis' on the nature of war and peace. The shield of Achilles, like much else, is entirely omitted from the film. Achilles has a shield, to be sure, but it has none of the features described by Homer. I believe this is symptomatic of the fact that the film's makers simply did not have any but the flimsiest sense of the moral and historical significance of the legend of the Trojan War – whatever awe Bana and Pitt may have felt when they read *The Iliad*.

Bobbitt's book is significant for a second reason, also related to the Trojan War. For the actual sack of Troy was a matter of history, not only of legend. And, as the epigraphs at the beginning of this essay indicate, it stood at the beginning of a 'tragedy without precedent' in human history up to that time – the devastation of the Bronze Age world by anarchic warriors and sackers of cities. This history has only in the past century been reconstructed out of the ruins of the deep past. Yet it is a gripping story – incomparably more dramatic than Wolfgang Petersen makes it seem in his film.

Imagine, for a moment, that Petersen's research team had had something other than a fancy dress melodrama in mind. They might have recreated the world of the late Bronze Age – the last three centuries of the second millennium BCE in more detail and at least hinted at the catastrophe that overwhelmed it between 1220 and 1170 BCE.[22] This is the context for the real Trojan War.[23] Invoking that real context could have had a powerful resonance in our time, whereas Petersen's thin context provided almost none.

A   See Essay 4 in Part I, 'On the Origins of Warfare'.
B   See Essay 2 in Part III, 'On Challenges Laid Down by the Islamists'.
C   See Essay 1 in Part III, 'On Strategy in the Twenty-first Century World'.

Troy as we call it, was a trading city on the western periphery of the Hittite Empire, known to the Hittites as Taruisa, the leading city of a kingdom called Wilusa, whence the classical names Troia/Troy and Ilios (Wilios)/Ilium.[24] Diplomatic exchanges occurred between the great king of the Hittites, in the centre of what is now Turkey, the great king of Ahhiyawa (the Hittite name for Achaea, or Greece), the Pharoah in Egypt and lesser principalities – including Wilusa. It was a world of high culture, with a history extending back many centuries.[25]

Within half a century, at the end of the thirteenth and the beginning of the twelfth century BCE, this whole world came crashing down. As Robert Drews puts it, 'almost every significant city or palace in the eastern Mediterranean world was destroyed, many of them never to be occupied again.'[26] Literacy disappeared from much of this hitherto highly civilised region and a dark age ensued in which much was forgotten or passed into legend. What Homer inherited was an oral tradition handed down over half a millennium. He was the equivalent of a Saxon bard telling tales of the Roman conquest of Britain.[27]

Only now, based on scientific and painstaking research, can we tell something like the true story and see it in its dramatic and historical context. Some few of those who watched Petersen's *Troy* will have known of this richer context. Most viewers will not have and perhaps, for that reason, did not miss it. One might wish, however, that Petersen had done more – as he could have done – to have opened their eyes to the terrors and awesome depths of the past, so that they might have felt more fully alive in our own time: the world of *The Shield of Achilles*.

1.   *Endnotes*

1   Homer, *The Iliad*, Book II, lines 355-6, trans. Robert Fagles, with an introduction and notes by Bernard Knox, Viking, 1990, p. 109.

2   Robert Drews, *The End of the Bronze Age: Changes in Warfare and the Catastrophe Circa 1200 BC*, Princeton University Press, 1993, pp. 3-4, 42.

3   Michael Wood, *In Search of the Trojan War*, BBC Books, rev. edn., 1998, preface, p. 4.

4   In his classic study of tragic theatre, literary critic George Steiner was concerned with the fundamental outlook on the world which underlies tragic drama. His contrast between the Hebraic and Homeric world views is still worth reflecting on. See *The Death of Tragedy*, Faber & Faber, London, 1974, pp. 4-5.

5   Walter Kaufmann captured quite well the enduring power of Homer's epic. The great tragedians, he argued, inherited from Homer the forms and themes of tragic drama and also a profound sense of humanity, in which the terrors of human existence are dwelt on, the glory and anguish of human suffering and the grief of one's enemies, as well as one's own. *Tragedy and Philosophy*, Chapter 5, 'Homer and the Birth of Tragedy', Princeton University Press, 1979, pp. 160-2.

6   Wood, *In Search of the Trojan War*, p. 143.

7   The Agamemnon who sacrifices Iphigenia in classical legend is not the coarse thug of *Troy*, but a man of his time, torn between paternal love and the ruthless demands of war and auguries. It was Calchas the priest who demanded the blood sacrifice to Artemis, who had sent a wind contrary to Greek hopes and aims. In Aeschylus' *Agamemnon*, his demand on behalf of the goddess was 'a thought to crush like lead the hearts of Atreus's sons [Agamemnon and Menelaus], who wept, as weep they must, and speechless ground their sceptres in the dust.' Agamemnon then responded, 'What can I say? Disaster follows if I disobey: surely yet worse disaster if I yield.' That, at least, is Philip Vellacott's translation (*Aeschylus: The Oresteian Trilogy*, Penguin, 1959, p. 49). The translations of *Agamemnon* by Robert Fagles, *Aeschylus: The Oresteia*, Penguin 1979, is stronger, as is that by Ted Hughes, a poet in his own right, *Aeschylus: The Oresteia*, Farrar, Straus and Giroux, New York, 1999. Hughes renders the passage in question: 'At that point, Calchas the seer spoke for heaven. He told us what had to be done to shift that wind – When they heard what Artemis demanded the warlords cried out, incredulous. But Agamemnon, Agamemnon, when he heard it, roared with anguish, sudden as the wound of a night-arrow. They took it in, those chieftains, with a jabbering of grief. Their royal staves pounded the earth. Then Agamemnon, our general for good reason, mastered himself with painful words: If I obey the goddess, my own daughter has to die. If I deny the goddess, this whole army has to dissolve ...'

8   Aristotle, *Poetics* #6, 'A tragedy, then, is the imitation of an action that is serious and also, as having magnitude, complete in itself; in language with pleasurable accessories, each kind brought in separately in the parts of the work; in a dramatic, not in a narrative form; with incidents arousing pity and fear, wherewith to accomplish its catharsis of such emotions.' Richard McKeon (ed.), *The Basic Works of Aristotle*, Random House, New York, 1941, p. 1460.

9   Benioff does, however, borrow a brief line from Homer for the climactic scene in which Hector confronts Achilles before the gate of Troy and asks that whoever wins their combat respect the corpse of the conquered. 'There are no pacts between lions and men,' Achilles (Pitt) tells Hector (Bana), in *Troy*. In Homer, Achilles speaks as grimly but at greater length: 'Hector, stop! You unforgivable, you ... don't talk to me of pacts. There are no binding oaths between men and lions – wolves and lambs can enjoy no meeting of the minds – they are all bent on hating each other to death. So with you and me. No love between us. No truce till one or the other falls and gluts with blood Ares who hacks at men behind his rawhide shield. Come, call up whatever courage you can muster. Life or death – now prove yourself a spearman, a daring man of war! No more escape for you – Athena will kill you with my spear in just a moment. Now you'll pay at a stroke for all my comrades' grief, all you killed in the fury of your spear.' Homer, *The Iliad*, trans. Robert Fagles, with an introduction and notes by Bernard Knox, Viking, 1990, p. 550. Book XXII, ll, 307-321.

10   Walter Kaufmann's remarks on this matter are worth noting: 'Nothing has obstructed a sensible reading of the *Iliad* more than the frequent failure to understand the role of the gods in Homer. Gods, one assumes, are supernatural; and Homer was a polytheist ... But the concept of the supernatural is out of place in Homer; it involves an anachronism, a reference to a wholly uncongenial vision of the world, and precludes an understanding of the experience of life in the *Iliad* ... The most crucial point about the gods in Homer is that belief is out of the picture ... Preoccupation with beliefs belongs to a far later stage in religion ... the whole antithesis of nature and the supernatural belongs to a post-Homeric climate of thought ... it has no place in the *Iliad* ... Polytheism suggests belief in many gods, as opposed to monotheism, which signifies belief in one god only. But Homer differs from monotheism in two ways. First, confronted with the reality of a cult of many gods, he does not oppose this diversity with any polemic; on the contrary, he turns it to poetic use. Secondly, belief is out of the picture. Polytheistic language is especially well suited to the description of war. No other poet has ever been able to capture so perfectly the confusion of war, the changing fortunes and the apparent cross-purposes.' *Tragedy and Philosophy*, pp. 168-178.

11   Euripides, *The Bacchae and Other Plays*, trans., Philip Vellacott, Penguin, 1954, pp. 129-31. The translation here, however, is my own, revising Vellacott, from *The Women of Troy*, lines 1213-16, 1236 and 1280-1.

12   Aristotle, *Poetics* #13, in Richard McKeon (ed.), *The Basic Works of Aristotle*, Random House, New York, 1941, p. 1467.

13   William Shakespeare, *Troilus and Cressida*, act 1, scene 1, lines 96-7.

14   Ernst Junger, *Storm of Steel*, trans. Michael Hofmann, Allen Lane, Penguin, 2003, p. 5. Consider, also, the passage on p. 58 in which Junger reflects on the ethics of war: 'Throughout the war, it was always my endeavour to view my opponent without animus, and to form an opinion of him as a man on the basis of the courage he showed. I would always seek him out in combat and kill him, and I expected nothing else from him. But never did I entertain mean thoughts of him. When prisoners fell into my hands, later on, I felt responsible for their safety, and would always do everything in my power for them.'

15   Robert D. Kaplan, *Warrior Politics: Why Leadership Demands a Pagan Ethos*, Vintage Books, Random House, New York, 2002, p. 118.

16   Ibid., p. 119.

17   Philip Bobbitt, *The Shield of Achilles: War, Peace and the Course of History*, Penguin, 2002, pp. ix-xiii; *The Iliad*, Book XVIII, lines 558-/20.

18   Ibid., p. xxxi.

19   Ibid., p. xix.

20   Ibid., p. 822.

21   There are, nonetheless, many passages in Bobbitt that are worth reflecting on against the background of the catastrophe at the end of the Bronze Age since, like Drews, he looks at changes in the nature of warfare as the harbingers of conflicts and possible catastrophes to come. Towards the end of his book, he writes almost as if the danger were from 'Trojan horses' – gifts for Poseidon, the god of the open sea and of international commerce,

one might say: 'We are entering a period ... when very small numbers of persons, operating with the enormous power of modern computers, biogenetics, air-transport, and even small nuclear weapons, can deal lethal blows to any society. Because the origin of these attacks can be effectively disguised, the fundamental bases of the State will change.' (p. 811.) Or again, 'Will we lay a long siege against ourselves or master the craft of the armourer when shields are made of secrets and not of bronze?' (p. 807.)

22   Fernand Braudel described an even earlier catastrophe, more than a thousand years before the Trojan War, in which the archaic Bronze Age world was overwhelmed by Indo-European invaders. This was the antecedent to the sack of cities right around the eastern Mediterranean littoral and throughout Anatolia and Syria in the late thirteenth and early twelfth centuries BCE – and archaic Troy was among the cities burned then, as well as later. *The Mediterranean in the Ancient World*, Allen Land, Penguin, 2001, pp. 129-131.

23   'We do care about the authenticity of the tale of Troy,' wrote Byron in his diary in 1821. 'I venerate the grand original as the truth of history (in the material facts) and of place, otherwise it would give no delight. Who will persuade me, when I reclined upon a mighty tomb, that it did not contain a hero? Men do not labor over the ignoble and petty dead – and why should not the dead be Homer's dead?' Benita Eisler, *Byron: Child of Passion, Fool of Fame*, Hamish Hamilton, London, 1999, p. 257.

24   Wood, *In Search of the Trojan War*, pp. 206-7: 'If there is anything at all in the legend, it must be tested against the only reliable sources for the history of the thirteenth century BCE in Asia Minor – archaeological finds, Linear B names, Hittite diplomacy – and it holds up surprisingly well.'

25   Ibid., p. 169: 'Remarkable discoveries in central Turkey have led to the decipherment of the Hittite language and have revealed the hitherto unsuspected existence of a great empire which stretched from the Aegean to the Euphrates valley at precisely the time when tradition places the Trojan War. In the Hittite archives ... we have "real" historical texts to interpret: diplomatic letters, treaties, annals and royal autobiographies, in which the characters of the Hittite kings and queens come to life in the most vivid way. Most exciting of all is the claim that Troy and the Trojan War are to be found in these files of the Hittite "Foreign Office".'

26   Drews, *The End of the Bronze Age*, p. 4.

27   'A deserted, ruined and overgrown site in a sparsely populated area of northwestern Anatolia, with no visible links with Greece, surely cannot have been selected as the setting for the Greek national epic, unless it had at some time in the past been the focus of warlike deeds memorable enough to have been celebrated in song. The simplest explanation is that the tale of Troy owed its central place in later epic tradition to the fact that it was the *last* such exploit before the disintegration of the Mycenean world ...' Wood, *In Search of the Trojan War*, p. 144.

My work is not a piece of writing designed to meet the taste of an immediate public, but was done to last forever.
- Thucydides (c. 400 BCE)[1]

Thucydides is one who, though he never digress to read a lecture, moral or political, upon his own text, nor enter into men's hearts further than the facts themselves evidently guide him, is yet accounted the most politic historiographer that ever writ.
- Thomas Hobbes (1628)[2]

We are still waging Peloponnesian Wars ... Or to put it in the terms of the tragic design drawn by Thucydides: our fleets shall always sail toward Sicily, although everyone is more or less aware that they go to their ruin.
- George Steiner (1961)[3]

2. On Writing Classical History

What is a classic? A book that is not ephemeral, that has endured across generations. A book that has set a standard to which others perennially aspire. A book that draws serious readers back, again and again, to reflect on what is has to offer. A book characterised not simply by the factual or narrative material in it, but by the power of the author's insights into that material. A book that is, finally, inimitable and cannot be displaced merely by someone covering the same material.

By all these criteria, the history of the Peloponnesian War by the fifth century BCE Athenian general Thucydides is certainly one of the great classics. The Peloponnesian War was a great war between coalitions led by Athens and Sparta that tore the Greek world apart between 431 and 404 BCE and ended in the downfall of the Athenian Empire. Indeed, Thucydides' history is my own master classic. I hold no other book in quite the same regard as *The Peloponnesian War* – though I hold many in very high regard and value countless others. This book possesses a universal relevance to human affairs, a gravity, a sombreness, an originality, a quality of humanity without illusions and a concern with critical objectivity that, together, set it apart from any other book I can think of.

I deeply admire what Peter Levi once called 'the dark thunder of Thucydides' prose'. Above all, I am awed by the fact that such a work was written so long ago and yet even now seems unsurpassed in its intelligence, realism and intellectual restraint. More than any book of moral or political theory, or any modern work of history that I can think of, I consider this book to be the greatest primer in war and human affairs, not simply in the Western canon, but in the world. Nothing in the Chinese canon, for instance, comes close; certainly not Sun Tzu's gnomic book on strategy, *The Art of War*.

One of the reasons for this is that the Peloponnesian War, though it was believed by Thucydides to be the greatest war ever fought up until his own time, was a very small affair compared to the colossal wars of later history. Thucydides was, therefore, able to relate it both in detail and yet within the bounds of an almost theatrical dramatic structure, in which the key characters are as active and articulate as those in a Shakespearean tragedy.

Yet they were not theatrical characters. They were actual, historical individuals. Pericles, Alcibiades, Nicias, Agis, Brasidas, Hermocrates, Cleon, Gylippus and Demosthenes, among others, remain alive in the pages of Thucydides as individuals clearly drawn, who are not the pawns of either tyrants or fate. They are shown making choices, making commitments, facing challenges and addressing their contemporaries. They occur as the articulate and purposeful citizens of polities with which we can still empathise, since they were ancestral to our own.

Thucydides self-consciously set himself a new project: to *get history right* and in doing so to understand the true nature of human affairs. He freely took on this project as an individual and spent some thirty years working on it. History, according to a tired and cynical saying, is merely the propaganda of the victors. But Thucydides embarked on his project shortly after the Peloponnesian War began and when he was writing his final draft he already knew that his side had *not* been the victors. He attempted accurate understanding and explanation so that it would be truly useful to those who came after him – forever. This, as he knew, was an original idea.

Few things mark Thucydides out more clearly as the world's first really serious historian than his introductory remarks about evidence and method. From the very beginning of the war, he decided to write its history, 'believing that it would be a great war and more worthy of relation than any that had preceded it.' This was a belief which, he claimed, 'was not without its grounds.' The war looked likely to be the greatest upheaval yet known, not only among the Greeks, but even in the 'barbarian' world around them, he thought.

But his dominant design was to be more accurate and incisive than the poets and chroniclers who preceded him. His scepticism and interest in evidence are registered in his very first paragraph, with the remark that 'though the events of remote antiquity, and even those that more immediately precede the war, could not from lapse of time be clearly ascertained, yet the evidences which an inquiry carried as far back as was practicable lead me to trust, all point to the conclusion that there was nothing on a greater scale, either in war or in other matters.'[4]

The depth of historical awareness which an educated person may, in our time, take somewhat for granted, was simply unavailable to Thucydides. There were almost no reliable records, as he testifies. What he meant by remote antiquity was no more than a thousand years or so before his own time, back to the era preceding the more or less legendary Trojan War. He makes no mention of Egyptian antiquity, or even of the rise of the Persian Empire, to say nothing of the earlier empires of Assyria, Babylon, Hatti or Akkad, of which he would seem to have been unaware. Of China he was certainly wholly ignorant.

Yet regarding the Trojan War he exhibits a critical reflectiveness that is striking. He does not doubt that the war took place under the leadership of Agamemnon of Mycenae. He is sceptical, however, of 'the exaggeration which a poet [Homer] would feel himself licensed to employ' and after reviewing the meagre evidence concludes that the expedition must surely have been 'inferior to its renown and to the current opinion about it formed under the tuition of the poets.'[5]

'Mycenae', he nonetheless observes, 'may have been a small place ... but no exact observer would therefore feel justified in rejecting the estimate given by the poets and by tradition for the magnitude of the armament [in the war against Troy]. For I suppose that if Sparta were to become desolate, and only the temples and the foundations of the public buildings were left, that as time went on there would be a strong disposition with posterity to refuse to accept her fame as a true exponent of her power.'[6]

He reflects more broadly on earlier history, but then makes a highly characteristic remark, which is the kind that has always – and increasingly – appealed to me. 'Having now given the result of my inquiries into early time, I grant that there will be a difficulty in believing every particular detail. The way that most men deal with traditions, even traditions of their own country, is to receive them all alike as they are delivered, without applying any critical test whatever ... So little pains do the vulgar take in the investigation of truth, accepting readily the first story that comes to hand.'[7]

His own conclusions, he argues, were, however, likely to prove reliable and 'Assuredly will not be disturbed either by the verses of a poet displaying the exaggerations of his craft, or by the compositions of the chroniclers that are attractive at truth's expense; the subjects they treat of being out of reach of evidence, and time having robbed most of them of historical value, by enthroning them in the realm of legend.'[8]

He claims, as regards the narrative of events in his history, 'far from permitting myself to derive it from the first source that came to hand, I did not even trust my own impressions, but it rests partly on what I saw myself, partly on what others saw for me, the accuracy of the report always being tried by the most severe and detailed tests possible. My conclusions have cost me some labor from the want of coincidence between accounts of the same occurrences by different eye-witnesses, arising sometimes from imperfect memory, sometimes from undue partiality for one side or the other.'[9]

Coming as they did in the context of a long and bitter war and in a culture still only partly literate and steeped in legends and auguries, these are stunning passages. They show Thucydides to have been a worthy contemporary of Democritus of Abdera (460-357 BCE), one of the world's first true natural scientists and the chief proponent of atomism in the ancient world.[10] He was also a close contemporary of Socrates (469-399 BCE) and of the great, sceptical tragedian Euripides (480-406 BCE), of Hippocrates of Cos (469-399 BCE), the pioneer of Western medicine whose work was not to be surpassed until after 1800 CE,[11] and the brilliant comic playwright Aristophanes (450-385 BCE).

One aspect of Thucydides' introduction seems to depart from the high standard he set himself – the case of the speeches in his history. There are over one hundred and forty of them, a remarkable number and an extremely rich part of his narrative. Yet he allows that he did not have access to written records of them and comments, 'some I heard myself, others I got from various quarters; it was in all cases difficult to carry them word for word in one's memory, so my habit has been to *make the speakers say what was in my opinion demanded of them by the various occasions*, of course adhering as closely as possible to the general sense of what they really said (emphasis added).'[12]

No historian could get away with doing this now, but what else was Thucydides to do? In any case, the speeches are extraordinary in their economy and power of expression and one would not want to be without them, or have substituted for them what may have been inferior speeches actually delivered by the historical individuals in question. In this respect, *The Peloponnesian War* has something about it of the appeal of Shakespeare. The Funeral Oration of Pericles, the single most notable example, belongs in any anthology of great speeches; regardless of how precisely the historian recorded what the great Athenian statesman actually said.

Indeed, George Steiner, forty years ago, suggested that the speeches in Thucydides may have been among the first demonstrations that literary prose was even possible. There had, of course, been literature before Thucydides, but it was in verse, not prose. Greek lyric and epic poetry were advanced art forms well before Thucydides and Greek tragedy flourished as an art form for decades before, as well as during, the Peloponnesian War. But it, also, was written in verse – being 'a convergence of speech, music and dance.' Thucydides, Steiner speculates, may have been the first to masterfully conceive 'that prose could aspire to the dignity and "apartness" of literature.'[13]

Nowhere is this more evident than in the famous Melian Dialogue, in Book V of *The Peloponnesian War*. This passage is unique in being set out by Thucydides in dialogue form, with the alternate remarks labelled, as if in a stage drama, 'Athenians' and 'Melians'. And the dialogue is completely worthy of a tragedy by Euripides – except that there is no *deus ex machina* at the end, as in some of Euripides' plays, to save the Melians from their grim fate.

The dialogue is about justice and 'realism' and is breathtaking in its lapidary clarity and hardness. The Athenians had come to the island of Melos to demand its submission and its participation in the war against Sparta. When the Melians declined, the Athenians destroyed the place, slew all the adult males and enslaved the women and children.

There can be few passages in historical or political writing as memorable or reflected upon as Book V:89 of *The Peloponnesian War*: 'Athenians: For ourselves, we shall not trouble you with specious pretenses – either of how we have a right to our empire because we [defeated the Persians seventy years ago], or are now attacking you because of wrong that you have done us – and make a long speech which would not be believed; and in return we hope that you, instead of thinking to influence us by saying that you did not join the Spartans, although their colonists, or that you have done us no wrong, will aim at what is feasible, holding in view the real sentiments of us both; since you know as well as we do that right, as the world goes, is only in question between equals in power, while the strong do what they can and the weak suffer what they must.'[14]

This passage itself, but even more the dialogue as a whole, is wonderfully representative of those qualities of Thucydides' history which led Clifford Orwin to remark, 'What Thucydides offers, through the unsurpassed artfulness of his narrative, is a vicarious experience of the events that he describes, for which no dogmatic presentation of the truths of political life could substitute.'[15] I believe this is so and that, for this reason, the study of *The Peloponnesian War* is more likely to impress upon reflective minds the enduring nature of human politics and conflict than anything in Plato or Aristotle.[16]

Yet Thucydides' history is not complete. Indeed, it ends abruptly, in mid-sentence, in the middle of the year 411, seven years before the war ended. We know that Thucydides lived to see the end of the war, for he speaks of it at various points in his history. We can even see emendations to his history which have plainly been written by him after the war ended with Athens' defeat and surrender to Sparta in 404. In short, he has left us a kind of grand, unfinished symphony.

It is, surely, a remarkable tribute to the faithfulness of 2,400 years of copyists, scribes and commentators that the text of *The Peloponnesian War* has come down to us – unfinished. It is greatly to be regretted that Thucydides died before completing his masterpiece – and that we do not have whatever notes he had accumulated for completion of the work. It is poignant, really – a testimony to the vulnerability of human endeavours. It lends an almost Romantic glow to this most severely classical of monuments.

Modern historical scholarship, however, has finally enabled us to round out the history Donald Kagan, Sterling Professor of Classics and History at Yale University, in particular, has laboured, over four decades, to produce a complete history of the Peloponnesian War. In four volumes – *The Outbreak of the Peloponnesian War* (1969), *The Archidamian War* (1974), *The Peace of Nicias and the Sicilian Expedition* (1981) and *The Fall of the Athenian Empire* (1987) – he drew together the findings of more than a century of Thucydides' scholarship and created a monumental history. In 2003, he finally published a condensed version in one volume, simply titled *The Peloponnesian War*.

Whereas Thucydides could look back on little serious historical scholarship of any description and but scant records, Kagan was heir to an enormous and institutionalised tradition of scholarship – the scrupulous, painstaking work of translation, documentation, archaeology and epigraphy that have made it possible for post-Enlightenment scholars to reconstruct the remote past in astonishing detail.

Jacqueline de Romilly's *Thucydides and Athenian Imperialism* and A. W. Gomme's *Historical Commentary on Thucydides*, written in the mid-twentieth century, laid the groundwork, but as Kagan remarks, the literature of specialised studies on the war and related subjects is vast. This is the work of Western universities. It's what they are for and it can only be accomplished when advanced standards of higher education are upheld. It is far from clear that they are being upheld in many Western universities at the beginning of the twenty-first century.

Donald Kagan, however, is of an old school, having been educated half a century ago. He has produced an impressive body of work in the great tradition, of which his history of the Peloponnesian War has been his greatest labour and is his crowning achievement.[17] When the four volume history had been completed, Bernard Levin remarked of it, 'although the great shadow of Thucydides must have been looking over his shoulder throughout, when Professor Kagan wrote The End, the shadow [would have] nodded in admiration.'

This would have to be said, also, of the one volume edition, since it completes the work Thucydides dedicated his life to, while exhibiting in the highest degree that relentless attention to detail and cross-checking of facts on which the Greek historian prided himself. Kagan even observes that restraint in digressing to 'read a lecture, moral or political, upon his own text', for which Hobbes praised Thucydides 380 years ago. In his own words, 'I have avoided making comparisons between events in the Peloponnesian War and later history, although many leap to mind, in the hope that an uninterrupted account will better allow readers to draw their own conclusions.'[18]

What is most striking, though, is the tribute Kagan pays to Thucydides himself. He declares of the great original, which he has laboured to update, that it 'is justly admired as a masterpiece of historical writing and hailed for its wisdom about the nature of war, international relations and mass psychology ... [Thucydides had] an extraordinary and original mind and, more than any other historian in antiquity, he placed the highest value on accuracy and objectivity.'[19]

Has Kagan's history finally displaced Thucydides, though? From the point of view of scrupulous accuracy and completeness, yes, it has. Perhaps that is, in part, what Levin meant, in writing of the 'great shadow' nodding in admiration. Yet the original is more powerful. Kagan's writing, though lucid, seems to lack the characteristic gravity of the original, conceivably because Thucydides was a general and a contemporary of the war, whereas Kagan is an academic at a very great remove from the blood and fire of what he describes. But what diminishes his version most is his omission or extreme condensation of the speeches that we find in Thucydides.

Even the Funeral Oration of Pericles and the Melian Dialogue are reduced – in both the four volume and the one volume histories – to bare summaries. He does not explain why he has done this, yet it is a major alteration in the character of the narrative. The effect is to excise the resonant voices of the figures of antiquity from the history, very much weakening its dramatic impact   the way it is 'heard' in a reader's mind.

Did Kagan omit the speeches precisely in order to leave the classic its most inimitable strengths – the reconstruction by Thucydides of one hundred and forty memorable and instructive public speeches on war, politics and strategy? Whether he omitted them deliberately or not, out of homage or strange tone deafness, I imagine the 'great shadow' raising an eye-brow, then smiling and resting in peace, secure in the knowledge that his work is honoured and its stature undiminished by Kagan's labours.

The decline of the humanities in our schools and universities has been widely heralded for quite some time. There are those who have argued that the displacement of print media by electronic media is a good part of the problem. Alexander Stille, for example, in *The Future of the Past*, wondered recently whether our sense of history itself – a product of the technology of writing – was in danger of being reduced to a televisual 'flat world in which everything occurs in a consumer present.'[20]

Whether that is so is a subject for another time, but what is indubitable is that no one individual made a more original or striking contribution to creating the sense of history, using the technology of writing, than did Thucydides of Athens. When, therefore, with Henry Petroski, I contemplate the books upon my shelves, none stand out as powerfully as my various editions of Thucydides.[21] His history is, for me, the master classic that has indeed become, as its author purposed, a possession for all time and a work to last forever.

2.   *Endnotes*

1   Thucydides, *The Peloponnesian War,* Book I:22, trans. Rex Warner, Penguin, 1954, p. 48. Hobbes, translating the same passage in the early seventeenth century, rendered it: 'And it is compiled rather for an everlasting possession than to be rehearsed for a prize.' *Thucydides: The Peloponnesian War – The Complete Hobbes Translation*, David Grene (ed.), University of Chicago Press, 1989, p. 14. Richard Crawley, translating Thucydides in the nineteenth century, rendered the same sentence: 'In fine, I have written my work, not as an essay which is to win the applause of the moment, but as a possession for all time.' *The Landmark Thucydides: A Comprehensive Guide to the Peloponnesian War*, Robert B. Strassler (ed.) with an introduction by Victor Davis Hanson, Touchstone Books, New York, 1998, p. 16.

2   Thomas Hobbes, 'To the Readers', in *Thucydides: The Peloponnesian War*, p. xxii.

3   George Steiner, *The Death of Tragedy*, Faber and Faber, London, 1961, pp. 6-7.

4   Thucydides I:1, Strassler, *The Landmark Thucydides*, p. 3.

5   Thucydides I:10-11, ibid., p. 9.

6   Thucydides I: 10, ibid., p. 8.

7   Thucydides I:20, ibid., pp. 14-15.

8   Thucydides I: 21, ibid., p. 15.

9   Thucydides I:22, ibid., p. 15.

10   Among the many works from classical antiquity that have not come down to us are, in addition to many of the lost works of Aristotle and most of the tragic dramas composed by the great masters Aeschylus, Sophocles and Euripides, the entire corpus of the writings of Democritus. According to the reliable Diogenes Laertius, his work was voluminous and of the highest quality. He wrote books on ethics, physics, geology, mathematics, music, literature, biology, medicine, agriculture, painting, law and tactics in war – an astonishing range, fully at the level of Aristotle. Why then did it all perish beyond recall? One cause was the hostility of the idealists, who detested his materialism. Plato, it is recorded, 'wished to burn all the writings of Democritus that he could collect', but was dissuaded by Pythagorean friends from doing so. Others, however, venerated him as 'the prince of philosophers' and 'the guardian of discourse.' Diogenes Laertius, *Lives of Eminent Philosophers,* vol. II, Loeb Classical Library, 1925, reprinted 2000, pp. 449-51 and 455-61.

11   'What unites [the Hippocratic writings] is the conviction that, as with everything else, health and disease are capable of explanation by reasoning about nature, independently of supernatural interference. Man is governed by the same physical laws as the cosmos, hence medicine must be an understanding, empirical and rational, of the workings of the body in its natural environment. Appeal to reason, rather than to rules or supernatural forces, gives Hippocratic medicine its distinctiveness.' Roy Porter, *The Greatest Benefit To Mankind: A Medical History of Humanity From Antiquity to the Present*, HarperCollins, 1997, p. 56.

12   Thucydides I:22, Strassler, *The Landmark Thucydides*, p. 15.

13   George Steiner, *The Death of Tragedy*, Faber and Faber, London, 1961, p 239.

14   Thucydides V:89, Strassler, *The Landmark Thucydides*, p. 352.

15   Clifford Orwin, *The Humanity of Thucydides*, Princeton University Press, 1994, p. 4.

16   In the 1880s Nietzsche wrote: 'My recreation, my preference, my cure from all Platonism has always been Thucydides ... For the deplorable embellishment of the Greeks with the colors of the ideal, which the "classically educated" carries away with him into life as the reward of his grammar school drilling, there is no more radical cure than Thucydides. One must turn him over line by line and read his hidden thoughts as clearly as his words: there are few thinkers so rich in hidden thoughts ... Courage in the face of reality ultimately distinguishes such natures as Thucydides and Plato: Plato is a coward in face of reality – consequently he flees into the ideal; Thucydides has *himself* under control – consequently he retains control over things ...' *Twilight of the Idols*, 'What I Owe to the Ancients' #2, Penguin, 1968, pp. 106-7.

17   His other books include: *The Great Dialogue: A History of Greek Political Thought From Homer to Polybius; Pericles of Athens and the Birth of Democracy;* and *On the Origins of War and the Preservation of Peace.*

18   Donald Kagan, *The Peloponnesian War*, Viking, New York, 2003, pp. xxvi-xxvii.

19   Ibid., p. xxvi.

20   Alexander Stille, *The Future of the Past*, Farrar, Straus and Giroux, New York, 2002, p. 338.

21   Henry Petroski, *The Book On the Book Shelf*, Vintage Books, New York, 1999.

... the best and most useful aim of my work is to explain to my readers by what means and by virtue of what political institutions almost the whole world fell under the rule of one power, that of Rome, an event which is absolutely without parallel in earlier history.
- Polybius (c. 135 BCE)[1]

Others, I have no doubt,
will forge the bronze to breathe with suppler lines,
draw from the block of marble features quick with life,
plead their cases better, chart with their rods the stars
that climb the sky and foretell the times they rise.
But you, Roman, remember, rule with all your power
The peoples of the earth – these will be your arts:
To put your stamp on the works and ways of peace
To spare the defeated, break the proud in war.
- Virgil (c. 25 BCE)[2]

The principal conquests of the Romans were achieved under the republic; and the emperors, for the most part, were satisfied with preserving those dominions which had been acquired by the policy of the senate, the active emulation of the consuls, and the martial enthusiasm of the people.
- Edward Gibbon (c. 1776)[3]

9. On the Rise of the Roman Empire

Modern ideals of political freedom, law and critical thinking have their foundations in the Greek and Roman republics of the ancient world. The great Renaissance and Enlightenment thinkers, from Machiavelli and Montesquieu to Hume, Smith and Gibbon, admired what they saw as ancient pagan civic virtue corrupted by empire, then overthrown by barbarism and religion. They sought to recover this virtue from the feudal and ecclesiastical morass into which, they believed, Europe had descended with the collapse of the Roman Empire. But they saw the Empire itself as the beginning of the corruption – because it succumbed to oriental despotism and eastern religion before it fell.

As John Pocock reminds us in his remarkable study, *Barbarism and Religion*, this was the core Enlightenment narrative.[4] If you are a modern rationalist, secularist and democrat (whether liberal or socialist), this is the foundation of your world view. It is this that has been under attack by Romantics and religious conservatives, deconstructionists, cultural relativists and postmodernists, from the nineteenth century to the present. It is attacked, also, by those who claim that these very Western ideals are the seeds of modern imperialism and 'orientalism', in Edward Said's sense.

Warwick Ball's *Rome in the East: The Transformation of an Empire* which appeared at the turn of the century, is a classic case of these anti-classical, post-Enlightenment perspectives.[5] 'We look to Rome for our own European roots', Ball writes, 'but Rome itself looked east.' True enough: Rome looked to Greece for political, legal and philosophical models. Romans also came to see themselves as descended from the survivors of the sack of Troy. Ball claims a lot more than this, though. It was the very acquisition of the eastern provinces of Asia Minor, Syria, Phoenicia, Judaea and Egypt, he argues, that 'led directly to the rise of Rome and of Roman greatness, the first period of greatness in European history.' Rome may have civilised the Celtic barbarians of Europe, but it was 'on the receiving end of civilisation' in its Asian provinces.

Far from Romanising the east, Ball asserts, the Roman Empire witnessed the 'triumph of the East'. This triumph began with the Romans gradually abandoning their republican institutions and opting for an imperial monarchy in emulation of the Iranian Great Kings. It was manifested by the abandonment of the crude and poor West to the barbarians and the removal of the seat of Empire to the East, from the fourth century until the final downfall of the Eastern Empire in the fifteenth century. It was manifested, also, by the rise in Europe itself of Christianity – an amalgam of Judaic, Zoroastrian and other Levantine religious cults. Christianity, he writes, was a truly 'revolutionary' phenomenon in Europe – and it came from Asia. In short, Western civilisation owes its spirit and culture to a fascination with the riches and religions of the East, not to the raw provincialism of the old Romans.

Ball's book is a passionate response to Fergus Millar's *The Roman Near East 31 BC – AD 337*.[6] Millar, in 1993, had asserted rashly that the Romans found only two cultures in the Near East, the Greek and the Jewish, and put a seal of unity and greatness upon them. This is lambasted by Ball, who remarks that Millar worked only with Greek and Latin texts, while the material evidence tells a very different story. Ball, himself an archaeologist, remarks that, in Anatolia, Syria, the Levantine coastlands and the Arabian interior, to say nothing of Egypt and Mesopotamia, Rome encountered civilisations that were thousands of years old and which outlasted its domination.

This polemic part of the book is strongly argued and beautifully illustrated. It fails, however, to provide a sufficient foundation for Ball's more sweeping argument. His pivotal claim is that the acquisition of the eastern provinces 'led directly to the rise of Rome and of Roman greatness, the first period of greatness in European history.' This is pivotal for two reasons. It is fundamental to his sustained argument that 'the East was the real heart of the Roman Empire' and to his further claim that Western civilisation 'as we know it' came out of the East in the form of Christianity. On the basis of these claims Ball implicitly overturns traditional Western historiography. His pivotal claim, however, cannot be sustained.

Acquisition of the eastern provinces certainly did *not* lead to the rise of Rome. Rather, the rise of Rome *culminated* in its acquisition of the eastern provinces. This is a matter of incontrovertible historical fact and it is astonishing that Ball's enthusiasm can have got the better of him to such an extent that he would have denied it. Moreover, 'the first period of greatness in European history' has, since classical times, been considered the Athenian fifth century BCE. What precipitated *that* period of greatness was, of course, the defeat of the Great Kings of the East by the free Greeks.

Finally, while Christianity certainly came from the East, it brooded during the Dark Ages, in Hobbes' famous words, 'among the ghosts of the deceased Roman Empire, sitting crowned upon the grave thereof.' Catholicism might owe much to the Levant, but it would have lacked its most formidable characteristics had it not draped itself in the mantle of Rome. And yet, above all, it was the recovery and enlargement of the republican and philosophical ideas of Greece and Rome by thinkers increasingly sceptical of Christianity that gave modern Western civilisation its defining character.

In writing what he did about Rome's rise, Ball disregards even the classical historians – as cavalierly as Millar disregards the Egyptian, Semitic and Iranian cultures of the Roman East. It would have struck

Polybius, for example, as very strange indeed to assert that the acquisition of the eastern provinces 'led directly to the rise of Rome'. In his second century BCE *Universal History*, Polybius argues that 'in all political situations ... the principal factor which makes for success or failure is the form of a state's constitution: it is from this source, as if from a fountainhead, that all designs and plans of action not only originate but reach their fulfilment.' It was the Roman constitution, he believed, that enabled the republic to master the whole Mediterranean world.

It was this constitution which Machiavelli,[7] Montesquieu,[8] Gibbon and Mommsen[9] alike, over four centuries of modern historiography, saw as the admirable thing about Rome. Not its empire, but its institutions. Not its riches, but its virtues. What Ball pointed to as the 'triumph of the East', all these thinkers saw as the decadence of the institutions they admired and the *failure* of the republican virtues to survive the acquisition of the eastern empire. This failure was foreshadowed by Antony's follies in Egypt and came to a head when the very classically minded Antonine emperors of the second century CE were succeeded by the eastern-derived Severan dynasty at the beginning of the disastrous third century.

According to Livy in the thirty-third book of his first century BCE *History of Rome from its Foundation*, when the Romans defeated Philip of Macedon in 197 BCE, they did so in the name of the *liberty of the Greek cities*. When they then confronted Antiochus the Great – heir to both Alexander and the Great Kings of Persia – they challenged him, also in the name of Greek liberty. The Greeks responded with joy that, 'There really was a nation on this earth prepared to fight for the freedom of other men, and ... even prepared to cross the sea in order to prevent the establishment of an unjust dominion in any quarter of the globe, and to ensure that right and justice and the rule of law should everywhere be supreme.'[10]

Was this merely the notorious 'propaganda of the victors'? No. Their own historians excoriated the crimes and follies of the Romans in the east and in the civil wars that ruined the republic, as have the moderns. Few historical phrases ring more clearly through the annals of Western history than the words Tacitus (56 – c. 120 CE) put into the mouth of a Scottish barbarian leader by the name of Calgacus. 'They [the Romans] create a desolation and call it peace.'[11] Yet none of those crimes and follies, nor the decline and fall of the Empire itself ever obliterated the memory and importance of the institutions and ideals which had been at the foundation of the rise of Rome. Ball, however, seems to have been strangely unacquainted with this whole historiography in writing his book.

His reflections on the architectural wonders of the ancient world aside (for this is his true metier), Ball is at his best in pointing out how vital the commercial networks of the East were before, during and after Roman domination. He writes of Phoenician commercial penetration of the western Mediterranean, centuries before the rise of Rome. He points to Indian and even Chinese merchants sailing up the Euphrates over many centuries. He argues that the

Achaemenid Empire in Persia between the sixth and fourth centuries BCE was a great cultural bridge between the West and the Indo-Iranian worlds. He observes that 'evidence for Palmyrene merchant houses has been found as far apart as Bahrain, the Indus delta, Merv, Rome and Newcastle-upon-Tyne, in England.'

He is perfectly correct in arguing that all this has a significance which the narrower views of historians like Fergus Millar tend to obscure. In wanting to go further, however, and insist that it was the East that 'made' Rome, he overreaches himself. His often vehement argument reminds one of Andre Gunder Frank's passionately anti-Western assertions in *Re-Orient: Global Economy in the Asian Age* (1998). Frank asserts that, in the second millennium CE, 'Europe did not pull itself up by its own economic bootstraps, and certainly not thanks to any kind of European "exceptionalism" of rationality, institutions, entrepreneurship, technology, geniality, in a word – of race.' Rather, he declares, 'Europe climbed up on the back of Asia, then stood on Asian shoulders – temporarily.' Both authors go astray where Polybius was on the right track. Institutions were in both cases crucial to the Western ascendancy. They are now equally crucial to Asian growth and the future of world order – for reasons that have nothing to do with 'race.'[12]

'Pagan architecture', Ball writes, 'has ... an arrogant self-confidence, an unquestioning belief in the superiority of its own culture that Christian architecture cannot match ... [late imperial Christian] construction was often second rate, even shoddy.' The same, unfortunately, is true of *Rome in the East* compared with the great works of Roman history and historiography. For a student of architecture, Ball exhibits a notable lack of architectonic sense in the construction of his argument and an often sloppy lack of attention to detail. He remarks at one point, for example, that 'the Parthians had little conception of the Romans as a great power, even less as a civilised one'. Yet he later writes that 'the Tigris may have been Rome's eastern frontier ... but the presence of so great a power as Rome was felt far beyond.'

He declares that Roman rule in the east was brutal and exploitative, but elsewhere that 'Roman oppression, such as it was, seems to have kept a remarkably low profile.' He describes Alexander the Great at Tyre as bringing 'Dunsinane Wood to Cawdor' – a confused allusion to Shakespeare's *Macbeth*, in which Birnam Wood is brought to Dunsinane. He confuses, at a crucial point, the great sixth-century king, Khusrau Anurshivan, who fought Emperor Justinian, with the seventh century Khusrau Parviz, who fought Emperor Heraclius. He describes Constantine as leaving 'a crumbling, ramshackle Rome' and moving to the East, but Rome in the early fourth century was very far from being either crumbling or ramshackle and Constantine further embellished it. He declares, at the very end of his narrative, that with Constantine's move to the shores of the Black Sea, 'Troy had achieved its greatest triumph.' But this is vacuous rhetoric. Constantine did not choose Troy; he chose Byzantium, for very specific strategic reasons. Yet he and all his successors, down to the fifteenth century, called themselves *Roman* Emperors and certainly not Trojan epigones.

The greatest of Ball's solecisms is his failure to acknowledge the modern Western preoccupation with how the classical world lapsed into barbarism and religion. As Nietzsche exclaims in his polemic *Antichrist* in 1888, 'The whole labour of the ancient world was in vain: I have no word to express my feelings about something so tremendous ... Wherefore Greeks? Wherefore Romans? All the presuppositions for a scholarly culture, all scientific *methods*, were already there ... Is this understood? ... the methods, one must say it ten times, are what is essential, also what is most difficult, also what is for the longest time opposed by habits and laziness. What we today have again conquered with immeasurable self-mastery ... had already been there once before ...'[13]

Ball shows no more sense of this understanding of the classical world than Frank before him had of the importance of *methods and institutions* in the modern world. This strange and, I suspect, wilful myopia vitiated their historical and economic arguments in each case. It is not that there is nothing to be said for the deepened perspective on universal history that they demand. On the contrary, such depth is, in general, what both classical circumspection and modern enlightenment have been about. It is precisely what inspired Herodotus and Polybius. Ball asserts, however, that the East was the 'heart' of the Roman Empire and that Christianity was the triumph of the religion of the East over the *provincialism* of Rome. In doing this, he is behaving like some intellectual barbarian, blundering into Rome as a would-be ruler without any grasp of what had made the city great.

Why does all this matter? Because, at the beginning of the twenty-first century, the whole human world faces the most profound challenges in consolidating free political institutions, the rule of law and the predominance of critical thinking over dogmatism and superstition. These things were not the fruit of the East, however much cultural relativists and Sinophiles like Andre Gunder Frank may for some reason resent this. They were overwhelmingly the fruit of classical Greco-Roman civilisation and the modern Western enlargement of it. Being in awe of Eastern temple ruins or palaces, like Ball, or of the ill-fated splendours of old China and India, like Frank, risks blinding one to this reality.

Montesquieu wrote *The Spirit of Laws* with Greek and Roman republican constitutions as his beacon. If, nearly four hundred years after Montesquieu, we seek to enlarge the sway of the spirit of law in the world, then, in irreducibly important ways, we must start with the *problem* of Rome in the East – the corruption of its republic by the East. If, with Warwick Ball, we were to see Rome as 'on the *receiving* end of civilisation' in the East, we might lose sight of why we need to extend political freedom and the rule of law in the emerging global culture. Rome failed to extend republican institutions to the east and its long decline can be measured precisely by its slide into oriental despotism and oriental mysticism. We face comparable challenges now and this is not the time to throw in our hand in deference to 'the East'.

3.   *Endnotes*

1   Polybius, *The Rise of the Roman Empire*, trans. F. W. Walbank, Penguin, 1979, Book VI, p. 302.

2   Virgil, *The Aeneid*, Book VI: 976-984, trans. Robert Fagles, with an introduction by Bernard Knox, Penguin, 2006, p. 210.

3   Edward Gibbon, *The Decline and Fall of the Roman Empire*, Everyman's Library, New York and London, 1963, vol. 1, p. 1.

4   J. G. A. Pocock, *Barbarism and Religion* , vol.1, *The Enlightenments of Edward Gibbon 1737-1764* and vol. 2, *Narratives of Civil Government*, Cambridge University Press, 1999.

5   Warwick Ball, *Rome and the East: The Transformation of an Empire*, Routledge, 2000.

6   Fergus Millar, *The Roman Near East 31 BC – AD 337*, Harvard University Press, Cambridge MA and London, 1993.

7   Nicolo Machiavelli, *The Discourses*, Bernard Crick (ed.), Penguin, 1974. 'The republic to him was the best of all possible worlds ... Not merely does he have a coherent theory of the preconditions for republican rule, but in many vital respects it is more fully worked out than even Aristotle's – and there is no rival in between,' Crick commented. *The Discourses* were reflections on the origins of the Roman republic, as recorded in the first ten books of Livy's history of Rome from its foundations.

8   Charles de Secondat, Baron de Montesquieu (1689-1755), *The Spirit of Laws* (1748), was directly preceded by his own *Considerations on the Causes of the Greatness of the Romans and Their Decline* (1734).

9   Theodor Mommsen (1818-1903) was a specialist in Roman law who writes, almost as a recreation from his chief scholarly labours, a monumental and acutely perceptive history of the Roman republic as well as a masterful, but unfinished one of the Empire. See *The History of Rome*, J. M. Dent, London, 1911 (4 vols.) and *A History of Rome Under the Emperors*, Routledge, London and New York, 1992.

10   Livy, *Rome and the Mediterranean*, Book xxxiii:33, Penguin, trans. Henry Bettenson, 1976, p. 127.

11   Tacitus, *Agricola* #29, Penguin, 1970, p. 81.

12   A. G. Frank, *Re-Orient: Global Economy in the Asian Age*, University of California Press, 1998, p. 4.

13   Friedrich Nietzsche, *The Antichrist*, #59, Penguin, 1968, p. 182. See also on this point Bryan Ward-Perkins, *The Fall of Rome and the End of Civilisation*, Oxford University Press, 2006.

Cui dono lepidum novum libellum/arida modo pumice expolitum?
(To whom should I dedicate my clever new book,
fresh, sharp and polished as it is?)
- Caius Valerius Catullus (c. 55 BCE)[1]

Catullus was the leading figure of the new poets of the late Republic, breaking with the tradition of Rome's past and finding his models in Greek poetry, both in the polished Alexandrian style and in the direct lyricism of Sappho. His style is immensely versatile and whatever he writes is his own, so that he is one of the greatest of all poets ...
- Betty Radice (1971)[2]

In the period immediately following his death, Catullus' literary impact was enormous ... Both Virgil and Horace show his influence again and again ... Martial, whose ideal was to rank second after Catullus ... especially fancied ... the kiss and sparrow poems, thus setting a fashion that is still with us today.
- Peter Green (2005)[3]

4. On the Fate of Classical Erotic Poetry

In Umberto Eco's *The Name of the Rose*, a single copy of the long-lost treatise on comedy by Aristotle is found in the library of an obscure abbey – with dramatic consequences. One wonders why Eco did not, instead, have his monks stumble upon a codex of the long-lost poems of Catullus – passionate, cheerfully obscene, kaleidoscopic in their variety of metre and subject matter, free spirited and brilliantly learned. Have you heard of Catullus? You should have. He wrote a single book of poems, no more, and died at the age of thirty in 54 BCE; but if you read his verse now you open a doorway into the history of the West.

Catullus was best known in antiquity for the brilliant variety of his metrics, but has been most famous, since the recovery of his poetry in the Italian Renaissance, for his passionate love poems to Lesbia:

> Vivamus, mea Lesbia, atque amemus
>
> Rumorseque senum severiorum
>
> Omnes unius aestimemus assis!
>
> Soles occidere et redire possunt:
>
> Nobis cum semel occidit brevis lux
>
> Nox est perpetua una dormienda.
>
> Da me basia mille, deinde centum,
>
> Dein mille altera ...
>
>
> (Let's live, Lesbia mine, and love – and as for
>
> scandal, all the gossip, old men's strictures,
>
> value the lot at no more than a farthing!
>
> Suns can rise and set ad infinitum –
>
> For us, though, once our brief life's quenched, there's only
>
> One unending night that's left to sleep through.
>
> Give me a thousand kisses, then a hundred,
>
> Then a thousand more ...)[4]

If you never studied Latin, or never discovered its power and beauty due to poor tutoring or lack of the opportunity, revel in these lines from two millennia ago – the passionate individualism and vigorous self-expression of the West in a nutshell. For myself, I was taught Latin indifferently for five years at a Catholic secondary school, then told that, with everyone else at the school abandoning the subject, it was not worth taking it any further, even by correspondence. It would have been worth it – if only in order to have been able to drink deeply of Catullus (and Ovid) at a tenderer age.

Caius Valerius Catullus was a wealthy and witty wastrel of the last decades of the Roman Republic, a kind of Lord Byron of his time. His family owned a good deal of land in northern Italy. He was a gentleman of Verona, where he was born in 84 BCE and grew up. Rome, however, was a magnet to the likes of young Caius and he gravitated to it at the age of twenty, not to seek his fortune but to expend his wit and energy. It was the age of Catiline and Cicero, Caesar and Pompey, Crassus and Spartacus. The young poet rubbed shoulders with the wealthy and powerful, mastered the musical properties of Greek and Latin, wrote his verse and died, like Keats, of tuberculosis, when still young.

The Lesbia, to whom the above poem like many of his others, was addressed, was herself a wealthy woman who rubbed a good deal more than shoulders with the wealthy and powerful, as Cicero famously testified in his oration *Pro Caelio* – defending in court one of her aggrieved lovers. She was Clodia Metelli, one of the sisters of a notoriously populist aristocrat, tribune of the people and rabble rouser, Publius Clodius. She married a powerful senator, her cousin Quintus Metellus Celer, cuckolded him all over Rome, then poisoned him to get him out of the way altogether. She had large, luminous eyes and a quickness of both mind and hand that made her one of the great *femmes fatales* of all time.

Ten years older than Catullus, Clodia dazzled him, became his muse, but infuriated him with her relentless promiscuity, occasioning first his most intimate and then some of his most vitriolic verse:

> Passer, deliciae meae puellae
>
> Quicum ludere, quem in sinu tenere,
>
> Cui primum digitum dare appetenti
>
> Et acris solit incitare morsus,
>
> Cum desiderio meo nitenti
>
> Carum nescio quid lubet iocari,
>
> Et solaciolum sui dolores,
>
> Credo ut tum gravis acquiescat ardor:
>
> Tecum ludere sicut ipsa possem
>
> Et tristis animi lavare curas!

(Sparrow, precious darling of my sweetheart,

Always her plaything, held fast in her bosom,

Whom she loves to provoke with outstretched finger

Tempting the little pecker to nip harder

When my incandescent longing fancies

Just a smidgin of fun and games and comfort

For the pain she's feeling (I believe it!),

Something to lighten that too heavy ardour –

How I wish I could sport with you as she does,

Bring some relief to the spirit's black depression).[5]

This poem, overflowing with tender wit and a teasing play on 'sparrow' – a bird sacred to Aphrodite and its name slang for the penis – has inspired endless variations and adaptations for more than five hundred years since Catullus's verse resurfaced after a thousand years of near oblivion. But its memorable counterpoint, for which Catullus is equally famous, is the bitter rebuke to his muse for spurning him and turning her life into a veritable 'salax taberna', a bordello, the front of which he declares he'll cover with obscene graffiti:[6]

Caelius, Lesbia nostra, Lesbia illa,

Illa Lesbia, quam Catullus unam

Plus quam se atque suos amavit omnes,

Nunc in quadriviis et angriportis

Glubit magnanimos Remi nepotes.

(Caelius, Lesbia – our dear Lesbia, that one,

That Lesbia whom alone Catullus worshipped

More than himself, far more than all his kinsfolk –

Now on backstreet corners and down alleys

Jacks off Remus's generous descendants.)[7]

His affair with this Carmen of the late republic went through every variation in mood and each found expression in verse written some time between 62 and 54 BCE. Early, he longs to be able to cast in her path something like that 'golden apple' which 'long ago', thrown in front of 'the maiden runner' (Atalanta), 'freed, at last, a girdle too long knotted.'[8] He is magnanimous himself, at first, about her other loves. It is only after falling thoroughly in love and being spurned that his mood becomes dark and his verse bitter.

One late Lesbia poem is especially beautiful in this context and appears to suggest the poet had gotten over his bitterness, though not his sense of loss. Written in Sapphic mode, it is also Sapphic in mood, echoing the legendary poetess of Lesbos in singing the praises of Lesbia. Indeed, it seems to be an actual translation or appropriation of a poem by Sappho herself.[9] It must be a very god, or one who eclipses godhead, who is able, uninterrupted, 'spectat et audit', to look at you and listen to you 'dulce ridentem', sweetly laughing. But the thought of this:

> ... misero quod omnis
> eripit sensus mihi: nam simul te,
> Lesbia, aspexi, nihil est super me
> Vocis in ore.
>
> Lingua sed torpet, tenuis sub artis
> Flamma demanat, sonitu suopte
> Tintinant aures, gemina tenguntur
> Lumina nocte.
>
> (... sunders unhappy me from
> All my senses: the instant I catch sight of
> You now, Lesbia, dumbness grips my voice
> It dies on my vocal chords.
>
> My tongue goes torpid, and through my body
> Thin fire lances down, my ears are ringing
> With their own thunder, while night curtains both my
> Eyes into darkness).[10]

This is the poet who wrote the longish (408 hexameter lines) poem about Theseus's abandonment of Ariadne on Naxos,[11] two thousand years before Hugo von Hofmannsthal and Richard Strauss laboured to turn that Grecian tale of erotic betrayal into their semi-burlesque of an opera. It is not difficult to appreciate, given much else in Catullus's poetry, that he felt able to empathise with Ariadne, as with Sappho. He was no-one's idea of a dour and soldierly Roman.

'Graecia capta ferum victorem cepit et artis intulit agresti Latio,' Horace famously wrote in his epistle to the first emperor, Augustus, a generation after the death of Catullus.[12] Conquered Greece led captive its own conqueror and introduced the arts into rustic Latium. Catullus embodied that introduction, imbibing the pre-classical lyric poets and the Hellenistic masters like Callimachus in his youth. There seems to be no clear account of how he received his 'classical education' in Verona, but he so absorbed it as to be able to make it his own. In one poem, he writes to a boyfriend, Licinius, of what fun they had had the previous day, writing impromptu light verse, 'playing around with every kind of metre'.[13] Clearly, whoever his teachers were, they did not put him off his Latin, or his Greek – or poetry.

The late republic was a profligate time, the close study of which is endlessly illuminating in our own time, and Catullus captured its spirit in much the same way as, say, Karl Kraus, captured the somewhat different spirit of late imperial Vienna. I have always associated his brief and incandescent life with that of his contemporary, the extravagant and brilliant political entrepreneur, Caius Scribonius Curio, who died in Africa in 49 BCE, serving Julius Caesar in the civil war. Of Curio, the great historian Theodor Mommsen wrote, 'We may regret that this exuberant nature was not permitted to work off its follies and to preserve itself for the following generation ...'[14]

When a talented individual is cut off in the prime of life, or when on the verge of possibly outstanding achievements, it lends a kind of romantic aura to his or her life.[15] The trio of English Romantic poets, Byron, Shelley and Keats, is famous in this respect, Keats not least, since he died at just 25. The death of Franz Schubert at 31, or of Mozart at 35, are other famous cases, though the enormous and extraordinary creative achievement of both of those composers makes it difficult to imagine what they might have left undone. In the case of Catullus, it seems possible that he would have gone on to write works of the calibre achieved by his successors, Virgil and Horace. Instead, he left behind only a single book of poetry.

That book of poetry however, marked him out, in the opinion of the finest Latin poets of the following generation, as a master to emulate. Yet, while their work survived the downfall of the ancient world, six centuries later, the book of Catullus's poetry almost did not. The history of that little book, over the two millennia since its author died, takes us through a virtual history of books, as such, of education and of the art of poetry. The hazards of copying and editing, censorship and emulation through which it has passed remind us just how precarious was the life of books before the invention of printing and how powerfully that technology has reinforced our collective capacity to preserve and propagate our cultural inheritance.

Given its impact and reputation for two hundred years after Catullus's untimely death, it is plain that the book must have been copied and kept in the great public libraries of Rome during that time. The first such library was, in fact, built in the 30s BCE, a generation after the poet died, by Asinius Pollio, an author himself and a friend of poets. A decade later, Augustus himself erected the Library of the Temple of Apollo on the Palatine Hill, some fragmentary ruins of which still remain.[16] These libraries had separate Greek and Latin sections and Catullus's book will have lodged in the Latin sections, in all probability, for as long as the libraries endured.

However, it was not, as Peter Green observes, the kind of book that conservative schoolmasters or the censors of the Empire would have looked upon with favour. Two generations after Catullus's death, the great erotic poet Ovid was banished to the Black Sea coast by Augustus because his book of poems, *The Art of Love*, was believed to be contributing to the degeneration of public morality. This was roughly the equivalent of *Lady Chatterley's Lover* causing a scandal and a tussle over freedom of expression, in the early 1960s. Catullus is unlikely to have enjoyed greater favour, even under the Empire, and so copies of his work, if they were not removed from libraries, are not likely to have multiplied, save in a few private collections.

The rise of the Catholic Church and the downfall of the Empire, between them, nearly saw the end of Catullus's little book. It is easy to overlook how very much, in general, was lost at that time and, given the enormous superfluity of our current holdings, not to feel the loss. It is also easy to overlook the significance of those unusual books, like Catullus's poetry, that only just survived – representative of a good deal that did not. Peter Green puts Petronius's *Satyricon* and 'the puzzling extracanonical plays of Euripides' in this category. 'What unites all these,' he comments, 'is their oddness, their unpredictability, their deviation from the norm – which suggests that, if our literary heritage from the ancient world were more complete, our view of it might be radically different.'[17]

Lost for centuries, seemingly beyond recall, the Catullan corpus was rediscovered in the form of a single copy by the Bishop of Verona itself in 965 CE. That copy, the so-called *Codex Veronensis*, then vanished for another 300 years, only to be found again in 1290, still in Verona – being used as a bung in a wine barrel. Nothing at all is known of its fate in the interim. The *Codex Veronensis* was then lost, apparently for good, but not before a copy had been made. That copy, too, was lost, but not before it had itself been copied twice. One of *those* copies belonged to the great Italian sonneteer Petrarch, but has since been lost.

Another, made in 1370, resides still in the Bodleian Library at Oxford and is known as the *Codex Oxoniensis*. Petrarch's copy was itself copied and, after immense labour by Renaissance scholars in the fourteenth and fifteenth centuries, was printed for the first time in 1472 – along with the poetry of Tibullus, Propertius and Statius. This was the high Renaissance and 'Graecia capta' and fallen Rome alike were surging back into European life and letters so vigorously that, within half a century, Christian 'Wahhabists' like Martin Luther would denounce the Catholic Church itself as pagan and repudiate its magisterial tradition in order to concentrate theological authority in the Bible alone.

Catullus was never again, however, in danger of extinction – thanks to the printing press. His influence on English poetry and letters has been enormous, from Wyatt to William Carlos Williams, from Herrick to Tennyson, from Jonson and Pope to Yeats and Pound – and the Lesbia poems have always been especially loved. It was Montaigne's tutor, Marc-Antoine de Muret, who, in 1552, is said to have identified Catullus's Lesbia as the wanton Clodia, but most of what we know of Catullus's life and the way in which it is reflected in his poetry was not pieced together until, in 1862, Ludwig Schwabe painstakingly reconstructed the great love affair between the poet and his muse.

A great labour of archaeology, such scholarship has excavated the long dead and given us back one of the treasures of antiquity. Peter Green's masterful edition, *The Poems of Catullus*, published in 2005, brought both his own life's work as a classicist and the collective work of many other scholars over half a millennium to a burnished completion. The richness of his commentary, the passionate care expended in working Catullus's metrics into a form of English that does not jar on the ear, the great gift of a bilingual edition so that we can enjoy the originals alongside the faithful translator's renditions of them, make this a book to be prized.

Truly, such an edition invites the reader, as Green remarked in his preface, 'to study and enjoy an ancient poet who can be, by turns, passionate and hilariously obscene, as buoyantly witty as W. S. Gilbert in a Savoy opera libretto, as melancholy as Matthew Arnold in "Dover Beach", as mean as Wyndham Lewis in *The Apes of God*, and as eruditely allusive as T. S. Eliot in *The Waste Land*.'[18] Yet it is to an earlier translator, Peter Wigham, that I turn for a piece of verse which tells the extraordinary story of Catullus and his recovery in a manner Catullus himself might have enjoyed:

> Wine stains the verse;
> The curse of time obliterates the arrogant line.
>
> Then, in Verona, Campesani knows
> The 'Roman hand':
> 'One woman could command this song.'
>
> He sang
> And fourteen hundred years
> Later, it reappears –
> In the barrel's bung
> (the hand that Campesani knows)
> codex from wine-bung springing,
> as from the dung –
> the rose.[19]

In this spirit, prompted by the caustic and talented poet of 2,000 years ago, we should all of us feel free to claim as our own and enjoy the riches of the classical tradition of the West.

4.   *Endnotes*

1   These are the opening lines to the first poem in the Catullan corpus. The translation is my own. Peter Green, in his new edition of the poems, renders them 'Who's the dedicatee of my witty new booklet, all fresh-polished with abrasive?' Peter Wigham, forty years ago, translated them 'To whom should I present this little book so carefully polished?'

2   Betty Radice, *Who's Who in the Ancient World*, Penguin, Harmondsworth, 1971, pp. 85-6.

3   Peter Green, *The Poems of Catullus*, University of California Press, 2005, p. 45, Poem #1.

4   Ibid., pp. 48-9, Poem #5.

5   Ibid., p. 45, Poem #2A.

6   Ibid., p. 85, Poem #37.

7   Ibid., p. 105, Poem #58A.

8   Ibid., p. 45, Poem #2B (a fragment).

9   Ibid., p. 37. In his notes on Catullus's use of the Sapphic strophe, Green remarks that, in the labour of rendering this Greek and Latin metric into passable English verse, he owes much to Swinburne, 'a poet not much in favour these days, but he was a master metrist, and I am glad to acknowledge what I have learned from him.'

10   Ibid., p. 99, Poem #51.

11   Ibid., pp. 133-157, Poem #64.

12   David Ferry (trans.), *The Epistles of Horace: A Bilingual Edition*, Farrar, Straus and Giroux, New York, 2001, p. 122. Ferry renders the lines 'Captive Greece took its Roman captor captive; invading uncouth Latium with its arts.'

13   Green, *The Poems of Catullus*, pp. 96-8, Poem #50.

14   Theodor Mommsen, *The History of Rome*, vol. 4, p. 370, J. M. Dent and Sons, London, 1911.

15   We are reminded of this, in midsummer 2008, with the sudden death of the brilliant young Australian actor Heath Ledger, at the age of 28, even as his career blossomed.

16   Lionel Casson, *Libraries in the Ancient World*, Yale University Press, 2001, pp. 80-82.

17   Green, *The Poems of Catullus*, p. 19.

18   Ibid., p. xiii.

19   Peter Wigham, *The Poems of Catullus*, Penguin, 1966, p. 11. This is also the note on which I concluded my own book of love poems, *Sonnets to a Promiscuous Beauty* (Barrallier, 2005).

Alors, pourquoi le ciel, les etres, la lumiere? A quoi bon l'univers? Thais va mourir! Ah! La voir encore! La
revoir, la saisir, la garder! Je la veux! Je la veux!
(Why, then, do heaven, living beings, or light exist? What good is the universe? Thais is going to die! Ah! To see
her again! See her, take hold of her, protect her! I want her! I want her!)
- Athanael, in Jules Massenet's *Thais* Act III, Sc 2[1]

Keep Ithaca always in your mind.
Arriving there is what you are destined for.
But do not hurry the journey at all.
Better if it lasts for years,
So you are old by the time you reach the island,
Wealthy with all you have gained on the way,
Not expecting Ithaca to make you rich.
- Constantine Cavafy,[2]

I have not lingered in European monasteries
... I have not worshipped wounds and relics,
or combs of iron,
or bodies wrapped and burnt in scrolls.
I have not been unhappy for ten thousand years.
During the day I laugh and during the night I sleep.
My favourite cooks prepare my meals,
My body cleans and repairs itself
And all my work goes well.
- Leonard Cohen[3]

5. On Classical Courtesans and Modern Opera

etween conservative religion and the lugubriously self-titled 'sex industry', militant feminist critiques of 'the beauty myth' and frankly erotic advertising, we are forever being confronted by the need to make judgements about the commercialisation of sexual desire. Fortunately, we have the fine arts to turn to as a defence against both puritanism and soulless hedonism. Jules Massenet's beautiful 1890s opera *Thais* is a case in point. It offers a rich reflection on the figure of the courtesan and the cult of Venus in late classical paganism through the prism of early Alexandrian Christianity and in *fin-de-siècle* Paris.

In Massenet's opera, the ascetic Athanael declares to his fellow monks that he will go to Alexandria and put an end to the shame and scandal being spread in the city by the famous courtesan Thais, 'a foul priestess of the cult of Venus.' Suppressing his passion for her extraordinary physical beauty and overcoming her taunts, he converts her to Christianity, then leads her to the desert for a life of penitence in a narrow cell as the bride of Christ. Back in his own monastic community, however, he finds himself haunted by her. 'In vain have I scourged my flesh!' he tells his spiritual master, Palemon. 'A demon possesses me! ... I see only Thais, Thais, Thais! Or rather, it is not her I see, but Helen and Phryne, Venus, Astarte, all the splendours and all the voluptuous delights in a single being!' The conversion had been mutual: she to Christ, he to Venus.

Helen and Phryne, Venus, Astarte are what Athanael sees in Thais. Of these, only one was an actual historical woman. Helen, of course, is the legendary adulterous beauty of Mycenae, whose face 'launched a thousand ships' in war against Troy. Venus and Astarte are different names, Latin and Levantine, for the goddess of erotic love, whom the Greeks called Aphrodite. Phryne was a real woman, in the Grecian world, more than 600 years before Thais. Both in her day and ever since, she has been celebrated as the greatest of all courtesans in the Western world, the closest there has been to a mortal incarnation of Aphrodite. It was, in a sense, *she* who defeated Athanael's asceticism, which had itself defeated Thais.

Phryne had as lovers and clients many of the most talented men of her time, including the great sculptor Praxiteles and the great painter Apelles – between them, the Michelangelo of classical Athens. She was the model for Praxiteles' breathtaking female nude, *The Venus of Cnidus*, which spread her image throughout the classical world. She was also the model for Apelles' famous painting *The Birth of Venus*, which inspired Botticelli's celebrated Renaissance painting of the same name. Geoffrey Grigson's *The Goddess of Love* has on its cover a picture of the Benghazi Aphrodite which was itself inspired by Apelles' painting of Phryne as Aphrodite.[4] Reportedly, Phryne became so rich that, after the Macedonians razed the city of Thebes, she offered to pay for the walls to be rebuilt. She asked only that the citizens put up an inscription reading, 'Alexander may have made it fall, but Phryne the hetaera got it back up again' – a cunning pun on the work of Aphrodite.

James Davidson tells us, in *Courtesans and Fishcakes: The Consuming Passions of Classical Athens*, that Phryne was put on trial for impiety. The great lawyer Hyperides, who was also one of her lover/clients, defended her with his greatest speech. Yet even this failed to win over the male jury. He then led Phryne herself into view, tore off her tunic, exposing her breasts, and appealed to the jury not to condemn to death for impiety 'Aphrodite's representative and attendant'. The jury were apparently so overawed at the sight of Phryne's perfect breasts that their resistance collapsed and they acquitted her at once.[5]

It is a great pity, surely, that we do not have a body of writing by Phryne: diaries, letters, memoirs or reflections on the society and politics of her time, on the characters of her lovers and clients and on the nature of her sexual experience. *The Memoirs of Phryne*, I think, would be at least as interesting as the diaries of Anaïs Nin. A study of her based on her own letters and reflections would be intrinsically even more interesting than Margaret Rosenthal's biography of the sixteenth-century Venetian courtesan Veronica Franco.[6] Marguerite Yourcenar wrote a splendid, semi-fictional biography of the Emperor Hadrian,[7] but no-one, to my knowledge, has seriously attempted to do the same for this astonishing icon of female beauty and sexual freedom. Living, as she did, one hundred years after Aspasia, the great courtesan companion of the Athenian statesman Pericles, a generation after Plato and at the same time as the aging Aristotle, she might have written the most fascinating of social and sexual reflections. Alas, she did not, apparently, do so. We have her image, therefore, sculpted and painted by her great male contemporaries, but we do not know what it was like to *be* her.

Can we, however, on the basis of what we *do* know, condemn her as 'a foul priestess of the cult of Venus'? Athanael, in the end, could not condemn Thaïs in such terms, and the classical world celebrated the glorious Phryne as the very epitome of beauty. It was, perhaps, a reflection of Massenet's *fin-de-siècle* humanism and his infatuation with his American soprano, Sybil Sanderson, that he could enter into Athanael's psyche as he did. But the courtesan was a major figure in the France of Massenet's era. There were many famous courtesans in Paris, the capital of the nineteenth century, as Walter Benjamin called it. Among them, at least two inspired major works of art: Marie Duplessis (1824-47) whose early death led Giuseppe Verdi to compose *La Traviata* and Apollonie Sabatier (1822-1889) whose beauty inspired poems by Theophile Gautier and Charles Baudelaire.[8]

The Catholic Church, on the other hand, did and does condemn the cult of Venus.[9] The Catholic bishop and ecclesiastical historian Eusebius, writing in the fourth century CE when Massenet's opera is set, denounced what he called 'the foul demon known by the name of Aphrodite'. Yet eight hundred years earlier, one hundred years before the birth of Phryne, the philosopher Empedocles had hailed Aphrodite as the true deity of the human Golden Age. She was, he thought, the Queen of Life and Love, whom human beings worshipped not with blood sacrifices but 'with sacred presents, with painted images and with subtle perfumes' – before the descent of humankind into the Iron Age of domination and war. What led to the inversion of Empedocles' judgement, such that Aphrodite could be described as a foul demon and her courtesan incarnations as her foul priestesses?

It would be playing to the gallery merely to say that Christianity was to blame. Similar shifts can be found in other cultures without Christianity having anything to do with it. Various forces, both of darkness and of light, are at work in the age-old repressive reactions to both female sexual freedom and the commercialisation of carnal desire. In Massenet's opera, the heart of the matter is dealt with exquisitely in terms of the soul of Thaïs and the psychology of her dramatic conversion from priestess of Venus to bride of Christ. When she first comes on stage, she is the 'Rose of Alexandria', the idol of the pagan crowd, whose performances in mime of the loves of Aphrodite drive the crowd to delirious idolatry. Yet she greets her client Nicias, who has blown his entire fortune on her and written her three books of elegies, with words filled with irony and a hint of bitterness. 'Thaïs the frail idol comes for the last time to sit at your flower-strewn table', she tells him. 'Tomorrow I shall be for you no more than a name.' Our clue lies in the expression 'frail idol.'

When Nicias introduces her to Athanael, the 'frail idol' plays the haughty priestess. She taunts Athanael about the fire in his eyes and tells him, 'You have not tasted the cup of life … Sit down among us, wreathe yourself in roses; the only truth is love, stretch out your arms to love.' When he declares that he is the servant of a stronger deity and has come to save her soul, she mocks him, 'Dare to come, you who defy Venus!' After the feast, when she is alone, however, the extent and nature of her frailty are revealed. 'Ah! I am alone, alone at last!' she cries. 'All these men are unfeeling and gross. The women are spiteful and the hours weigh heavily … My soul is empty … Where can I find rest? And how can I preserve happiness? O my faithful mirror, comfort me! Tell me that I am fair and that I shall be fair for ever! For ever!' How many such sighs might have found their way into the diaries of Phryne, had they been written?

The anxieties that prey on Thaïs are of a peculiarly self-conscious, human kind. In the idolatry of her clients and admirers she experiences only an infatuation without depth of feeling for a beauty which cannot endure. She dreads what may come after and finds the hours unsustained by deeper love tedious and hollow. It is as if she was the soul of Phryne imprisoned within the beautiful statue carved by Praxiteles. It is the statue that they adore, not me, she thinks – and what if the statue should crumble with age? Yet, when Athanael returns and promises her 'I love you in the spirit. I love you in truth. I promise you more than the flowery abandon and the dreams of a brief night! The happiness that I bring you today will never end!' she is initially defensive and even mocking. 'You come too late to teach me anything about love, my friend', she tells him, 'since I already know every pleasure.'

Aroused by her mockery, he bursts out in anger, playing unknowingly on her deep anxieties, cursing 'the death which is within you.' Terrified by his ascetic severity, she throws herself tearfully at his feet, pleading, 'Do not harm me! ... it is no fault of mine if I am beautiful. Mercy! Do not let me die!' You will not die, he promises, but will have eternal life – a moment of poignant dramatic irony, since it is precisely her death by which he will later be shattered. She hesitates, steps back into defiance, telling him 'No! I remain Thais. Thais the courtesan.' Overnight, however, she has a change of heart and, the next morning, goes to him and tells him 'Your words have lodged in my heart like a heavenly balsam.'

Beset by conflicted longings himself, Athanael has touched the soul of Thais as Nicias and her other clients had never done. He has addressed in her the thirst for something deeper and more profound than the role of courtesan/supermodel/priestess of Venus. In an oasis en route to the desert convent where she will be confined, there is a sublime moment where this something deeper is briefly realised. Having driven her to the destruction of her possessions and the near collapse of her body, he relents out of compassion. They pause for rest. He kisses her bleeding feet, brings her fresh fruit and refreshes her lips and hands with cool water. In soliloquy, while he is fetching the fruit and water, she sings, 'Oh messenger of God, so harsh and yet so good, bless you, for you have opened heaven to me. My flesh is bleeding, but my soul is full of joy ... and my spirit, free from the earth, now soars in that vastness ...' They sing together, 'Bathe my hands and lips with water ... My life is yours, God has entrusted it to you ...' Thais declares 'Everything intoxicates me!' and Athanael is himself filled with a 'supreme sweetness' and a profound sense of spiritualised erotic intimacy.

The moment when Athanael leaves Thais with the White Sisters of the desert and realises that he will never see her again has aptly been described as one of the most powerful dramatic moments in nineteenth-century opera. Back in his hermit's cell, besieged by erotic imaginings of her, he suddenly has a vision of her dying. 'Thais va mourir!' he cries out, in despair: 'Thais is going to die!' This sets the scene for the truly extraordinary finale, act 3, scene 3. As Thais dies, remembering their 'radiant journey' through the desert and seeing the angels and saints coming to bring her to God, Athanael renounces all belief in God and Heaven and declares that he remembers 'only your mortal beauty' and pleads with her to live and be his. What was Massenet offering us? A celebration of Christian spiritual redemption or a brilliant denunciation of it as a cruel delusion? Neither and both.

The theme of erotic passion and its spiritual tumults runs through opera from Monteverdi's *Orfeo* and *Poppaea* to Berg's *Lulu*. As Peter Conrad argues in his reflection on the meaning of opera, *The Song of Love and Death*, music is able to convey the immense range of feelings involved more powerfully than any non-musical drama. This is because music reaches beyond the possibilities of coherent or credible verbal utterance in the very way that erotic longing reaches far beyond the bounds set by coherent or credible moral sanction or physical reality. The 'foul priestess of Venus' is, consequently, the single most notable presence in opera. Yet the genre does not merely constitute the cult of the 'frail idol', the beautiful body. Rather, it explores musically the drama of the pagan soul, both male and female, in its quest for passionate intimacy, freedom and meaning in a world of fickle desires, physical frailty and mortality. More completely than almost any other opera, Massenet's *Thais* embodies the profound insight of Stendhal that, where love overcomes mere lust, 'the soul envelops the body.'

Carole Pateman asks in *The Sexual Contract*, 'what's wrong with prostitution?' One possible answer, remembering Phryne and Thais, could be: 'The same thing, in sober fact, that is wrong with all too many marriages and *de facto* relationships: the soul does not envelop the body, but is its more or less anxious prisoner.' 'Loving Thais' might be a useful metaphor for recognising this and living by its wisdom of care and imaginative sympathy in any erotic encounter one has, whether with an avatar of Phryne, an incarnation of Eros, a true Muse or a marital partner. The anxieties of Thais and Athanael's shattering loss indicate the boundaries within which the music drama of the soul must be played out.

It is the arts, from painting and sculpture to poetry, drama and music, but opera in particular, which show us the way to this Empedoclean world. It might be called, to coin another metaphor, 'the Corinth of the soul'. For Corinth was the luxury city and courtesan capital of Greece in the age of Phryne. Many libertines think that love is merely a matter of carnal gratification, while the religious and neurotic claim that it should be 'pure'. Some of the latter still go so far as to 'curse the flesh' with the ascetics rather than blessing it. To them one can only say, in the words of Horace, '*non cuivis homini contingit adire Corinthum*' – 'Not everyone makes it to Corinth.'[10]

5.   *Endnotes*

1   Jules Massenet (1842-1912), *Thais*, Choeur de l'Opera de Bordeaux, Orchestra National de Bordeaux Aquitaine, conducted by Yves Abel, with Renee Fleming and Thomas Hampson, Decca, 2000.

2   C. P. Cavafy, *Collected Poems* (rev. edn.), edited by George Savidis, trans. Edmund Keeley and Philip Sherrard, Princeton University Press, 1992, pp. 36-7.

3   Leonard Cohen, *Stranger Music: Selected Poems and Songs*, Vintage Books, Random House, New York, 1993, p. 18.

4   Geoffrey Grigson, *The Goddess of Love: The Birth, Triumph, Death and Return of Aphrodite*, Stein and Day, New York, 1977.

5   James Davidson, *Courtesans and Fishcakes: The Consuming Passions of Classical Athens*, Fontana Press, HarperCollins, 1997, p. 134.

6   Margaret Rosenthal, *The Honest Courtesan: Veronica Franco, Citizen and Writer in Sixteenth Century Venice*, University of Chicago Press, 1992. Franco (1546-1591) was the most famous and sophisticated courtesan of her time.

7   Marguerite Yourcenar (1903-1988), *Memoirs of Hadrian* (1951), Penguin, 1986. Born Marguerite de Crayencour in Brussels, Yourcenar was brought up and educated by her French father. She was reading Racine and Aristophanes (in French) by the age of eight and both Latin and Greek by the age of twelve. Her first book of poetry was published when she was eighteen – paid for by her father. He died when she was twenty-four and she spent the rest of a long life travelling and writing. She was a fine essayist. See especially *The Dark Brain of Piranesi and Other Essays*, Farrar, Straus, Giroux, New York, 1984; and *That Mighty Sculptor, Time*, Noonday Press, Farrar, Straus, Giroux, New York, 1993.

8   Joanna Richardson, *The Courtesans: The Demi-Monde in Nineteenth Century France*, Phoenix Press, London, 1967, p. 107.

9   For a particularly scholarly and fascinating reflection on sexuality and the first centuries of the Catholic Church, see Peter Brown, *The Body and Society: Men, Women and Sexual Renunciation in Early Christianity*, Columbia University Press, New York, 1988.

10   Quoted in Grigson, *The Goddess of Love: The Birth, Triumph, Death and Return of Aphrodite*, p. 111.

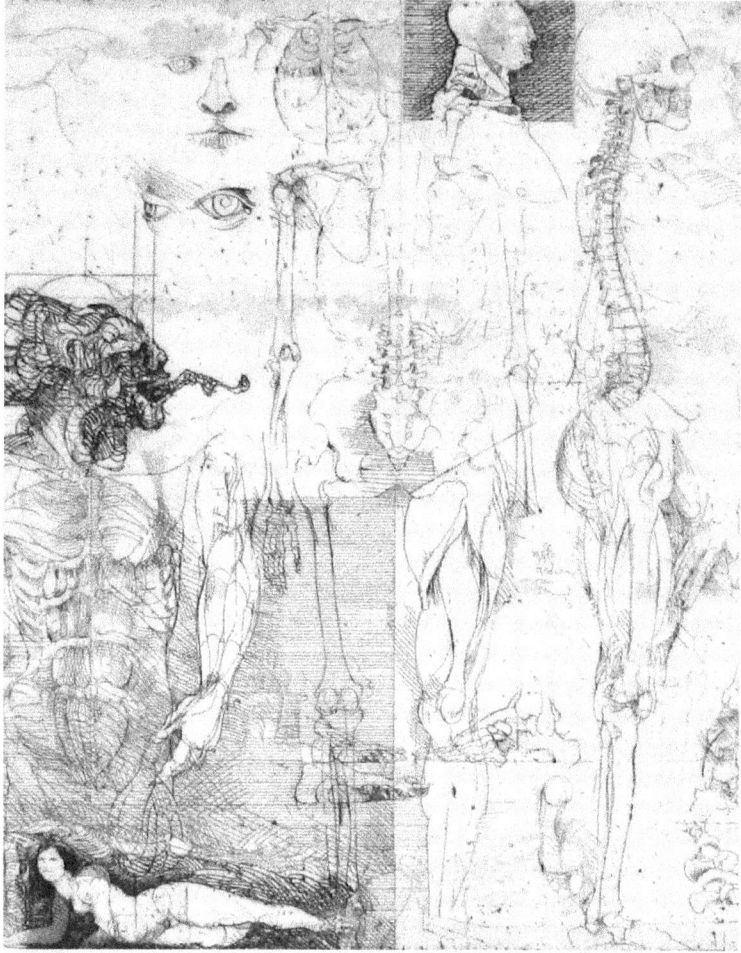

JS 110, *Sketches Also After Dürer* (detail)
Jörg Schmeisser, etching, 1974.

Boxing ... was especially favoured in Italy along with bloody combat in the arena. As the most brutal among popular sports, it appealed to the robust taste for violence which the Romans shared with the Etruscans. Boxing naturally lends itself to simple violence, and in the ancient world its object was to inflict damage as directly as possible. Since the head, not the body, was the principal target, the chance for a quick kill (literally, at times) which made rest periods between rounds unnecessary was enhanced by the caestus or brass knuckles (small metal dumbbells held in the fist).

- Paul Plass[1]

In England (in the 18th and 19th centuries) there evolved some new and broadly based attitudes toward games and competitions and athletes and their performances. These new notions favouring equal (sporting) opportunity, fair play, codified rules, training, transregional leagues and referees had striking analogues in English social and economic life, which were being transformed ...

- Richard Mandell[2]

Sunday had a feel of its own. It was a special day: the day we went to church and the day we watched 'World of Sport'. These two institutions kept the earth spinning safely on its axis.

- John Harms[3]

6. On Sport, Ancient and Modern

'You have to be born somewhere', remarks John Harms in a delightfully folksy rumination on Australian Rules football as a way of life published in 2002. The book is called *Loose Men Everywhere* and, from a decidedly provincial, Australian perspective, it offers an unusual point of entry to reflecting on the nature of sport as such and its role in the cultural and psychological life of complex human societies, ancient and modern.

'I could have been born into some comfortable bluestone villa in Hawthorn', Harms comments, '... or I could have been born into a housing commission fibro in Footscray ... I wasn't. I was born in a little town cut out of the brigalow scrub on the rich black soil of the western Darling Downs. The town of Chinchilla would have been fine if I'd been born into a Catholic rugby league family, or was the son of some sun-spotted rugger-loving grazier who'd been to school at Churchie in Brisbane. Things might have made sense then. But I was born into a life of confusion and frustration, a life of upset and despondency, a life of tears. I was born a Geelong supporter. Not only was I born a Geelong supporter but I was born in exile; a Geelong supporter in the Queensland bush, where people had no idea of what footy was or meant.' He was also born a Lutheran. More importantly, he was lovingly nurtured as both a Geelong supporter and a Lutheran. And both were matters of childhood faith. 'We knew. We didn't believe: we *knew*', he remarks of his Lutheran upbringing. 'And we belonged. We always belonged.' And Geelong? One knew, likewise, and belonged.[4]

Reading *Loose Men Everywhere* was a complete delight.[5] At countless points I was able to identify with exactly the childhood and adolescent experiences Harms recalls so fondly and warmly: listening to 3KZ radio broadcasts, collecting Scanlan's footy cards, watching black and white TV football replays, keeping scrapbooks of clippings from the *Sun* and *Sporting Globe* ... on and on. The only differences are that I was born a Catholic and a Collingwood supporter. Confusion and frustration, upset and despondency were every bit as much my experience as they were those of Harms. Of course, unlike the innocent Harms, the Lutheran and the Geelong supporter, I doubtless *deserved* my sufferings. After all, I was a Catholic. And a supporter of Collingwood – the black and white club that all others hate with a particular vehemence.

Harms relates that he was born in the ninth round of the 1962 season of the old Victorian Football League. I was born out of season in the midst of Norm Smith's reign as coach of Melbourne during which he won them six premierships, three of them at Collingwood's expense. Norm Smith, along with his less famous brother Len, was one of the shapers of the modern game of Australian Rules football. The Smith brothers were mentors to Ron Barassi and Tom Hafey, two of the greatest football coaches of the 1960s and 1970s, who won eight premierships between them, several of them at Collingwood's expense.

These men brought a revolutionary physical mobility, tactical flexibility and athleticism into the game and, in the process, made my young life miserable. Harms thinks it was frustrating in those years seeing Geelong seldom make the finals. As a Collingwood supporter, one had the far more devastating experience of seeing them almost invariably make the finals, only to be conquered by teams coached by Barassi and Hafey.

In both 1969 and 1973, Collingwood finished on top of the league ladder, only to lose first to Carlton in the second semi-final and then to Richmond in the preliminary final. In 1970, famously, Collingwood, having again finished on top of the league ladder, defeated Carlton in the second semi and led them by 44 points at half time in the grand final. I was there. This was going to be a great and glorious victory!

But Barassi's Carlton came from nowhere in the third quarter and took the game away, winning by ten points. The heavens fell in. How could this be? 'Quiet confidence is a wholly inappropriate feeling when you are a Geelong supporter', Harms concludes, watching his team lose several grand finals in the 1990s. The same, discomforting sense of the fragility of hope and the treacherous nature of reality overtook me as a bewildered boy that September day in 1970.[6]

In 1977, it appeared that the world would be set to rights when the marvellous Tommy Hafey came to Collingwood as coach. The results were immediate and impressive. Having been wooden spooners in 1976, suddenly, under Hafey's disciplined guidance, Collingwood was beating everyone. The match with Carlton, halfway through the season, set the seal on this revival.

In front by 44 points at half-time, which was ominously reminiscent of the 1970 grand final, Collingwood came out in the third quarter and absolutely trampled Carlton, to extend their lead to 98 points by three-quarter time. The Collingwood crowd was jubilant. Sweet, sweet revenge! 'Kill! Kill! Kill!' they chanted, as Collingwood piled on the goals.

The illusion of revenge was short-lived, of course. Collingwood was beaten in the 1977 grand final by North Melbourne – coached by Barassi. It was then beaten in both the 1979 and 1981 grand finals – by Carlton – and, for good measure, in the 1980 grand final, by Richmond. Hope, it turned out, had become even more fragile and reality more treacherous than ever!

I've never forgotten, however, that barely sublimated bloodlust in the winter of 1977. At this remove, it prompts a reflection on what I think is the most significant passage in *Loose Men Everywhere*. That passage comes towards the end of the book where Harms reflects, very briefly, on European sports compared with Australian Rules football.

'In Rome', he writes, 'I walk. As I head down from the Forum, I get my first glimpse of the Colosseum. It looks just like the MCG when you walk from the city along Brunton Avenue. I go inside and sit on a broken column which has been smoothed on top at about Bay 13. I hear the screams for Gary Ablett.'[7] What has changed?'[8]

It doesn't take a lot of imagination, of course, to see the Colosseum as a sort of MCG of the ancient world. The famous Roman stadium had only half the crowd capacity of the MCG – about 50,000 spectators – but its structure invites comparison with the great Melbourne stadium. The really awesome Roman stadium was the Circus Maximus, which truly was colossal, having a peak capacity, reportedly, of 385,000 spectators at its height in the fourth century CE. While that figure seems hardly credible, the arena was indeed gigantic: 600 metres in length and 118 metres in width, with three great tiers of seats around its circumference.[9]

When I first visited Rome, in 1985, I stood on the edges of the Campus Martius where the Circus Maximus used be – before it was dismantled by the medieval Popes to provide masonry for churches. Here, I reflected, crowds gathered for almost a thousand years to witness sporting contests, most famously *Ben Hur*-style chariot races. I had the uncanny experience of 'hearing' the ghostly sound of crowds roaring. The hair stood up on the back of my neck. This was the greatest 'sporting' arena of the ancient world, perhaps of all time. But, Harms asks, what has changed? A great deal, actually. And this change is of the greatest significance and importance.

While the Collingwood crowd, in the winter of 1977, chanted 'Kill! Kill! Kill!', there was, in fact, no killing that day at the football. The Colosseum, by contrast, was an arena dedicated precisely to killing. Anyone raised as a Christian, as both Harms and I were, will have heard tales of Christians thrown to the lions by the pagan Romans. Christians, however, were the least of it. Killing itself, not the sublimated idea of it, was mass entertainment in imperial Rome for some four or five hundred years.

When the Colosseum was first opened, the Emperor Titus celebrated by providing what the Roman historian Suetonius described as 'a most lavish gladiatorial show ... and a wild beast hunt, 5,000 beasts of different sorts dying in a single day.'[10] Gladiatorial contests had begun on a small scale centuries earlier as grim rituals associated with funeral rites. But as the Empire grew, they became ever more brutal and extravagant in nature until thousands of human beings and animals could meet their deaths in monstrous public spectacles.

The Romans were notoriously addicted to 'bread and circuses' but their circuses were singularly bloodthirsty, entailing less screams *for* than the screams *of* the Gary Abletts of the era. In a brilliant study of this phenomenon, Paul Plass has argued that the brutality of the Roman games grew in direct proportion to the deepening of arbitrary imperial rule. It was also, he suggests, a reflection of the seething psychological tensions in a society based on war and slavery in which violent death was an endemic phenomenon.[11]

While the Roman excess of violence was stunning, ritual games in the ancient world were, in general, more violent than modern sports. This seems to have been due, in part, to the chronic and grim nature of warfare in the Iron Age. 'Warfare', as Michael Poliakoff points out, 'was an inescapable reality of the ancient world, and it exercised profound influence on civic and political life; it is impossible to understand antiquity in general, to say nothing of its predilection for violent games, without considering what war meant in day to day concerns.'[12]

The bloody excesses of the Roman circus were not, however, merely a reflection of grim social utility. They were a cancerous development, shot through with anomaly and irrationality. There were atavistic elements of human sacrifice involved and a sort of sickness or demoralisation, which led the first century CE philosopher Seneca to remark that he came away from among human beings at the arena feeling less human. The cruelty and depravity of these games was a constant target of Christian criticism and it is astonishing that the Christian Harms was not alive to this in his visit to the Colosseum in the 1990s.

It is not only cruelty and excess, though, that set the Roman games apart from modern sports in general and Australian Rules football in particular. It is the very idea of sport itself as a rule-governed and fair proceeding with referees and equality between contestants. This is a distinctively modern development. Indeed, it is a distinctively *English* development, arising in the eighteenth and nineteenth centuries and spreading from England to the rest of the world.[13] The now universally used word 'sport' is English in derivation.[14]

'Games in general are rule-bound procedures reflecting deeper social patterns and performing several complementary functions', Paul Plass remarks. But the emergence of fastidiously *fair* rules, of the Marquis of Queensbury variety, designed to limit brutality and put a premium on 'sportsmanship' are what make modern sports distinctively different from earlier ritual or competitive activities. The transition or mutation is epitomised by no single sport so much as by cricket; but even tennis bears the marks of the transformation, as James Joyce indirectly testified in his swipe at the verbal gentility of 'Alfred Lawn Tennyson'.

Harms shows himself, in his book, to be very much heir to this transformation and a 'sport' to a fault. Nothing indicates this more clearly than his reaction to the dramatic opening of the (among Australian Rules fans) famous and exceedingly vigorous 1989 grand final. At the first centre bounce,

Geelong's Mark Yeates ran straight at Hawthorn's magnificent champion, Dermott Brereton, and laid him out with a crunching hip and shoulder blow. 'In an instant, all sense of fun and play disappear from the match,' Harms exclaims, looking back. 'This isn't how I like footy. This isn't Geelong ...' he remembers thinking, even at the time.

This isn't how he likes footy? What? The Hawthorn ace was cut down and a stupendous goal kicked in the first minute of play by Gary Ablett. But, thinks the gentle Harms, it isn't *sporting* to do that sort of thing – take champions out of the play by shirt-fronting them off the ball. 'What is happening here? This isn't how we play,' he laments. 'We're not barbarians, head-hunters; we're footballers.' Not barbarians? Not Romans either! Not gladiators or those who enjoyed watching them kill one another. The MCG is not, after all, the Colosseum. Much has changed.

The 1989 grand final was, surely, one of the very greatest of Australian Rules contests, for all Harms' sportsmanlike concerns. As far as the opening is concerned, I was always struck less by Dermott Brereton being felled than by his tremendous courage in taking an overhead mark moments later and, with determination written all over his youthful face, kicking a goal that set his team on the way to victory.

But the most unforgettable feature of that classic game was the nine goals kicked in peerless fashion by Gary Ablett, as Geelong fought back from a 40 point quarter-time deficit to get within six points at the final siren. That day, Ablett – widely regarded as the most gifted footballer in the game's history – gave an extraordinary demonstration of sporting prowess, the kind that legends are made of. Each goal was a gem, a sublime blend of power, poise and skill. Like Harms, I have long kept a video recording of the game. I watch it again and again because it's simply inspirational.

Sporting legends are an important part of Australian culture and of all cultures that have fallen under Anglophone and more broadly Western influence over the past two centuries. What makes Donald Bradman an enduring legend in Australia and around the world, wherever cricket is played, encapsulates the very idea of sport in the modern sense. It is not mere brute power, nor success alone. It is something more refined and interesting, something bound up with the whole modern concept of sport. Skill, discipline, character, fairness, leadership are all part of this. Above all, a certain existential graciousness, in which victory and defeat are *played at* bloodlessly and according to scrupulously observed decencies. That's cricket. That's *sport* in the modern, very English sense.

The Bradman legend is forever backlit by the brief, but luminous career of his contemporary, Archie Jackson. Jackson was to Australian cricket, in the 1920s and early 1930s, what John Keats was to early nineteenth century English poetry: the brief blooming genius, snuffed out by tuberculosis in his youth. In 1929, Bradman and Jackson were both at the very beginning of their cricket careers. Both were prodigiously talented. Yet Jackson, at nineteen, was already 'the finished batsman, the batsman who knows the one stroke for every ball ...[and] executes that stroke with an artistry that has no parallel', one contemporary journalist wrote.

He was, however, 'the shooting star of cricket: he died four years later of tuberculosis at the age of twenty-three.'[15] Bradman, by contrast, went on to a glorious career, interrupted, but not cut off by the Second World War. If Bradman, therefore, has almost the status of a cricket deity, Jackson has almost that of a sacrificial victim, even though it was disease not human violence that cut him down in his prime. Taken together, the two figures embody almost the whole psychological content of modern sport and the deeper, darker ritual roots which underlie it.

'Footy is about life in the face of death; hope in the face of despair. Footy is about the things at the very centre of existence, like religion is,' Harms concludes.[16] He is entirely correct, but it isn't just Australian Rules football that has this nature. It is sport itself. 'Sport is the ritual sacrifice of physical energy,' David Sansone tells us.[17] Its connections to religion are numerous and well documented. In some respects, it has even displaced religion as a ritual organisation of human experience, in the late twentieth century and the beginning of the twenty-first.

The Australian Rules grand final is an Australian high mass. It is a festive rite in which the fierce, but rule-governed, fair and peaceful competition, on which our modern, market-oriented civilisation is based, is enacted and celebrated as deep play.[18] The game of death in ancient Rome was a far darker ritual, for all its superficial similarities. Even when the Romans cheered ferociously for the Blues or the Greens in the Circus Maximus chariot races, their notions of competition were altogether less refined than are those of our football fans – even in Bay 13. We *insist* that the rules be applied with scrupulous and consistent fairness – however much we may hope that that will favour our particular team. That is what the *game* is all about.

Looked at in this perspective, sport is a kind of litmus test of civilisation itself. We look back, in so far as we still get any kind of classical education, on Athens and Rome as the progenitors of our civilisation; but the changes along the way have been vital. Harms, for some reason, overlooks, or omits to mention, the revolution through which Christian moral revulsion against the brutality of the circus and the passion for courtesans and fishcakes gave way to an era of asceticism and veneration of the Eucharist. What this deprived him of was the opportunity to celebrate the remarkable evolution of sport, from an often bloody and generally coarse spectacle into a festival of exuberant *restraint* and fair competition – under the aegis of English imperialism and industrialism. In its own way, quite as much as the verse of Catullus, our contemporary sports, too, represent the West in a nutshell.

6. *Endnotes*

1   Paul Plass, *The Game of Death in Ancient Rome: Arena Sport and Political Suicide*, University of Wisconsin Press, 1995.

2   Richard Mandell, *Sport: A Cultural History*, Columbia University Press, New York, 1984, p. xv.

3   John Harms, *Loose Men Everywhere*, Text, Melbourne, 2002, p. 31.

4   In 2007, Geelong dominated the Australian Rules football season and won a crushing victory in the grand final, defeating Port Adelaide by a record margin of over 100 points. Thus the tears for Harms' childhood were at last dried and being a Geelong supporter again became a matter of incredible pride.

5   Harms' little book is full of the most wonderful drollery which is, perhaps, typified by this victory song of the University of Queensland Aussie Rules team circa 1980.

> *Drunk last night, drunk the night before,*
> *Gunna get drunk tonight like we've never been drunk before,*
> *Cos here we are, happy as can be,*
> *We are the boys of the var-sitt-ee.*
> *Glorious, victorious, one jug of beer between the four of us,*
> *Cos one of us could drink the bloody lot without his pants on.*
> *One of us could drink the bloody lot.*
> *Roll over, Mabel, your navel's on the other side.*
> *You beauty, you beauty, you beauty.*

Ibid., pp. 122-3.

6   The most memorable single moment of that famous game was surely Alex Jesaulenko soaring over the shoulders of Collingwood's big man, Graeme Jenkin, in the second quarter, to take one of the all-time great 'speccies'. For years afterwards, we schoolboys would leap up on one another's backs, whether or not we were playing football, and yell out 'Jesaulenko!!'

7   A brilliant Geelong footballer of the 1980s and 1990s who is widely regarded as possibly the greatest football champion of all time.

8   Harms, *Loose Men Everywhere*, p. 206.

9   Samuel Ball Platner and Thomas Ashby, *A Topographical Dictionary of Ancient Rome*, Oxford University Press, London, 1929, pp. 114-120, authoritatively reports that the Circus Maximus measured a staggering 600 metres from end to end and could accommodate, at its peak, between 140,000 and 385,000 spectators. Pliny, in his *Natural History* (xxxvi: 102) in the first century CE, claimed the arena could hold 250,000 spectators while the fourth century CE official record, the so-called *Notitia*, claims it could contain 385,000. Both figures have been questioned on various grounds and the latter is highly implausible. The most scholarly estimate would seem to be in the vicinity of 200,000, still an immense number. According to the Byzantine historian Procopius, the last recorded games were held in the great arena in 550 CE, during the rule of the Gothic King of Italy, Totila. After that, the Circus Maximus began to fall into ruin and was dismantled by the Popes to build churches, though a considerable amount of it lasted for a thousand years. Now almost nothing remains.

10   Suetonius, *The Twelve Caesars*, trans. Robert Graves, Penguin, 1957, p. 290. According to Josephus, Titus organised repeated, massive spectacles of this nature in Judea during the Jewish War in 69-70. To celebrate his brother's birthday, for example, 'in the grand style,' he held 'games' at Caesarea in which 2,500 men 'perished in combats with wild beasts, or in fighting each other or by being burnt alive.' *The Jewish War*, Penguin, 1974, p. 364.

11   Plass, *The Game of Death in Ancient Rome: Arena Sport and Political Suicide*.

12   Michael B. Poliakoff, *Combat Sports in the Ancient World*, Yale University Press, 1987, p. 93. It was Plutarch (AD 46-125) who pointed out that the games staged by Achilles to mark the funeral of Patroclus in the twenty-third book of Homer's *Iliad* were all – boxing, wrestling, foot racing – 'imitations and exercises of war'.

13   Allen Guttman, *From Ritual to Record: The Nature of Modern Sports*, New York, 1978, p. 57.

14   David Sansone, *Greek Athletics and the Genesis of Sport*, University of California Press, 1992, pp. 3-5. Sansone points out that the Ukrainian word is sport, the Greek spor, the Gaelic spors, the Turkish spor, the Japanese supotsu and so on. In short, 'the English word sport refers to something for which many languages simply do not have a word.'

15   Charles Williams, *Bradman: 1908-2001*, Abacus, Little Brown & Co., London, 2001, pp. 37-8.

16   Harms, *Loose Men Everywhere*, p. 235.

17   Sansone, *Greek Athletics and the Genesis of Sport*, p. 37.

18   In a charming passage, Harms even shows how Australian Rules incorporates the football equivalent of both the rigour of economic models and the unpredictability of market fluctuations. It has to do with the shape of the ball. 'Footies were amazing things. They alerted you to a sense of perfection. What was truer and more beautiful than a sweetly timed drop kick? They also alerted you to a sense of chaos ... Why did they bounce end on end: two low forward bounces and then on the third they'd sit up on their point and bounce high? Often they gave you no chance. Why did they sometimes bounce any which way and no matter how closely you watched them you could never pick it?' Harms, *Loose Men Everywhere*, p. 37.

There is nothing preserved of this great genius which is worth knowing. Nothing which might inform us what education, what company, what accident turned his mind to letters and the drama.
- John Adams (on visiting Stratford, 1786)[1]

I am haunted by the conviction that the divine William is the biggest and most successful fraud ever practiced on a patient world.
- Henry James (letter to a friend, 1903)[2]

I think Oxford wrote Shakespeare. If you don't, there are some awful funny coincidences to explain away.
- Orson Welles (to an interviewer, 1954)[3]

7. On Shakespeare and Authorship

No author has a better claim to having captured the spirit of the West in a nutshell than Shakespeare. But who was he? Do you really know who *wrote* the works attributed to Shakespeare? What is the basis of your knowledge: what primary evidence, chains of inference, webs of belief? Shakespeare, after all, is to literature what Mozart is to music. The exquisite documentary *In Search of Mozart* drew on copious letters and records in reconstructing the life of the great composer. Various myths or corruptions of the biography popularised in *Amadeus*, twenty years earlier, were corrected. Mozart's music, astonishing in its precocity, range and beauty, was set in the context of his life, and its development from his childhood closely analysed.

But no documentary about Shakespeare can do any such thing with his life and work because the evidence is simply lacking. Michael Wood's 2003 BBC documentary, *In Search of Shakespeare*, is only superficially the equivalent of the Mozart film. The book of the series gave this away with the admission that 'Almost 400 years after his death, William Shakespeare is still acclaimed as the world's greatest writer, but the man himself remains shrouded in mystery.' What Wood actually did was to recreate 'the turbulent times through which the poet lived.' He did not give us the biography of the poet, because he could not.

The paucity of documentary detail about the life of Shakespeare is so remarkable that Mark Twain was prompted to comment wryly, 'as far as poverty of biographical details is concerned, the only parallel in history, romance or tradition to Shakespeare is Satan.'[4] What do we know that is documented and not disputed? The standard account goes roughly as follows. William Shakespeare was born in 1564 in Stratford-on-Avon, married Anne Hathaway in 1582, had three children in five years, spent much of his time between 1587 and 1604 in London, was first mentioned as a minor actor in 1592, made money in business, retired to Stratford in 1604 at a time when plays and poems were being published in his name (some of which are no longer attributed to him) and died there unremarked in 1616. Very little else exists, save scattered references to him which either are irrelevant to his presumed writing career or are subject to contentious interpretation.[5]

There is no record whatsoever of his first eighteen years and thus of his education; nor of his first five years in London and thus of his discovery of theatre.[6] There is no evidence that he ever went to university, studied English history or law, worked in any capacity at the royal court, learned French, Spanish or Italian, or travelled abroad, though the plays attributed to him powerfully suggest that their author had done all of these things. *The Sonnets* appear to be the love poems of a man who had nothing in common with what we know of the man from Stratford, whether in terms of age, physical condition, social class or sexual inclinations.[7] His wife and children appear to have been and remained illiterate and his will, notoriously, was a rudimentary inventory of property which made no mention of any books, literary manuscripts, poems, collections of letters, musical instruments, or anything else suggestive of the life of the greatest poet and dramatist in the realm.[8]

Not only was his will devoid of evidence that he had been a writer but, in strenuous searches for more than two hundred years starting with those by the Reverend James Wilmot in 1780, not a scintilla of direct evidence has ever been found that William Shakespeare ever wrote a single poem or play, owned a single book or had intellectual interests. As far as anyone knows, he received only one letter during his life and it was of a business nature.[9] When he died, his passing stirred no word that has been recorded of eulogy or mourning or even recognition. Yet the deaths of other notable literary contemporaries such as Francis Bacon, Ben Jonson, Edmund Spenser and Walter Raleigh drew such things.

All this is, surely, very strange. It is the root of the 'Shakespeare Authorship Question': if William Shakespeare of Stratford did *not* write the poems and plays long since attributed to him, then who did?[10] And how did Shakespeare's name become affixed to them? Why did the true author (to whom I shall refer as 'the Author') obscure his own identity? How could he have done so successfully? Though vaguely aware of the old canard, dating from the mid-nineteenth century, about Francis Bacon having been the Author, I took none of this seriously until a few years ago and, I confess, I still find it more a bemusing puzzle than a vital preoccupation.[11]

But in 2004-05, at least four books were published purporting to be biographies of the Author.[12] One of them, Peter Ackroyd's *Shakespeare: The Biography*[13] maintains it was the man from Stratford; one, Rodney Bolt's *History Play*[14] that it was Christopher Marlowe; one, Mark Anderson's *'Shakespeare' By Another Name*[15] that it was Edward de Vere, the seventeenth Earl of Oxford; and the other, Brenda James and William Rubinstein's *The Truth Will Out: Unmasking the Real Shakespeare*[16] that it was a hitherto entirely overlooked figure – Sir Henry Neville. All these books were greeted with acclaim by well-credentialled figures, but at least three of them must be in error as regards their central claim.

If you are disposed to presume that it must have been the man from Stratford who wrote the complete works of Shakespeare you will, perhaps, declare that it's pretty obvious which three must be in error. The beauty of the case, though, is that, actually, it isn't *obvious* at all. There is a genuine problem with the case for Stratford. That is why such luminaries as Charles Dickens, Henry James, Mark Twain, Walt Whitman, Ralph Waldo Emerson, Sigmund Freud, Charlie Chaplin, Daphne Du Maurier, John

Galsworthy, Orson Welles, Paul H. Nitze and the great Shakespearean actors John Gielgud, Derek Jacobi and Kenneth Branagh,[17] have all declared that they cannot believe the man from Stratford was the Author.[18]

Jacobi wrote the foreword to Mark Anderson's book. In England, Jacobi is a leading advocate of the view that Edward de Vere was the author of the Shakespearean canon. He notes that 'when one advocates that de Vere wrote under the pen-name "Shakespeare",' one courts 'charges of the wildest eccentricity, outrageous snobbery, and downright heresy.' However, he declares that Anderson's book 'demonstrates the intense intellectual energy and attention to factual detail that are required to unravel what, to an honest mind, is an obvious mystery.' The book, he continues, 'presents the logical, valid and excitingly precise arguments for recognizing that de Vere, like all writers, drew from his own experiences, interests, accomplishments, education, position and talents.'[19]

'An actor's instincts and the evidence of a growing body of research convince me,' Jacobi writes, 'that de Vere was – along with being a scholar, patron and author *par excellence* – an actor. The troupe kept by Edward de Vere's father had influenced his early childhood. De Vere's own troupe had nurtured those interests, and acting and stagecraft became intrinsic to his talents. Hence the precise and very special observation of the mechanics and meaning of the world of theater are everywhere expressed in the plays ...'[20] This is, of course, the kind of observation which should be true of the Author, whoever he was; but none of it can be said of the man from Stratford, because there is simply no evidence to support the claim.

Leaving aside, for a moment, the question of what evidence there is that de Vere was the Author, it is worth amplifying this point about the formation of *an* author. If we read the biography of Charles Dickens, Leo Tolstoy or Fyodor Dostoevsky, Thomas Mann, Tennessee Williams or Graham Greene, to take a few well-known names almost at random, we see the roots of their writings in their education and experience. The puzzle with the man from Stratford is that we cannot do this. In much orthodox Shakespeare scholarship, however, this absence of a plausible biographical seedbed for the flowering of his work tends to be waved away as a sign of his sublime 'genius.'[21]

In 1949, Northrop Frye, for example, remarked that Shakespeare 'was an expert in keeping his personal life out of our reach'[22] and in 1994 Harold Bloom stated that Shakespeare's 'personality always evades us, even in the sonnets. He is everyone and no-one.'[23] But whoever the Author was, he most certainly was neither everyone nor no-one, but a specific individual, whose background will have surfaced in his poems and plays. As it happens, at least *plausible* cases can be made for any one of de Vere (Oxford), Marlowe, or Neville that his background education and experience enable us to make sense of the possibility that he was the Author, other things being equal. If the man from Stratford was the Author, surely his background would trump theirs in this regard. The puzzle is that it clearly does not.

A plausible case, however, is not necessarily a true one. To be a true account of who the Author was, a case would have to explain how the name of an obscure actor and moderately successful businessman called William Shakespeare came to be accepted as that of the Author. The career of the Author would also have to be shown to have been consistent with his having written the plays and poems, given all we know about their composition, performance and publication, and with the various references to William Shakespeare, admittedly not numerous, that were made by his contemporaries.[24] There would have to be an explanation for why Ben Jonson, for example, a well-known contemporary and peer of the Author, would refer to him, seven years after his death, when the First Folio of his collected works was published, as 'the sweet swan of Avon', if in fact he was someone other than William Shakespeare.

Above all, the case for the Author should be *economical*. It should provide the clearest and simplest argument consistent with the known evidence, rather than requiring us to embrace elaborate or specious arguments. This is the principle known to philosophers as Ockham's Razor. In the case of the Shakespeare Authorship Question, it provides a fairly simple rule of thumb by which to gauge the relative plausibility of one case or another as to who the Author was. Simply ask how many things need to be *explained away*, if we are to accept any given argument as true. Were it not for the paucity of direct evidence that the man from Stratford was a writer at all, this would likely present an insurmountable obstacle to arguments that someone else wrote the poems and plays. The puzzling paucity of such evidence is the source of the Question.

Such direct evidence, however, is not the only kind of evidence that has a bearing on the matter. One of the simplest and most telling pieces of evidence is chronology. William Shakespeare's life began in 1564 and ended in 1616. The standard account is that his poems and plays were written between 1593 (*Venus and Adonis*) and 1612 (*The Tempest*).[25] How, then, can they have been written by Christopher Marlowe, who was murdered at Deptford in 1593, just before *Venus and Adonis* was published? How could the later works have been written by Edward de Vere, who died in 1604? If you are to make a case for either man having been the Author, you have a lot to explain away, especially in the case of Marlowe. Brenda James and William Rubinstein accept this. They then argue that no such problem exists in the case of Henry Neville, because his life (1562-1615) overlapped almost exactly with that of the man from Stratford.[26]

Rodney Bolt, like other Marlovians, argues that the young playwright's alleged murder at Deptford was faked, after which he fled abroad and wrote the poems and plays in exile, many of them in Italy.[27] Bolt had to invent the link with William Shakespeare from whole cloth. He has, therefore, to explain away almost everything. The astonishing thing is that this did not deter him. In an afterword to his book, he wrote: 'This book has not been an attempt to prove that Christopher Marlowe staged his own death, fled to the Continent and went on to write the works attributed to

Shakespeare. It assumes that as its starting point ... By assuming the seemingly preposterous, I have hoped to shake up our notions of the possible, or at the very least to look a little more sharply at how we construct truth.'[28]

'How we *construct* truth' is an interesting turn of phrase. He did not write 'how we *discern* truth'. His disclaimer is epistemologically provocative: 'This book is, of course, an exercise of purest (or most impure) conjecture. But then so is the work of countless other writers of lives of Shakespeare and Marlowe. This story differs only in the degree to which invention has played a role in the outcome, and in the method by which it was told ... The book is grounded in fact, but has the courage of its own (con-)fictions.'[29] It is, in fact, a highly readable book, a beguiling book, grounded in an impressive knowledge of the Author's times; but it does, indeed, *construct* its 'truth' and this must, at the very least, leave us a little uncomfortable. Were there no Authorship Question, of course, we would rightly dismiss it out of hand.

Mark Anderson, conversely, claimed that he is *discerning*, not *constructing* the truth, in his argument that Edward de Vere was the Author. He points to William Shakespeare as a convenient cover because he was obscure, and to Jonson as a de Vere confederate who kept the secret.[30] He acknowledges the problem of chronology and argues that, closely considered, it actually *proves* de Vere's authorship. There are several grounds on which he develops this line of argument. The first is that the Author drew on no source published or scientific discovery made after 1603. The second is that, starting in 1593, new plays or poems attributed to Shakespeare appeared on average twice a year, but 'in 1604 Shakespeare fell silent.' The third is that newly corrected, augmented or amended editions of the plays stopped appearing after 1604.[31] His argument concerning the dating of the 'last' of the Author's plays, *The Tempest* and *Henry VIII*, is especially interesting in this regard and plainly serious.[32]

Anderson added to these considerations several indirect grounds for believing that the manuscripts of the Author's works were left in the possession of the de Vere family after 1604. First, that three plays (*Pericles, King Lear* and *Troilus and Cressida*) and *The Sonnets* first appeared in print in 1608-09 when Elizabeth Trentham de Vere, Edward's last wife, was preparing to move out of the home she had shared with him during his final years.[33] Second, that the publication in 1623 of the Author's complete works, including eighteen plays never printed before then, was supervised by the Earls of Pembroke and Montgomery, who were very close to the de Veres – Montgomery (William Herbert) having been married to de Vere's youngest daughter, Susan, since 1604 and Pembroke having been proposed as a husband for another of de Vere's three daughters, Bridget, in 1597.[34]

'The Herberts were the premier literary aristocratic family in the early seventeenth century,' Anderson remarks, and Susan de Vere had inherited her father's love of letters and learning (contrast her education and love of theatre and literature with the illiteracy of the daughters of William Shakespeare, one of whom was called Susanna). The Herberts engaged the London bookseller and printer William Jaggard to publish the complete works of the Author in 1623. Anderson argued that politics had been the reason for de Vere using the pen-name of the bit-part actor from Stratford in the 1590s and because of politics in 1623 his family 'stuck to the cover story they'd inherited. Their own lives and fortunes too clearly hung in the balance for them to play games with their father's compromised identity.'[35]

This is a much more interesting, because much less 'constructed' case than the one Bolt makes for Marlowe having been the Author. Consider, further, de Vere's family background, as scion of the oldest aristocratic family in the realm; the fact that his uncles Henry Howard and Arthur Golding invented the 'Shakespearean' sonnet and translated 'Shakespeare's' most beloved Latin poet, Ovid, into English; the fact that de Vere was raised to courtly life, surrounded by theatre from infancy, educated in law, spoke several foreign languages and travelled extensively on the Continent, especially in Italy, in his youth.

Add to these considerations that, after his father's death, he was the rebellious young ward of William Cecil, Lord Burghley, on whom Polonius is widely supposed to have been based; that he married the fifteen-year-old Anne Cecil, Burghley's daughter, and had a relationship with her which bears striking parallels to the relationships between Hamlet and Ophelia, Othello and Desdemona; and that, like Hamlet, he was kidnapped by pirates in the English Channel. Consider, finally, that he plainly was bisexual and was intimate with the young Henry Wriothesley, Earl of Southampton, so that he could well have written the sonnets to him. The case for him being the Author surely begins to look quite impressive. The more so, it should be underscored, because there is no evidence of *any* of these things being true of the man from Stratford.

What, then, of Brenda James and William Rubinstein's Henry Neville? James and Rubinstein begin by rehearsing the case against the man from Stratford as the Author, then explain why no other candidate proposed to date is actually convincing.[36] They allow that the two perennially most favoured such candidates have been Francis Bacon and Edward de Vere. They observe that Bacon's candidature has been ruled out on various compelling grounds.[37] They allowed that de Vere looks much more plausible, but argued that he has to be excluded on chronological grounds because he died in June 1604. 'The single greatest stumbling block … is plainly that he died in 1604 and around eleven of Shakespeare's plays appeared after that date.' But there is also the problem that, by 1589-90 when the Author apparently first began to write, de Vere was already almost forty.

James and Rubinstein suggest that it is simpler to accept an orthodox chronology for the plays than to try to demonstrate that they were all written before 1604. They argue specifically that *The Tempest* was inspired by William Strachey's 1610 account of the wreck of the English ship *Sea Venture* in the Bermudas in 1609, which would automatically rule out de Vere as the Author.[38] They argue that de Vere's extant writing, which includes early poetry, though talented, lacks the complexity

of the poems attributed to the Author and that the chronology of de Vere's life does not seem
to match the demonstrable chronology of the 'evolution' of the Author's theatrical interests and
style, most especially the shift from comedy and history to tragedy after 1601. They argue, finally,
that the case for de Vere ultimately runs into the same impasse as that for Stratford – the lack of a
'smoking gun' after some eighty years of searching for direct evidence.[39]

Their case for Neville pivots, then, on the argument that none of these things is true of him.
Assuming that Stratford was not the Author and that no other candidate was, there must be a
hitherto unsuspected candidate. He must have had the kind of background necessary to have
written the poems and plays. They argue that Neville did. His chronology must be unproblematic
as regards the composition and evolution of the poems and plays. They argue that it is. He must
have had both occasion and opportunity to write as 'William Shakespeare' and reason to keep
his identity secret. His life must be consistent with that of the Author of *The Sonnets* and with the
circumstances in which they and the First Folio were published. They argue that all these things
were the case. Finally, they believe they have their 'smoking gun' – direct evidence that Neville
was the Author. Apparently as swayed by their reasoning as Derek Jacobi was by Mark Anderson's,
Mark Rylance, Artistic Director of Shakespeare's Globe Theatre since 1996, acclaimed their book
as 'pioneering' and 'historic'.[40]

Their 'smoking gun' is a recently discovered manuscript, 196 pages in length, known as the *Tower
Notebook*, compiled in the Tower of London by Henry Neville in 1601-02 when he was a prisoner
of the Crown there because of his involvement in the Essex Rebellion. In it there are notes for what
appear to be draft scenes for the play *Henry VIII*. The prisoner, they argue, was preparing a play to
flatter the Queen about the origins of her family dynasty. To be sure, 'so far as anyone knew or knows,
Sir Henry Neville was not a playwright. Yet here he was, in 1602, writing sketches which found their
way into a ['Shakespeare'] play [in 1613], in a notebook which also proclaimed itself to be principally
concerned with "Pastime".'[41] 'Almost all authorities regard *Henry VIII* as having been co-authored with
John Fletcher. In 1613, (when the play was finally produced) Fletcher – and his collaborator, Francis
Beaumont – were certainly close friends and political supporters of Sir Henry Neville ...'[42]

They argue that both the writing of *The Sonnets*, between 1589 and 1608, and their publication
in 1609, are best explained in terms of Neville's long relationship with Southampton. Even more
intriguingly, they argue that the Strachey Letter, reputed source for *The Tempest*, was a confidential
manuscript, unpublished until 1625, to which Neville had access as a member of the council of the
London Virginia Company, but which would not have been available to William Shakespeare, who
had no association of any kind with the Company. These are genuinely interesting arguments and do
much to buttress Mark Rylance's acclaim for the book as pioneering. Its authors, of course, have to
contend with two sets of enthusiasts: those who are convinced that the man from Stratford must have

been the Author and those who are wedded to the idea that de Vere was the Author. Neither set is likely to yield ground easily or gracefully.

Having read the three iconoclastic books, I turned to Peter Ackroyd's book to find, I hoped, a triumphant demonstration that the Question was a delusion and the man from Stratford was, after all, the real and true Author. Alas, he disregarded the entire Question with contumely, addressing it not at all. What's worse, he wrote in a tedious manner which makes his the least engaging of the four books. Harold Bloom, a Stratford man through and through, when asked by *Harper's* to make the case for Stratford in a special issue on the Question in April 1999, at least wrote entertainingly, though he failed to acknowledge that there was a Question at all and entered no argument whatsoever either way.[43] Jonathan Bate, in 1997, in *The Genius of Shakespeare*, was also dismissive of the Question, but he at least made an effort to marshal a lucid argument for Stratford.[44]

Who, then, was the Author? Certainty would seem to be elusive. Does it matter? Not nearly as much as many other pressing things. What *does* matter is how you *think* about puzzles of this nature. What do you most *want* to believe? Argue against yourself and try not to confirm your belief, but to confute it. You will be among the cognitively commonplace if you stick with the conventional wisdom without arguing the toss, like Ackroyd or Bloom, or if you embrace a plausible confiction in the manner of Bolt. You will be a mere pupil if you find yourself beguiled by Anderson, or James and Rubinstein, into believing that the Author was de Vere or Neville – or if you bow to Bate's reputation and erudition. You will be a free master of your own mind if and only if you sift the evidence with fine discrimination, look for diagnostic evidence rather than confirming evidence and account scrupulously for your deductions. It would, perhaps, be interesting to know who the Author of 'Shakespeare' was, but it is surely far more important, ultimately, to know what 'knowing' really requires.

7.   Endnotes

1   John Adams, *Notes on a Tour of English Country Seats etc April 1786*, reprinted in Warren Hope and Kim Holston, *The Shakespeare Controversy: An Analysis of the Claimants to Authorship and Their Champions and Detractors*, McFarland and Co., Jefferson, N.C., 1992, pp. 61-5.

2   Letter from Henry James to Violet Hunt, 26 August 1903, in Percy Lubbock (ed.), *The Letters of Henry James*, Charles Scribner's Sons, New York, 1920, vol. 1, pp. 424-5.

3   Orson Welles quoted in Cecil Beaton and Kenneth Tynan, *Persona Grata*, Putnam, New York, 1954, p. 98.

4   Mark Twain (Samuel L. Clemens), quoted in the Foreword to Rodney Bolt's *History Play*, p. xiii.

5   Diana Price, *Shakespeare's Unorthodox Biography: New Evidence of an Authorship Problem* (Greenwood, CT, 2001), p. 150 remarks, 'The biography of William Shakespeare is deficient. It cites not one personal literary record to prove that he wrote for a living. Moreover, it cites not one personal record to prove that he was *capable* of writing the works of William Shakespeare. In the genre of Elizabethan and Jacobean literary biography, that deficiency is unique. While Shakespeare wrote over seventy biographical records, not one of them tells us that his occupation was writing. In contrast, George Peele's meager pile of twenty some biographical records includes at least nine that are literary. John Webster, one of the least documented writers of the day, left behind fewer than a dozen personal biographical records, but seven of them are literary ... If Shakespeare had acquired the education and cultural experiences to write the plays, he would have left a few footprints behind to prove it. Shakespeare's extant records are not only devoid of literary evidence, they point *away* from a literary career and toward other vocations.'

6   In his acclaimed book *1599: A Year in the Life of Shakespeare*, Faber & Faber, London, 2005, James Shapiro encapsulates this strange problem: 'None of the men who wrote plays for a living in 1599 was over forty years old. They had come from London and the countryside, from the Inns of Court, the universities and various trades. About the only thing these writers had in common is that they were all from the middling classes. There were about fifteen of them at work in 1599 and they knew each other and each other's writing styles well: George Chapman, Henry Chettle, John Day, Thomas Dekker, Michael Drayton, Richard Hathaway, William Haughton, Thomas Heywood, Ben Jonson, John Marston, Anthony Munday, Henry Porter, Robert Wilson, *and, of course, Shakespeare*. Collectively this year they wrote about sixty plays, of which only a dozen or so survive, a quarter of these Shakespeare's. Their names – *though not Shakespeare's* – can be found in the pages of an extraordinary volume called Henslowe's Diary, a ledger or account book belonging to Philip Henslowe, owner of the Rose Theatre, in which he recorded his business activities, mostly theatrical, from 1592 to 1609 [emphases added]. See p. 11.

7   'The Sonnets form a huge riddle that demands a solution', wrote Joseph Sobran in 1997. 'But our natural curiosity about it [the riddle] meets with the sophisticated scorn of commentators who regard such an interest as somewhat improper ... Try as we may, we can't banish the sense that something real lies behind the Sonnets, if only we could find it. C. S. Lewis says they tell "so odd a story that we find a difficulty in regarding it as fiction." ... The poet takes bold liberties with the youth [to whom the Sonnets are chiefly addressed], praising him in terms that would be incredibly presumptuous if he were a common poet addressing a man of Southampton's rank [as is supposed by orthodox scholars] ... The poet speaks continually of his "age", of being "old", "bated and chopp'd with tann'd antiquity", and laments the loss of "precious friends, hid in death's dateless night". His "days are past the best". He looks forward to his grave and obscurity ... But why would Mr Shakspere [sic], writing in the early 1590s, feel that he was old, looking death in the face, incurably disgraced, doomed to oblivion, and so forth? There are not the normal feelings of a man of thirty who is doing quite well for himself." *Alias Shakespeare*, Free Press, New York, 1997, pp. 82-9.

8   Ibid., Appendix 1, 'Mr Shakspere's Will', pp. 227-230, supplies the actual text of the will of Will.

9   Jonathan Bate takes issue with the claim that no letters by William Shakespeare survive with the remark that 'Letters addressed *by* William Shakespeare to Henry Wriothesley, the Earl of Southampton, may be read at the beginning of the texts of *Venus and Adonis* and *The Rape of Lucrece* in any complete edition of his works. The letter prefixed to *Venus and Adonis* is couched in the servile language which low born writers had no choice but to use if they aspired to the patronage of aristocrats ...' *The Genius of Shakespeare*, Picador, 1997, p. 73.

10   Diana Price's *Shakespeare's Unorthodox Biography* is widely regarded as the most decisive account of what Brenda James and William Rubinstein have called 'the insuperable difficulties involved in accepting the view that Shakespeare of Stratford wrote the works attributed to him.' *The Truth Will Out: Unmasking the Real Shakespeare*, p. 307.

11   Joseph Sobran's *Alias Shakespeare*, Free Press, New York, 1997, is the place to start.

12   See also Frank Kermode's *The Age of Shakespeare*, Weidenfeld & Nicolson, London, 2004; James Shapiro's *1599: A Year in the Life of William Shakespeare*, Faber & Faber, London, 2005; Stephen Greenblatt, *Will in the World: How Shakespeare Became Shakespeare*, Jonathan Cape, London, 2004; and Stanley Wells, *Shakespeare For All Time*, MacMillan, London, 2002, as well as David Crystal and Ben Crystal, *Shakespeare's Words: A Glossary and Language Companion*, Penguin, London, 2002.

13   Peter Ackroyd, *Shakespeare: The Biography*, Chatto and Windus, London, 2005.

14   Rodney Bolt, *History Play*, Harper Perennial, 2005.

15   Mark Anderson, *'Shakespeare' By Another Name*, Gotham Books, New York, 2005.

16   Jonathan Bate opens his chapter on the subject with the remark: 'There is a mystery about the identity of William Shakespeare. The mystery is this: why should anyone doubt that he was William Shakespeare, the actor from Stratford-upon-Avon?' *The Genius of Shakespeare*, p. 65.

17   The testimony of these three outstanding Shakespearean actors, as well as that of Mark Rylance, confutes the unsupported claim by Jonathan Bate, in 1997, that 'no major actor has ever been attracted to Anti-Stratfordianism.' Ibid., p. 67.

18   http://www.shakespeare-oxford.com/index.htm 'The Honor Roll of Skeptics', contains original quotes from the individuals mentioned and others besides.

19   Anderson, *'Shakespeare' By Another Name*, p. xxiv.

20   Ibid., p. xxiii.

21   Jonathan Bate summarises this position in his 1999 essay 'Golden Lads and Chimney Sweepers': 'The best response to skeptics who doubt that the Stratford man could have written the plays on the foundation of nothing more than a grammar school education is an education to read the complete plays of Ben Jonson. They are vastly more academic than Shakespeare's, yet they, too, were written on the foundation of nothing more than a grammar school education. The thing is, Elizabethan grammar schools were very good. They put our high schools to deep shame.' *Harper's* Folio 'The Ghost of Shakespeare', April 1999, p. 61.

22   Northrop Frye in Edward Hubler (ed.), *The Riddle of Shakespeare's Sonnets*, Basic Books, New York, 1962, pp. 26-7.

23   Harold Bloom, *The Western Canon: The Books and School of the Ages,* Harcourt, Brace & Co., New York, 1994, p. 90.

24   Jonathan Bate, *The Genius of Shakespeare* pp. 69-73, supplies the short list: Ben Jonson, Francis Beaumont, William Camden, John Davies, George Buc and Leonard Digges. He does not enter into any discussion of the various pieces of evidence which cast these few references into a different light. Most notable among these is the case of Henry Peacham who 'In a book on education published in 1622 ... made a list of poets, including Sidney and Spenser, who had made Elizabeth's reign "a golden age" of poetry. The list began with "Edward, Earl of Oxford". It made no mention of William Shakespeare, despite his popularity.' Sobran, *Alias Shakespeare*, p. 142. See also Mark Anderson, *Shakespeare By Another Name*, pp. 365-67. Peacham's book was The *Compleat Gentleman*. It went through multiple editions for forty years after 1622, but the leading role attributed to Edward de Vere was never altered, nor the name of William Shakespeare included.

25   Bate, *The Genius of Shakespeare.*

26   James and Rubinstein, *The Truth Will Out,* pp. 31, 37-41.

27   For the conventional life of Marlowe, see Park Honan, *Christopher Marlowe: Poet and Spy*, Oxford University Press, 2005, especially Chapter 10, 'A Little Matter of Murder', pp. 321-60 and Appendix, 'The Coroner's Inquest of 1 June 1593', pp. 376-7.

28   Bolt, *History Play*, pp. 313-14.

29   Ibid., pp. 314-15.

30   Anderson points out that Jonson was 'a friend to the Herberts and to Henry de Vere [Edward's son]' and 'was hired to edit and oversee the Folio.' *'Shakespeare' By Another Name*, p. 376.

31   Ibid., Appendix C, 'The 1604 Question', pp. 398-9.

32   Ibid., pp. 350-3 and 401-3.

33   Ibid., p. 398.

34   Ibid., pp. 314, 371.

35   Ibid., p. 374.

36   James and Rubinstein, *The Truth Will Out,* pp. 33-42.

37   Ibid., pp. 35-7. The grounds are that he lived too long, dying in 1626, for it to be plausible that he stopped writing plays in 1612; that his well-known prose works are written in a style altogether different from the works of Shakespeare; that he never visited Italy and that he had no close relationship to Henry Wriothesley, Earl of Southampton, that would have induced him to write two long poems for him, much less write sonnets expressing a homoerotic love for him, as the Author is believed to have done.

38   Ibid., pp. 40, 195-8. They also raise the question of *Henry VIII* (1613) and observe that *Macbeth* could not have been written before the 1605 Gunpowder Plot and that *The Winter's Tale* was licensed by Sir George Buc, 'who only began licensing plays for performance in 1610' (p. 66). Anderson, *'Shakespeare' By Another Name*, discusses at some length the autobiographical traces that link *The Winter's Tale* to de Vere and the question of its composition, performance and publication.

39   James and Rubenstein, *The Truth Will Out*, pp. 40-1.

40   Ibid., foreword, p. xi.

41   Ibid., p. 49.

42   Ibid.

43   *Harper's* April 1999 Folio 'The Ghost of Shakespeare' pp. 35-62. Bloom's style is one of blooming rhetoric, not systematic argument. He wrote: 'Oxfordians are the sub-literary equivalent of the sub-religious Scientologists. You don't want to argue with them, as they are dogmatic and abusive. I, therefore, will let the Earl of Sobran be and confine myself to the poetic power of Shakespeare's Sonnets ...' Instead of addressing any argument to the Oxfordians, he sought to cast ridicule on the whole debate by offering the hilarious suggestion that the works of Shakespeare were all written by the poet's 'dark lady', whom he identifies as Lucy Negro, 'Elizabethan England's most celebrated East Indian whore', following Anthony Burgess' fictive account of Shakespeare's life, *Nothing Like the Sun*. This assertion, he ventured, enables us to read Shakespeare with assured political correctness, 'since Lucy Negro was, by definition, multicultural, feminist and post-colonial.' *Harper's* Folio 'The Ghost of Shakespeare', April 1999, p. 56.

44   Bate, *The Genius of Shakespeare*, Chapter 3, 'The Authorship Controversy', pp. 65-100.

Beauty is the promise of happiness.
- Stendhal

My formula for happiness? A yes, a no, a straight line, a goal.
- Nietzsche

There is no simple formula for finding happiness. But if our great big brains do not allow us to go sure-footedly into our futures, they at least allow us to understand what makes us stumble.
- Daniel Gilbert

8. On Goethe and Human Happiness

What does it take to be happy? Is it important to be happy? What do we actually *mean* by happiness? Nietzsche once quipped that 'strength makes stupid'. Does happiness make us frivolous or banal? The utilitarians take as the central plank of their philosophy that public policy should generate 'the greatest happiness of the greatest number', but many a critic of modern 'consumer' society has derided such a philosophy as leading to vacuity, boredom and aimlessness.

There must, the argument goes, be something more than mere 'happiness' – a higher purpose, a lofty ideal or final goal, a spirit of sacrifice and a sense of meaning to make life truly fulfilling. Of course, these claims seem to imply both that 'happiness' and higher purposes are somehow mutually exclusive and, at the same time, that *true* happiness requires that one have a higher purpose, be it religious, ideological, or creative.

My earliest recollection of wrestling with or being troubled by these questions is reading Henri Daniel-Rops's *The Church of Apostles and Martyrs* as an adolescent Catholic and being stirred by his sonorous rhetoric about the depth of meaning and historical purpose that Christianity brought to the hedonistic, violent and decaying world of late antiquity. He seized my imagination with an arresting phrase about 'that abyss into which history hurls empires and civilizations with sublime indiscrimination.'

Even earlier, as a young boy, I was profoundly affected by J. R. R. Tolkien's *The Lord of the Rings*, chiefly because of the elegiac sadness and profound dignity of the High Elves and the Ents. One could not have said, in any ordinary sense of the word, that Elrond, for example, was 'happy'. His capacity, however, to remember long ages of history and ponder the meaning of things deeply, while keeping a quite musical sense of proportion, made me look for these things in Catholicism. Wisdom, knowledge, gravity, meaning – these things seemed deeply appealing by comparison with mere 'happiness'.

Those were the years immediately after Vatican II but, to my bewilderment, I failed to find what I was looking for in Catholicism. At the level of ritual and tradition, it seemed to be falling apart. Morally, it seemed divided and confused. Culturally and cognitively, it seemed stranded. Vatican II notwithstanding, I found myself, as a very young man, reflecting that the Church resembled Tolkien's description of the High Elves in the Third Age – 'for long they were at peace', he wrote, 'wielding the Three Rings while Sauron slept and the one ring was lost, but they attempted nothing new, living in memory of the past.'

What is the use of long memory of the past, I thought, if it has lost its grip on the present and its way into the future? Besides, as I grew into my late teens, I simply could not profess to believe what I had recited by catechetic rote since childhood, nor could I get satisfying answers from my teachers when I asked them perfectly serious questions about doctrine and belief. Happiness was difficult in these circumstances, but it was not the point.

Any unhappiness I felt was derivative. My response was not to try to be happy as things were – to get along by going along, as the old saying has it – but to seek a true understanding of what life is about. Lots of us go through this, of course. The disturbing thing, looking back, is to contemplate how apparently random the search was and how very difficult it was to find any resting point: any point at which I could feel that my understanding and my well-being were in reasonable equilibrium, any point at which I felt content to ride at anchor.

It was only slowly that I realised I was going through a process that the whole world had been going through for quite some time. I had often felt terribly alone, but actually I was in abundant and very good company. Our species has been going through a staggering metamorphosis since the end of the last ice age. We have invented agriculture, cities and writing, mathematics, logic and physical science, monotheistic religion and complex music, industrial production and mass transport, telecoms and IT.

The drastic acceleration of this metamorphosis in the past few centuries has been overwhelming. How is any individual to find meaning and truth while we are all being swept along by this gigantic process? For several thousand years, most human beings have found such things, or at least sought them, through religion – folk religion mostly, of course, in rough symbiosis with more mundane concerns and pleasures. A relative few have sought to master the scheme of things cognitively, or at least to create something expressive of their sense of meaning in the world.

These few are philosophers and scientists, poets, composers, dramatists and artists. Being 'educated' enables us to draw upon the work of such people to enlarge our personal understanding of the world. Of those few, a very small number tend to stand out because of the unusual scope or originality of their conceptual or creative work. Individuals such as Moses, Confucius, Buddha and Jesus, Archimedes, Plato, Dante, Shakespeare, Newton, Mozart and Einstein are our beacons in the search for meaning and truth – though not necessarily for 'happiness'.

Johann Wolfgang von Goethe (1749-1832) is widely held to have been among these remarkable individuals. He is so regarded, however, not simply on account of his creative work or analytical insight, but because he seems to have been a kind of paragon of the fully rounded man of the world: poet, novelist and practical statesman, traveller and businessman. Goethe may also have been a better than usual guide in the search for happiness. This, at least, is the argument of John Armstrong in *Love, Life, Goethe: How to be Happy in an Imperfect World*. Goethe, he writes, while 'an outstanding intellectual and artistic figure ... was very much a "Weltkind" – to use one of his own favorite words: "a creature of this world".'[1]

This is the way Goethe has been regarded for 200 years. After his death, as industrialisation and secularism developed in both Germany and the West at large, Goethe came to seem like a beacon of human completeness and cheerful practicality in a world of increasing specialisation of labour and the erosion of old cultural traditions. When Max Weber wrote, at the conclusion of *The Protestant Ethic and the Spirit of Capitalism*, that his contemporaries in the early twentieth century had become a 'nullity', devoid of both heart and spirit, it was the counter-example of Goethe, as much as anything, that he had in mind.[2]

In the mid-twentieth century, the Hungarian Marxist literary critic Gyorgy Lukacs wrote of Goethe as the highest example of the old bourgeois artist and intellectual.[3] Lukacs saw Goethe as a kind of human ideal no longer possible under advanced capitalism, but somehow representative of what communism would make possible. Ironically, it was only in tenacious resistance to communism that literary humanism flourished in the twentieth century – whether in the form of Anna Akhmatova and Nobel Prize winner Boris Pasternak in Russia, or more recently Nobel Prize winner Gao Xingjian in post-Mao China.[4]

Armstrong, on the other hand, sees Goethe not as a monument of what capitalist society has lost, or of what a utopian society might make generally possible, but as an interlocutor in our own search for 'happiness' in an imperfect world. 'Through his representative life', he writes, 'Goethe invites us to connect things that are dislocated in our society as well as in ourselves – creative freedom and emotional stability; profundity and practicality; refined taste and power. To live well, we have to thrive in the imperfect world we have.'

Armstrong himself was born in Glasgow in 1966, of parents whom he describes as 'distressed bohemians and not obviously in control of life'. He sought meaning, truth and happiness through the study of art and philosophy. At one time a research fellow and director of the aesthetics program at the University of London, he has become, more recently, Associate Professor in Philosophy at the University of Melbourne, director of the aesthetics program at the Monash Centre for Public Philosophy and philosopher in residence at the Melbourne Business School.

Love, Life, Goethe was Armstrong's fourth book and, like the earlier ones, is written in a lucid, conversational style. His first book, *The Intimate Philosophy of Art* (2000), was his opening salvo in a campaign to re-educate the eye of the contemporary citizen, to draw us, as it were, into the picture of things. He extended this campaign with *The Secret Power of Beauty: Why Happiness is in the Eye of the Beholder* (2002), an inquiry into the psychology of aesthetic perception, and then *Conditions of Love: The Philosophy of Intimacy* (2004). The latter is a sustained argument against the endemic belief that 'love' consists in the intense feelings associated with 'falling in love'; that it is, rather, about the maturing of partnership as romantic intensity recedes.

Conditions of Love begins with a close reference to Goethe. 'A decisive moment in the history of thinking about love occurred in 1774. That was the year in which Goethe's first novel, *The Sorrows of Young Werther*, appeared and quickly achieved an overwhelming success throughout Europe. This short book presented a simple, and seductive, vision of the nature of love: love is a feeling.'[5] A bestseller in the 1770s, *The Sorrows of Young Werther* is not much read these days, but it deals with the subject that saturates our air waves. It is a reflection on the roots of unhappiness in love.

Longing, rapture, torment and obsession are the themes of young Werther's love for Charlotte, a woman both uninterested and unavailable. 'Goethe didn't invent romantic love,' Armstrong comments, 'he merely provided an unforgettable, exact rendition of the motions of the heart which constitute such love.' The problem was that Werther committed suicide when Charlotte rejected him. This set a romantic fashion that swept across Europe although the novel is actually, Armstrong argues, a reflection on why one should *not* be like Werther.

In reflecting on the true nature of love and its relation to 'happiness', Armstrong draws on Wittgenstein's observation in his *Philosophical Investigations* that we use a whole range of words of which we cannot actually give a satisfactory definition. Armstrong is thinking of love and happiness, but Wittgenstein used the word 'game' as his chief example. The word can't be adequately defined, he wrote, because games do not have any one set of common characteristics that are definitive. We are not just somehow failing to see the true definition. There simply is no single way or set of ways in which all games resemble one another. Rather, they exhibit various overlapping 'family resemblances'.

To develop an adequate conception of such a thing as love or happiness then, we have to explore the family resemblances of the various strands of meaning that attach to such a word. When we do that, Armstrong argues, we open up the possibility of a more nuanced and balanced, mature and creative appreciation of the domain in which the word is used. This sets us free, to some extent, from unbalanced perceptions, misjudgements, delusions and, in general, narrow ways of looking at the world. This is Armstrong's counsel in regard to love: that we widen our conversation about it to avoid both cynicism and delusion.

In his book on Goethe, he makes the same argument in regard to happiness. He styles the book as a kind of extended 'conversation' with Goethe about happiness, in the belief that a poet and man of the world who reflected so much on life and lived so richly still has much to offer us.

The conversation leads us through Goethe's singularly fortunate and long life as through a museum with Armstrong as the tour guide.

We learn about Goethe's comfortable family background, his education, his sexual awakening, his creative writing, his professional work, his travel and the maturing of his status as a sage of Enlightenment Europe in the era of Hegel and Napoleon. We discover how all of Goethe's creative works, including the famous *Faust* on which he worked for decades, his erotic poetry, his novels (*The Sorrows of Young Werther, Wilhelm Meister's Apprenticeship* and *Elective Affinities*), his stage dramas (*Goetz, Iphigenia, Tasso, Egmont*), his travel writing and autobiographical reflections, were meditations on the challenges of moral and practical life.

There are countless delightful moments in the conversation, both by virtue of the subtlety and variety of Goethe's writing and by way of Armstrong's analytically acute and luminous observations. Between them, they represent the broad humanity of Goethe, which is the pivotal reality that engages Armstrong and to which he seeks to draw our attention. Several of these moments, during Goethe's famous and extended travels in Italy in 1786-87, are especially representative of the way in which Armstrong saw 'happiness' in Goethe's life.

He describes Goethe in northern Italy visiting buildings by the sixteenth-century architect Andrea Palladio (1508-1580) and being tremendously impressed by the proportionality and tastefulness of Palladio's designs for living. 'In fact,' he writes, 'Palladio became the single most important role model for Goethe ... He wanted to do with his life what Palladio had done with stone.' Decades later, in his autobiography, *Poetry and Truth*, Goethe was to write, 'There is something divine about Palladio's talent, something comparable to the power of a great poet who, out of the worlds of truth and deception creates a third whose borrowed existence enchants us.'

Palladio's style was one of both nobility and gracious moderation. Goethe saw him as sublimely classical and as opening up 'the way to the glorious days of antiquity'. Realising that many of Palladio's buildings had never been completed, Goethe reflected on his own unfinished creative manuscripts and on the fact that the world is full of completed works that one might wish had remained unwritten or fragmentary. 'Oh kindly Fates, who favour and perpetuate so many stupidities,' he wrote, 'why did you not allow [Palladio's] work to be completed?' He went on to Venice and worked quietly on his play *Iphigenia*, a reworking of the classical drama by Euripides.

There is a famous painting by Johann Friedrich August Tischbein (1750-1812) of Goethe in the Roman *campagna* in a broad-brimmed red hat and cape. Here, Armstrong observes, we see 'the face of a man who has full control of his faculties, who is ripe for action or thought; it is an image of a man in his prime – the point of perfect ripeness. He is among the lovely remains of the classical world, a world of which he has taken inner possession; he is at ease and yet he is completely in earnest.' Albert Camus remarked that one must imagine toiling Sisyphus to have been happy rolling his rock up the mountain ever and again. It is surely more plausible to see happiness embodied in Tischbein's Goethe.

Goethe's Palladio and Tischbein's Goethe tell us much of Armstrong's ideal: the man of sound faculties with a profound sense of proportion who has taken inner possession of the great classical tradition of civilisation and who is ripe for thought or action. Armstrong shows us this Goethe en route from Sicily to Naples in the summer of 1787, involved in a near shipwreck and doing all he can to keep the other passengers calm in order to avoid panic and all the disasters that can come with it. 'Life is a very slow shipwreck,' he remarks. 'Goethe tried to communicate a *cheerful pessimism*: to see life as it is and yet to enjoy it as it is.'

Goethe the artist and intellectual, he suggests, set himself to 'control unhelpful anxiety'. Given the intractable challenges of life, we have too many voices clamouring that we are all going to die and someone is to blame and something urgent must be done. Modern artists and intellectuals tend, Armstrong remarks, to assume that we are all sleepwalking to disaster and that it is their appointed task to rouse us from our slumber with angry, shocking or disturbing voices. Goethe believed that hysteria rather than complacency was the greater danger. 'Therefore, a significant task for art and culture might be to calm us down, to bring order and harmony – so that we can do what we need to do.'

So addicted have we become to intensity of feeling, hysteria and cynicism that all this talk of order and harmony might make both Goethe and Armstrong seem rather like stuffed shirts or ivory tower types contemplating their navels, falsely aloof from the everyday disorder of the world and the ordinary passions of human life. But Armstrong's whole point was that Goethe was no such stuffed shirt. He emphasised, in this respect, Goethe's sexual appetites, his political work and his business acumen. His sexual appetites, in particular, were anything but contemplative.

Goethe's relationship with Christiane Vulpius – one of the many loves of his life, but the most stable of them all – illustrates this admirably. 'Everything we know about her,' Armstrong relates, 'bears witness to her pronounced sensuality: her fondness for wine, for lively company and louche jokes. She was erotically adventurous and Goethe talks of their mutual exploration of "all twelve books" of sexual experience.'[6] She, for her part, thought Goethe 'a god from another world. He was rich, funny, handsome, generous, famous, passionate, keen on drinking and by turns quietly domestic and outrageously dirty.'

It was during the early years of his life with Christiane after his return from Italy that Goethe wrote his famous cycle of erotic poems, the *Roman Elegies*. He had been fascinated by the stones of Rome, but then reflected that without an experience of love the stones did not come alive. 'Without love, the world is not the world, and without love, Rome is not Rome,' he wrote in the opening lines of *Roman Elegies*.[7] His *Venetian Epigrams* tell us even more. They enunciate, as Armstrong expresses it, 'a breezy, obscene and very witty erotic philosophy.'

Goethe was, in short, a man of exceptional accomplishments and worldly success. The question is, does this make him a model for the rest of us in our own search for meaning, truth or happiness? Not at the level of fortune or accomplishment. After all, if it is necessary for us to be born into wealthy families, get a first class education, enjoy ample leisure and much travel, write novels, erotic poetry, plays and memoirs, be ministers of state and have many lovers in order to be happy, there is little hope for the overwhelming majority of us.

We might, however, feel a little more at home in the world after seeing it through Goethe's eyes. We might seek something of the cheerful enjoyment of food and sex, the sense of aesthetic proportion, the attention to practical affairs that characterised Goethe. Above all, if we are the citizens of affluent societies, we might see in Goethe a certain kind of model of the integrated and balanced life, neither frenetic nor anxious, neither gloomy nor relentlessly acquisitive. We might also emulate his behaviour on the ship from Sicily and seek to keep others calm and rational, amid the possibilities for shipwreck that confront us all. By such means, we might actually find a reasonable kind of happiness in an imperfect world.

8.   *Endnotes*

1   John Armstrong, *Love, Life, Goethe: How to be Happy in an Imperfect World*, Allen Lane, Penguin, 2006, p. 4.

2   'Limitation to specialized work, with a renunciation of the Faustian universality of man which it involves, is a condition of any valuable work in the modern world; hence deeds and renunciation inevitably condition each other today. This fundamentally ascetic trait of middle class life, if it attempts to be a way of life at all, and not simply the absence of any, was what Goethe wanted to teach, at the height of his wisdom, in the *Wanderjahren* and in the end which he gave to the life of his *Faust*. For him the realization meant a renunciation, a departure from an age of full and beautiful humanity, which can no more be repeated in the course of our cultural development than can the flower of the Athenian culture of antiquity.

The Puritan wanted to work in a calling; we are forced to do so. For when asceticism was carried out of monastic cells into everyday life, and began to dominate worldly morality, it did its part in building the tremendous cosmos of the modern economic order. This order is now bound to the technical and economic conditions of machine production which today determine the lives of all the individuals who are born into this mechanism, not only those directly concerned with economic acquisition, with irresistible force. Perhaps it will so determine them until the last ton of fossilised coal is burnt. In Baxter's view the care for external goods should only lie on the shoulders of the "saint like a light cloak which can be thrown aside at any moment." But fate decreed that the cloak should become an iron cage.

... No one knows who will live in this cage in the future, or whether at the end of this tremendous development entirely new prophets will arise, or there will be a great rebirth of old ideas and ideals, or, if neither, mechanized petrification, embellished with a sort of convulsive self-importance. For of the last stage of this cultural development, it might well be truly said, "Specialists without spirit, sensualists without heart; this nullity imagines that it has attained a level of civilization never before achieved.'"

Max Weber, *The Protestant Ethic and the Spirit of Capitalism*, George Allen and Unwin, 1976, pp. 180-2.

3   Gyorgy Lukacs, *Essays on Thomas Mann*, Universal Library, Grosset Dunlap, New York, 1965.

4   Roberta Reeder, *Anna Akhmatova: Poet and Prophet*, St. Martin's Press, New York, 1994; Ronald Hingley *Pasternak: A Biography*, Weidenfeld and Nicolson, London, 1983; Gao Xingjian *One Man's Bible*, 1999, translated from the Chinese by Mabel Lee, HarperCollins, 2002. See also Vitaly Shentalinsky, *The KGB's Literary Archive*, The Harvill Press, London, 1997.

5   John Armstrong, *Conditions of Love: The Philosophy of Intimacy*, Penguin, 2002, p. 1.

6   Armstrong, *Love, Life, Goethe*, p. 201.

7   This is a theme developed also in my *Sonnets to a Promiscuous Beauty* (Barrallier, 2005), notably in sonnets v and xii, 'European Love Songs' and 'A Roman Elegy'.

The greatest accomplishment of past mankind is that we no longer have to live in continual fear of wild animals, of barbarians, of gods and of our own dreams.
- Nietzsche (1881)[1]

... I want to learn more and more to see as beautiful what is necessary in things; then I shall be one of those who make things beautiful. Amor fati: let that be my love henceforth ...
- Nietzsche (1882)[2]

The great epochs of our life are the occasions when we gain the courage to rebaptize our evil qualities as our best qualities.
- Nietzsche (1886)[3]

9. On Nietzsche and Amor Fati

On 15 October 1888, his forty-fourth birthday, Friedrich Nietzsche wrote, 'On this perfect day, when everything is ripening ... the eye of the sun just fell upon my life: I looked back, I looked forward, and never saw so many and such good things at once ... How could I fail to be grateful to my whole life?' He had fewer than three months of sanity left before complete mental collapse rendered him an invalid for the last decade of his life. The preceding decade had been one of illness, loneliness and lack of recognition. Why, then, the gratitude? In one word: creativity, and of the most remarkable kind. He was in the process of crowning two decades of stunning writing with five brilliantly iconoclastic books in one year. His gratitude struck the keynote of his whole philosophy: *amor fati*, the love of one's lot.

It was Walter Kaufmann, author of the marvellous *Nietzsche: Philosopher, Psychologist, Antichrist*, who remarked, half a century ago, 'I confess that I love best the five books Nietzsche wrote during his last productive year, 1888 – not least because they are such brilliant works of art. *The Case of Wagner, The Twilight of the Idols, The Antichrist* and *Ecce Homo* present a crescendo without equal in prose. Then comes the final work, *Nietzsche contra Wagner*, Nietzsche's briefest and most beautiful book ... The preface to that book was dated Christmas 1888. A few days later, during the first week of January, he collapsed on the street, recovered sufficient lucidity to dispatch a few mad but strangely beautiful letters – and then darkness closed in and extinguished passion and intelligence. He suffered and thought no more. He had burnt himself out.'

I am irresistibly reminded here of Friedrich Holderlin's poem, 'To the Fates':

> A single summer grant me, great powers, and
> A single autumn for fully ripened song
> That, sated with the sweetness of my
> Playing, my heart may more willingly die.
> The soul that, living, did not attain its divine
> Right cannot repose in the nether world.
> But once what I am bent on, what is
> Holy, my poetry, is accomplished:

> Be welcome then, stillness of the shadows' world!
>
> I shall be satisfied, though my lyre will not
>
> Accompany me down there. Once I
>
> Lived like the gods, and more is not needed.

It was in this spirit of pagan epiphany that Nietzsche concluded the year 1888. The Fates, according to ancient tradition, are Lachesis (who assigns one's lot at birth), Clotho (who spins the thread of life) and Atropos (who cuts that thread at the moment of death). In his *Lycidas*, Milton referred to Atropos as 'the blind Fury with the abhorred shears', but this was not the outlook of either Holderlin or Nietzsche. Death, in their poetics, could be embraced as a completion – like the plunge of Empedocles into the volcanic crater of Mount Etna, in 433 BCE, which Holderlin, in fact, celebrated in verse.

Nietzsche had known for years, before 1888, that his grip on both sanity and physical existence was tenuous. He therefore sought conditions under which he could bring into being the 'poetry' that was within him. That he did so at all was a triumph of the spirit. Like Vincent van Gogh who, in 1888, was similarly fending off madness while doing some of his most remarkable work, Nietzsche burnt himself out in passionate creativity. What he created, however, was rather more earthshaking than the work of Van Gogh.

In *Twilight of the Idols* and *The Antichrist*, summarising and re-emphasising themes first developed in *Human, All-Too-Human* (1879), *Daybreak* (1881), *The Gay Science* (1882), *Thus Spake Zarathustra* (1883-85), *Beyond Good and Evil* (1886) and *The Genealogy of Morals* (1887), Nietzsche radically challenged the whole Western philosophic, metaphysical and moral tradition since Plato and Aristotle. Above all, he indicted Christianity as a profound *corruption* of the human spirit. In *Ecce Homo*, which Kaufmann describes as 'one of the treasures of world literature' and something in exchange for which 'we should gladly trade the whole vast literature on Nietzsche', he wrote down his own reflections on his work and his life. It was a short life – but a body of work that rivals in richness and possibly even in insight that of Plato or Shakespeare.

In *Nietzsche In Turin: An Intimate Biography* (1996), Lesley Chamberlain attempts to get 'inside' Nietzsche's year of pagan epiphany. It is a thoughtful and often stimulating book, though Irvin Yalom's marvellously conceived fiction *When Nietzsche Wept* (1992) is both more entertaining and more profound. The odd thing about Chamberlain's book is the posture she takes. 'This book', she writes, 'is an attempt to befriend Nietzsche.' There is a strange sort of presumption in this statement. Imagine the author of a book on Shakespeare, focused especially on the metaphysical terrors and moral question marks of *Hamlet, King Lear* and *The Tempest*, declaring that he or she was making 'an attempt to befriend Shakespeare'. It would be almost as if the dramatist was himself Hamlet in soliloquy, or Lear on the heath, and the author was cheerfully offering to provide consolation.

One has the sense that Chamberlain saw herself in the role of Lou Salome, the only woman with whom Nietzsche appears (unsuccessfully) to have actually sought intimacy, venturing to attempt, in Salome's place, to accomplish intimacy with the proclaimer of the death of God. At times, she seems to have been suggesting that Nietzsche's iconoclasm can be explained largely as a compensation mechanism for his loneliness, failure as a musician and, above all, lack of a sexual relationship with 'a woman of his own class.' At one point, she quotes Wagner's quip to Nietzsche, at Tribschen in the early 1870s, 'Write an opera, Fritz, or get married.' She comments that he was incapable of doing either and *therefore* turned against Wagner, whose wife, Cosima, he was in love with.

Chamberlain's book was a plausible successor to Salome's own book on Nietzsche, concerning which her biographer Rudolph Binion wrote, in 1968, that it was 'an object lesson in reducing a philosophy to a personal dossier on its author.' Salome's liaisons between 1880 and 1930 included, to borrow Tom Lehrer's words about Alma Mahler-Gropius-Werfel, 'practically every top creative man in Europe.' In her youth, she was close to Nietzsche though she was half his age. Chamberlain, like Salome, comes across clearly, in Binion's words, as a woman of 'prodigious intellect and personality,' but not quite in Nietzsche's class. In venturing to Turin more than a hundred years after Nietzsche collapsed there, she seems to have been prompted, in part, by a desire to find out why Nietzsche had *not* formed a satisfactory sexual relationship with her *alter ego,* Salome.

Unfortunately, this distracted her from actually analysing the extraordinary books Nietzsche wrote in 1888. She constantly digresses from what Nietzsche was thinking and writing at the climax of his life to his relationships with Lou, his sister and mother, his father and the Wagners. Along the way there are quite a few interesting observations about his general cast of mind, but there is no clear judgement concerning the climactic pieces of work. She remarks, at one point, that Nietzsche was 'the first philosophic troubadour, concerned with love for the unattainable one.' The 'unattainable one' being 'the truth' philosophically, but woman – Lou Salome or Cosima Wagner – psycho-sexually.

Given Nietzsche's remark at the beginning of *Beyond Good and Evil* that one might suppose truth to be a woman whom philosophers had, so far, been too unsubtle and incautious in wooing, Chamberlain had a target of opportunity here. She concedes, however, that the link between the promptings of Eros and the quest for the truth had already been identified by Plato. One can later find a comparable longing for the 'unsayable truths of idealism' in the philosophical work of Wittgenstein, she notes.

So Nietzsche's 'troubadour' status becomes, surely, not much more than a poetic gloss on an activity which had always taken erotic longing as its *point of departure*. To suggest, as Chamberlain does, that sexual consummation would have saved Nietzsche writing iconoclastic books is to *miss* the point of departure. It was Binion's conclusion, in 1968, that Nietzsche, in fact, 'surmounted the sentimental casualty' of his 1882 encounter with Salome, 'more rapidly, more fully and more healthily' than she did and was then able 'to turn the dross of it into purest gold.'

The gold in question was his greatest books, starting with *Thus Spake Zarathustra*. Nietzsche, Chamberlain observes, was concerned with refashioning philosophy from a metaphysical rumination on the dissonance between mind and body into 'a most modern tool, cleansing consciousness of unnecessary encumbrances to living.' Truth, she paraphrased him as saying, had for millennia been obscured by humanity's greatest idols (what it *called* its truths). He, however, took his slogan from Ovid's *Amores* (III, iv:17) 'Nitimur in vetitum semper, cupimusque negata.' (We always strive for what is forbidden and desire what is denied us.)

Again, this demonstrates that the erotic undercurrent had been clearly acknowledged by Nietzsche himself. It is not 'unconscious' or repressed. Yet, having acknowledged that much of his most formidable work took this as its motto, Chamberlain barely touches on the content of his five culminating works. To read them, however, is to be dazzled by a fire-wielder who has set about dynamiting the accumulated prejudices, self-indulgences and self-delusions of European civilisation, from Socrates to Wagner and from Caesar to Bismarck. It is a pyrotechnic display almost without parallel, by turns exhilarating, disorienting and actually alarming. To *think through* all this enlarges, tempers and disciplines one's mind as few other exercises can.

In the first of his 1888 *tours de force*, *The Wagner Case*, Nietzsche engages in a delightfully malicious commentary on the theme of 'redemption' in Wagner's operas. He concludes of *The Flying Dutchman* that here Wagner 'preaches the sublime doctrine that woman settles even the most unsettled man – in Wagnerian terms, she "redeems him". Here we permit ourselves a question: Suppose this were true – does that also make it desirable? What becomes of the eternal "Wandering Jew" whom a wife adores and settles? He merely ceases to be eternal; he gets married and does not concern us any more ...' So much, one might think, for Lesley Chamberlain's desire to 'befriend' Nietzsche. He continues to concern us a great deal, because he was so *unsettled*. His analysis of human civilisation was unusually thought-provoking. It has become even more instructive looking backwards over the twentieth century – and looking forwards into the twenty-first.

In *Twilight of the Idols*, Nietzsche could almost be seen as having set the curriculum for the whole of twentieth century philosophy, but in dazzling prose that awakens the mind by its sheer pungency. His reflections on 'Reason in Philosophy' and his stunning summation of Western metaphysics since Plato, 'How the 'True World' at Last Became a Fable', anticipate an enormous amount of language philosophy, both analytic and Continental, but unlike almost all of it are written in a style that is a delight to read. His reflections on 'The 'Improvers' of Mankind' uncannily anticipated the horrendous atrocities and hypocrisies of the twentieth century's totalitarian movements, both major and minor. 'Neither Manu nor Plato nor Confucius nor the Jewish and Christian teachers have ever doubted their *right* to lie,' he comments acidly. He concludes: 'Expressed in a formula, one might say: *all* the means by which one has so far attempted to make mankind moral were through and through *immoral*.'

In a culture which continues to cling by its teeth to the notion that Christianity is the summation of virtue and spiritual truth, even if one does not actually *believe* in its ancient doctrines any more, Nietzsche's third book of 1888, *The Antichrist*, remains the most 'scandalous' of his writings. It is the one most prone to prompt lesser mortals to dismiss him as 'mad'. Yet this blistering polemic is a paean to human integrity. The spirit of it is, in fact, beautifully encapsulated in his remark in *Twilight of the Idols*, 'Raphael said Yes, Raphael *did* Yes; consequently, Raphael was no Christian.'

Given 'how much Plato there still is in the concept "church", in the construction, system and practice of the church', the best '*cure* from all Platonism has always been Thucydides; the unconditional will not to gull oneself and to see reason in *reality* – not in "reason", still less in "morality",' Nietzsche writes. In the preface to *The Antichrist* he declares, 'One must be honest in matters of the spirit to the point of hardness before one can even endure my seriousness and my passion.' If one can, however, *The Antichrist* makes bracing and profoundly challenging reading.

Nietzsche's anti-Christian philosophy, he believes, offers an 'exit out of the labyrinth of thousands of years.' 'Who *else* has found it?', he asks. 'Modern man, perhaps? "I have got lost: I am everything that has got lost", sighs modern man.' Who would deny that this 'sigh' of modern (and even more of so-called *postmodern*) man is distinctly more audible at the beginning of the twenty-first century than it was at the end of the nineteenth? *Did* Nietzsche, then, find 'the exit'?

This is precisely the question he ponders, with considerable irony and self-parody, in his fourth book of 1888, *Ecce Homo*. At the end of its preface he quotes himself from *Thus Spake Zarathustra*: 'You had not yet sought yourselves; and you found me. Thus do all believers; therefore all faith amounts to so little. Now I bid you lose me and find yourselves; and only when you have all denied me will I return to you.' This is, of course, an unmistakable reversal of the demands on our fidelity in discipleship by the Jesus of the Gospels. That was the seal on Nietzsche's realism. He had no desire to be a Pied Piper in the Hamlin of the modern world, either to children or to rats.

Neither, as he makes clear in his fifth and final book of 1888, *Nietzsche contra Wagner*, was he interested in any mountebank role as 'redeemer'. His critique of Wagner's *Parsifal* was withering, precisely because he saw Wagner in this opera turning his back on freedom of the spirit in favour of a murky mysticism. The role of a Shakespearean clown was the closest he came to defining his work as diagnostician and 'prophet'. Look reality in the eye, he dares us, don't wallow in delusion or humbug; *create* your own 'exit' - with cleanliness, courage, honesty and gratitude for your fate. Then, and *only then*, will you be a friend of his.

What a note to go out on! But there was a little more: Nietzsche's last few, strangely beautiful letters. For me, the most evocative and beautiful fragment from them has always been a postscript in the final letter to his old friend and mentor, Jacob Burckhardt, postmarked Turin, 5 January 1889. 'I go everywhere in my student's coat and, here and there, slap somebody on the shoulder and say, *Siamo contenti? Son dio ho fatto questa caricatura* (Are we happy? I am God who made this caricature).'

Surely that is a sentence worthy of a Shakespearean clown; worthy of a character in a Sophoclean satyr-play? Surely it sums up with remarkable brevity, Nietzsche's ultimately *humane* critique of Christianity and his tragic world view? It belongs, as an atheistic joke, alongside Stendhal's wonderful *bon mot*, 'God's only excuse is that He doesn't exist.' It was the last glow of Nietzsche's culminating year, his fiercely creative *annus mirabilis*; his year of pagan epiphany.

9.   *Endnotes*

1   Friedrich Nietzsche, *Daybreak: Thoughts on the Prejudices of Morality,* trans. R. J. Hollingdale, Cambridge University Press, 1982, #5, p. 9.

2   Friedrich Nietzsche, *The Gay Science*, Book IV, #276, trans. commentary by Walter Kaufmann, Vintage Books, Random House, New York, 1974, p. 223.

3   Friedrich Nietzsche, *Beyond Good and Evil,* trans. R. J. Hollingdale, Penguin, 1973, #116, p. 79.

We here in Cambridge all keep each other going by the unquestioned assumption that what we do is important, but I often wonder if it really is. What is important, I wonder? Scott and his companions dying in the blizzard seem to me impervious to doubt – and his record of it has a really great simplicity. But intellect, except at white heat, is very apt to be trivial.

- Bertrand Russell to G. L. Dickenson (1913)[1]

Sometimes, things inside me are in such a ferment that I think I'm going mad; then the next day I am totally apathetic again. But deep inside me there's a perpetual seething, like the bottom of a geyser, and I keep on hoping that things will come to an eruption once and for all, so that I can turn into a different person. Perhaps you regard this thinking about myself as a waste of time, but how can I be a logician before I am a human being? Far the most important thing is to settle accounts with myself.

- Ludwig Wittgenstein to Russell (1914)[2]

What's the good of sticking in the damned ship and haranguing the merchant pilgrims in their own language? Why don't you drop overboard? Why don't you clear out of the whole show? ... You said in your lecture on education that you didn't set much count by the unconscious. This is sheer perversity. The whole of consciousness and the conscious content is old hat – the millstone around your neck.

- D. H. Lawrence to Russell (1916)[3]

10. On Russell's Retreat from Pythagoras

'We are sometimes accused of being arrogant scientists,' wrote Alan Sokal and Jean Bricmont in their classic critique of postmodernism, *Intellectual Impostures*, 'but our view of the hard sciences' role is in fact rather modest. Wouldn't it be nice (for us mathematicians and physicists, that is) if Gödel's theorem or relativity theory did have immediate and deep implications for the study of society? Or if the axiom of choice could be used to study poetry? Or if topology had something to do with the human psyche? But alas, it is not the case.'[4]

Sokal and Bricmont were concerned with exposing intellectual charlatanism so that critical reason could be applied to our common social and contemporary concerns. How that is to be done has been a fundamental concern of inquirers into reason and truth at least since the time of the Greek philosophers. It is often called 'the Enlightenment project' as if it commenced in the modern era, but its roots are classical.

The problems with which this project has collided, in the classical and in the modern world, have again and again led to suggestions that it is inherently flawed, that either metaphysical reality or human nature are not accessible to the dictates of reason and that some other form of understanding is necessary. The practitioners of reason themselves, from Plato to Wittgenstein, have often espoused a kind of mysticism which leaves both reason and practical concerns behind it.

In the past century, 'the Enlightenment project' came in for round after round of criticism or outright rejection. The sources of such criticism were, broadly speaking, of three kinds: psychological, ideological and philosophical. The psychological critique stemmed from the claim that human behaviour was inescapably rooted in the drives of the unconscious, not in a free or rational consciousness. The ideological critique stemmed from assertions that human beings needed religious revelation, symbolic culture or group identity and that these made claims which trumped those of reason. The philosophical critique stemmed from the disconcerting sense that the claims of reason could not themselves be anchored in a bedrock of demonstrable truth.

All these critiques took place against the background of unprecedented advances in pure and applied physical science, quantitative social science and economic analysis. Yet this did less than one might have thought to dispel their appeal. Indeed, such advances were often described as either irrelevant to the fundamental problems in question, or as symptoms of the problems themselves. Just as nineteenth-century industrialism had led to Romantic critiques of the proverbial 'dark Satanic mills', even as those mills first generated unparalleled material prosperity and then underwrote ameliorative social reform; so twentieth-century science was often seen as ominous rather than exhilarating, because it was seen as creating a cultural and possibly ecological wasteland.

Perhaps the most famous psychological critiques of reason were those of the psychoanalysts, Sigmund Freud and Carl G. Jung; but they were anticipated by Friedrich Nietzsche and others in the nineteenth century.[5] Human sexual and aggressive drives, they argued, suffered repression under the constraints of civilisation, leading to discontent and neurosis. The ideological critique, though it came from many quarters, was perhaps most gloomily articulated by Theodor Adorno and Max Horkheimer in *The Dialectic of Enlightenment* in the 1940s. Reason, they reflected, seemed to be producing a reductionist, consumerist and militarist order of things which was itself a disaster.[6]

The philosophical critique of reason was the most subtle of all. Its roots, like those of 'the Enlightenment project' itself, are classical. Among the Greeks themselves, scepticism and atomism developed alongside geometry and formal logic. The Enlightenment itself consisted not in a unitary scheme for the rational reconstruction of the natural and human order, but of increasingly strenuous inquiries into the roots of reason, truth and meaning. Immanuel Kant, one of the towering figures of the Enlightenment epoch, at one and the same time, urged human beings to reason boldly and engaged in a searching *Critique of Pure Reason*.

The bedrock of reason as a means to truth has always been mathematics. The late classical Neoplatonist philosopher Iamblichus (245-325 CE) summed up an attitude first articulated by Pythagoras, when he wrote, at the beginning of the Christian era, 'The Pythagoreans, having devoted themselves to mathematics, and admiring the accuracy of its reasonings, because it alone among human activities knows of proofs ... deemed these (facts of mathematics) and their principles to be, generally, causative of existing things, so that whoever wishes to comprehend the true nature of existing things should turn his attention to ... numbers ... and proportions, because it is by them that everything is made clear.'[7]

The great philosophical quest of the modern era has been for the deepest foundations of truth – the bedrock of reason itself. The quest was self-consciously in a direct lineage from Pythagoras and Plato, Aristotle and the great mathematician and geometer Euclid.[8] The problem was that, the deeper the most relentless thinkers delved, the more elusive they found that bedrock. The philosophical critique of 'the Enlightenment project' derives from this disconcerting collective experience. The confusions of postmodernism have arisen against this very real background.

The quest might be seen, in shorthand, as extending from the discovery by Carl Friedrich Gauss and Janos Bolyai, in 1829, that the axioms of geometry articulated by Euclid and believed for more than two thousand years to have constituted a solid piece of the bedrock, were not as solid as they had been thought to be, to the demonstration by Kurt Gödel (1906-1978), in 1931, 'that, in various axiomatic systems for the foundations of mathematics ... there are arithmetical propositions whose truth cannot be settled by the axioms, one way or the other, assuming that the system in question is consistent.'[9]

What was at stake here? What the great twentieth-century philosopher Rudolf Carnap called 'the logical structure of the world'.[10] To even set out on the quest for such a 'structure' in earnest required a serious grasp of what reason and mathematics were capable of doing.[11] To demonstrate that the structure was 'incomplete' in the way that Gauss and Gödel did, required reasoning of the most exacting and exquisite nature. Yet it seemed to generate the paradox that reason at its most intense could not provide a secure foundation for itself.

Where, then, did this leave 'the Enlightenment project'? On what secure basis were the proponents of reason now to respond to the psychological and ideological critics of the project? What were the implications for practical judgement in regard to the study of society where such critiques were sheeted home, in relation to poetry and the life of the imagination, or in relation to the nature and emotional workings of the human psyche? Where did the acute reasonings of the reasoners leave them in relation to the claims of ideology and religion?[12]

To keep our bearings here, it may help to refer back to the classical beginnings of the great quest, with Pythagoras (582-507 BCE). Few aspects of human experience seem more expressive of the 'irrational' element in the human psyche than our enjoyment of music. Yet it was precisely in music that Pythagoras first saw the possibility of analysing reality in terms of numbers. What he discovered was that 'the differences in vibration that characterize the notes of a musical scale can be calculated. Each individual tone is determined by the number of times a particular sound-producing medium – such as the strings of a lyre or guitar – oscillates in a given time.'[13]

His remarkable insight was that numbers enabled us to grasp a hidden, but rigorous and demonstrable order. 'If in music, then why not in other matters?' he wondered. He and his disciples came to think that one could 'construct the whole universe out of numbers', as Aristotle put it two hundred years later.[14] One could now apprehend the order of things as a 'cosmos', a harmonious order, rather than as a chaos. One could tune one's mind to the very music of the spheres and grasp, through the mathematical structure of this cosmos, the very mind of God.

Here is the origin of the quest that Carnap was engaged in 2,500 years later. Yet Aristotle had already sounded a cautionary note, so to speak, in observing of the Pythagoreans that, being intoxicated by the idea of numerical harmony, 'if there was any gap anywhere, they readily made additions, so as to make their whole theory coherent. For example, as the number 10 is thought to be perfect and to comprise the whole nature of numbers, they say that the bodies which move through the heavens are ten, but as the visible bodies are only nine, to meet this they invent a tenth – the "counter-earth".'[15]

Whereas the Pythagoreans believed they had discovered in number the key to a complete theory of reality and the mind of God, others grasped logic, less portentously, as a method for clarifying the mind of man. How to ascertain something as true, rather than merely superstitious or plausible was their quest. Parmenides of Elea (circa 500 BCE) was, perhaps, the founder of this variant of the quest. It was his judgement that, provided one followed certain rules and techniques, mere opinion could be sifted from true belief by the detection of fallacies and contradictions.

This tradition did not apprehend or seek completeness or harmony in quite the same way as did Pythagoras. Rather, it rested on a sceptical sense that human beings are highly prone to error and that true opinion could only be separated from false opinion by a careful, logical and provisional testing of evidence and a clear use of language. To be free from an error was not, in itself, to have discovered the truth, but it was nonetheless an epiphany and a good thing in its own right.

This tradition was perhaps most famously represented in the twentieth century by the philosopher Karl Popper (1902-1994), with his insistence on the need for falsifiability criteria in serious claims and his abiding interest in how human beings can be set free to engage in conjectures and refutations. Seen in the context of this sceptical tradition, the work of Gauss and Gödel was not demoralising at all, but extraordinary and liberating.

Popper is possibly most famous for his critique of all those philosophers, beginning with Plato, who sought 'completeness' in their systems, since he saw this as constituting a leaning toward totalitarianism. His polemical mid-twentieth-century work, *The Open Society and Its Enemies*, directed at all myths of origin and destiny whether Platonist, Nazi or Marxist, remains a classic philosophical response to the critics of 'the Enlightenment project'.[16] It was in honour of Popper's work that George Soros, in the late twentieth century, created his Open Society Foundations around the world.[17]

One possible response, therefore, to the whole set of critiques of 'the Enlightenment project' is the 'Parmenidean' or Popperian one: that reason is best exercised not in discovering 'the mind of God', but in overcoming illusion and working free of prejudice, naiveté and fundamental error. The open society, then, is liberal in a classical sense, as defined by John Stuart Mill: customary ways and conventional opinion keep life workable and stable, but are open to debate and correction by careful, piecemeal, peaceful and rational means.

There may be no better case study in how all of these profoundly important matters played out during the last century than the life of Bertrand Russell (1872-1970).[18] Born to parents who came from a long line of Whiggish (English Reformation) aristocrats and were close personal friends to John Stuart Mill, Russell was given the best education available in his day. He quested strenuously for the bedrock of logic and mathematics in his youth and tried all his life to apply his philosophical ideas to practical reality.

The life of Russell stands out, in the context I have sketched, for several reasons. What made him famous was the monumental three volume work he co-authored with Alfred North Whitehead at Cambridge University, *Principia Mathematica* (1910-1913). Yet the key to understanding his life is what he himself called 'the retreat from Pythagoras' that he felt compelled to undertake in the wake of this great labour – because he did not succeed in his quest.[19]

He was, in any case, ambivalent about purely intellectual pursuits all his life because of unresolved inner conflicts about his emotions. Finally, he longed to make a practical difference in political affairs, but nothing that he wrote or did in this domain had the intellectual distinction that characterised the technical philosophic work of his youth.[20] Indeed, his political and geopolitical thinking were erratic and tended to veer toward extremes such as radical pacifism in the 1930s, pre-emptive nuclear war advocacy in the late 1940s and Guevarist revolutionism in the 1960s.

What makes it possible for his life to be seen as a case study in the relationship between logic and practical judgement, however, is not simply that he wrestled with that relationship for most of his very long life, but that he left behind an immense documentary record which enables the scrupulous scholar to explore in detail how he thought and how he exercised his formidable powers of reason in the study of society, the appreciation of literature and, above all, the life of his own psyche.

Ray Monk, professor of philosophy at the University of Southampton, examined this documentary record. Monk's biography of Russell is deeply absorbing, but disconcerting reading. It is disconcerting for anyone who would prefer to believe that the exercise of critical reason can readily deal with the enduring and intractable matters of the emotional life of the psyche and the tensions and troubles of political and geopolitical affairs.

Russell's youthful interest in the philosophy of mathematics and the quest for the bedrock of truth was inspired by the same ancient intoxicants that had inspired Pythagoras and Plato long before: the hunger for certainty and wonder at the sheer beauty of 'eternal forms'. He sought in such certainty and beauty both a refuge from the messiness of the social world and a defence against his own emotional needs and fears. He floundered when it came to applying his reason to politics and geopolitics.

By a curious parallelism, Russell's three volume autobiography, published at the end of his life, approximates in bulk the three volume magnum opus he co-authored in his youth. The two parallel one another in a second respect: *Principia Mathematica* fails to complete its task and the *Autobiography* does the same. Autobiographies, whether confessional or apologetic, tend to omit or bend much of the less flattering truth about the author. But only the very best biographies do much better. Ray Monk's biography of Russell does better than both Russell's autobiography and any earlier biography of the man.

It does so because Monk, himself deeply conversant with the philosophic issues with which the young Russell was engaged, set out to explore the relationship between Russell's philosophical development and his personal and public life. In the process, he created an illuminating account of how an enormously intelligent human being struggled without final success to both get his mind around the foundations of reason and to live a passionate, humane and politically responsible existence. His judgement is neither censorious nor sparing of Russell's failings, and in both respects it is a monument of rational inquiry.

It takes something to register the sheer mass of writing that Bertrand Russell left behind him when he died. Monk discovered some 40,000 letters, for example, written by Russell himself over some eighty or more years. Apart from the letters, there were the massive three volume works, the *Principia* and the *Autobiography* and, in between, more than two dozen books, from *German Social Democracy* (1896) to *Logic and Knowledge* (1956). Merely to have sorted through this huge archive was a monumental achievement. To have crafted from it a narrative at once dispassionate and deeply reflective is a remarkable accomplishment.

The concern at the centre of Monk's reflections is the implicit relationship between Russell's commitment to abstract reason and the way his mind actually worked psychologically and politically. Three kinds of concern run right through the biography: the lifelong fears Russell harboured of descending into madness; the deep conflicts he experienced between his passion for abstract reason and his emotional needs; and the floundering efforts he made to take his reason out of the cloisters of Cambridge University and apply it in the 'real world' of English politics and international affairs. Each is worth briefly illustrating.

The second volume of the biography is sub-titled 'The Ghost of Madness'. It centres on Russell's fear that, like his Uncle William, who suffered a nervous breakdown in 1874 and was confined to an asylum until he died in 1933, he would lapse into madness himself or have children who were mad. In addition, throughout his life, Russell oscillated between writing of hope, peace, enlightenment and progress, on the one hand, and feeling that both he himself and mankind in general were, beneath a veneer of civility, barbarous and cruel.

He admired the novels of Fyodor Dostoevsky and, even more, those of Joseph Conrad, because they explored the psychology of moral corruption, existential vertigo and madness with such insight and artistry. When he met Conrad, in September 1913, he thought he had met a rare kindred spirit. Immediately afterwards, he wrote to the great love of his life, Ottoline Morrell:

> At our very first meeting, we talked with continually increasing intimacy. We
> seemed to sink through layer after layer of what was superficial, till gradually both

reached the central fire. It was an experience unlike any other that I have known. We looked into each other's eyes, half appalled and half intoxicated to find ourselves together in such a region. The emotion was as intense as passionate love and at the same time all-embracing. I came away bewildered and hardly able to find my way among ordinary affairs.[21]

The contrast between this experience and view of the world and that of the abstract purity of the Pythagorean or Platonist world he inhabited at Cambridge could hardly be greater. And whereas the Platonist in him longed to see metaphysical truths as reality, in his encounter with Conrad it was as if such truths were merely a layer of the 'superficial', beneath which lay true reality – the 'central fire'.

Russell never came close to reconciling these discordant themes in his mental life. One can imagine them being reconciled in various ways, but he did not embrace any of them.[22] The chief way in which they have been reconciled by human beings has been in the form of religion, but Russell had a deep ambivalence toward religion and seems never to have been able fully to come to terms with it. Writing to another of his great loves, Constance Malleson, in 1916, Russell confessed:

> The centre of me is always and eternally a terrible pain ... a searching for something beyond what the world contains, something transfigured and infinite – the beatific vision – God ... I can't explain it or make it seem anything but foolishness ... I have known others who had it – Conrad especially – but it is rare – it sets one oddly apart and gives a great sense of isolation.[23]

The author of *Why I Am Not a Christian* (1927) and *Religion and Science* (1935) saw himself, in other words, as 'oddly set apart' by precisely those things which characterise the Christian view of the human condition.

Why was his reason so unable to work through all of this? The answers, Monk suggests, have to do with his psychological character – personality traits deeply ingrained in him and not readily amenable to being overcome by his own reason. But if this is so, if one of the most gifted and rational minds of the twentieth century was left stranded in this manner, must not significant concessions be made to the psychological critique of 'the Enlightenment project'? Must we not concede that human beings have psychological needs which, be they ever so 'irrational', are substantially ineradicable and need to be addressed on their own terms, rather than being dismissed or neglected?

Similar questions arise in regard to Russell's emotional life. All his life he felt a terrible sense of loneliness and a profound need for passionate intimacy. Both of these things are common enough of course. What is striking is that Russell's intense rationalism not only did not enable him to deal with them better than most others; it actually served to compound the loneliness and inhibit the intimacy.

The repressive sexual upbringing he went through cannot have helped and his desolate first marriage to Alys Pearsall Smith accentuated all the problems he brought into it and he appears never to have been able to find the kind of balance and happiness he craved.

This was exhibited in Russell's relationships with his first three wives, a number of lovers and various friends and colleagues, but it is especially well illuminated by his relationship with the greatest love of his life, the aristocratic Ottoline Morrell, and by his encounter with D. H. Lawrence. Morrell was both fascinated by Russell and repelled by him. They had known each other six years when she recorded in her diary, in July 1915:

> He gets dreadfully on my nerves, he is so stiff, so self-absorbed, so harsh and unbending in mind or body, that I can hardly look at him, but have to control myself and look away. And of course he feels this and it makes him harsher and more snappy and crushing to me. What can I do? I feel I must be alone and go my own way to develop my life, my own internal life. Bertie crushes it out; he would remake me and the effort of resisting him and of protecting myself makes me desperate. It is far better to be alone than to be false.[24]

This was not merely a late development or an experience peculiar to Morrell. It ran through Russell's life and caused him intense anguish.[25] He loved the poetry of Shelley, especially his great love poem *Epipsychidion,*[26] and Shakespeare's sonnets. Yet he could never overcome the tendency to coldness and self-absorption that Morrell saw in him.[27]

The encounter with Lawrence, brokered by Morrell as the encounter with Conrad had been, brought all this violently to the surface for Russell. He was immensely impressed by Lawrence and told Morrell, 'He is amazing. He sees through and through one.' For once it was Morrell who played the sceptic. 'But do you think he sees correctly?' 'Absolutely,' Russell replied. 'He sees everything and is always right.'[28] When this 'infallible' seer wrote to him denouncing his anti-war activism as arrant hypocrisy and telling him he was 'simply full of repressed desires,' Russell was stunned and depressed to the point where he contemplated committing suicide. Nor was this a passing mood. He remained convinced throughout his life that he was barely containing inner demons

The politics of D. H. Lawrence, of course, were barely coherent and verged on a phantasmagorical form of communism. They were ancestral to some of the wilder versions of New Leftism in the 1960s and after. Russell, perhaps in part because of his repressive and rationalistic character, never allowed himself to wallow in the kind of wild rhetoric in which the novelist freely indulged. He did, however, indulge in some remarkably errant political reasoning throughout his life which neither his Whig background, nor his wide and deep education, nor his formidable training in philosophical logic and mathematics corrected.

The most admirable moments in Russell's rather quixotic political life were, surely, his intelligent reflections on German social democracy in 1896; his recoil from anti-German racism during the First World War; the writing of *Principles of Social Reconstruction* during the same years; and his horrified reaction to Bolshevism during his 1920 visit to the new Soviet Union, which were captured in his book *The Practice and Theory of Bolshevism*, published that same year. After that, however, his brilliant capacity for formal logic was less and less converted into anything resembling sound, practical judgement.

Against his better efforts must be set his naïve reflections on China in 1921; his espousal of unconditional pacifism in the 1930s, even as the Nazi menace rose; his call for a pre-emptive nuclear war against the Soviet Union in the late 1940s, should it fail to bow to an American ultimatum to disarm; his espousal, conversely, of unilateral nuclear disarmament in Britain, in the 1950s; and his lapse into Guevarist anti-Americanism in his last decade. What is astonishing again and again is not simply the extravagance of Russell's opinions over the last forty years of his life, but the wild inconsistencies, morally and strategically, in his thinking.[29]

What are we to make of all this? Monk quotes John Maynard Keynes, a man who had much in common with Russell philosophically, but a far more practical cast of mind, as remarking, in his *Essays in Biography*, 'Bertie held two ludicrously incompatible beliefs: on the one hand, he believed that all the problems of the world stemmed from conducting human affairs in a most irrational way; on the other hand that the solution was simple, since all we had to do was to behave rationally.'[30] Instead of being corrected both by reflection and experience over the course of his life, these incompatible beliefs of Russell's seem to have become entrenched, with more and more lamentable consequences, in terms of the substantive foolishness and superficiality of his political and geopolitical judgements.[31]

But let us reflect a little more closely on the implications of Keynes' remark. For he was himself convinced that human affairs were often conducted in a most irrational manner and was committed, in intensely practical ways, to ameliorating this often appalling state of affairs by rational means. He simply viewed human behaviour in a more realistic way than did Russell. He was very much a man of the Enlightenment project' and stood in a grand tradition of practical economic reason going back via Alfred Marshall to Adam Smith. He, like Popper, saw reason as an instrument for solving tangible problems in imaginative new ways and not merely as a way to discover and contemplate eternal truths.

I have, throughout this essay, placed 'the Enlightenment project' between inverted commas. I have done so not because it is something of which I am sceptical, but because it is a somewhat ill-defined something. The truth is that it has never been a single or coordinated project. Rather, it is a broad label for a general ferment of inquiry which took hold in Europe in the seventeenth century and has since opened up to common understanding the natural world and the world of the human species to a completely unprecedented depth. In doing so, it has presented us with both astonishing opportunities and daunting challenges.

Perhaps the greatest of those challenges is how to *integrate* our social and psychological needs with what our reason tells us is so about the world.[32] How do we maintain even a semblance of cultural and social cohesion amidst the ferment of ideas? How do we avoid errant kinds of 'integration' in the form of religious fundamentalism or ideological fanaticism and, at the same time, avoid the kinds of disintegration implicit in 'postmodernism'? Surely through the application of both prudence and imagination to the challenges we face as evolved beings in a complex world. That much our reason surely tells us. As for precisely how to accomplish such ends, it tells us, also, that the hopes of Pythagoras were excessive and that incompleteness, rather than mathematical exactness will always attend our schemes. Should that dismay us? Not of any necessity. It keeps the horizon open and the day after tomorrow provocatively uncertain.

10.   *Endnotes*

1   Bertrand Russell writing to Goldie Lowes Dickinson, 13 February 1913. Ray Monk, *Bertrand Russell: The Spirit of Solitude*, Jonathan Cape, London, 1996, p. 292.

2   Ludwig Wittgenstein writing to Bertrand Russell, January 1914. Ibid., p. 340.

3   D. H. Lawrence writing to Bertrand Russell, 19 February 1916. Ibid., p. 452.

4   Alan Sokal and Jean Bricmont, *Intellectual Impostures*, Profile Books, London, 1998, preface, p. x.

5   The most representative works of Freud in this regard are *Civilisation and Its Discontents* and *Beyond the Pleasure Principle*. For a brief introduction to Jung's outlook on the problem, see his *The Undiscovered Self*, first published in German in 1957, published by Routledge Classic in 2002.

6   A classic philosophic response to such trends and critiques, written at the same time as *The Dialectic of Enlightenment*, is Friedrich A. Hayek's *The Road to Serfdom* (1944), Routledge Classics, 2001.

7   Quoted in Arnold Hermann, *To Think Like God: Pythagoras and Parmenides*, Parmenides Publishing, 2004, p. 101. Iamblichus was the chief representative of Neoplatonism in the last decades before the Emperor Constantine made Christianity the state religion in the Roman Empire. The main tenets of his belief can be worked out from his extant writings, chief among which are the surviving fragments of his ten-book treatise on the teachings of Pythagoras. Only a fraction of his books have survived, most of them having been destroyed during the Christianisation of the Roman Empire.

8   Euclid of Alexandria (325-265 BCE) was the most prominent mathematician of antiquity and is best known for his treatise on mathematics, *The Elements*. Proclus, the last major Greek philosopher, who lived around 450 CE recorded a famous anecdote about Euclid, to the effect that Ptolemy, the Greek ruler of Egypt in the early third century BCE, once asked him if there were a shorter way to study geometry than *The Elements*, to which Euclid is said to have replied that there was 'no royal road to geometry'. He considered himself a Platonist.

9   Anita Burdman Feferman and Solomon Feferman, *Alfred Tarski: Life and Logic*, Cambridge University Press, 2004, p. 84.

10   Rudolf Carnap (1891-1970) was powerfully influenced by Kant's *Critique of Pure Reason* and Alfred North Whitehead and Bertrand Russell's *Principia Mathematica*. He was a founding member of the so-called Vienna Circle in the 1920s which espoused the philosophy of logical positivism. He published two

important books in 1928 which established him as a major philosopher: *The Logical Structure of the World* and *Pseudo-problems in Philosophy*. Working at the University of Prague, he wrote *The Logical Syntax of Language* in 1934, then emigrated to the United States in 1935 because, as a socialist and pacifist, he dreaded what the Nazis were about to unleash on Europe.

11   For an engaging reflection on this question by two mathematicians, see Philip J. Davis and Reuben Hersh, *Descartes' Dream: The World According to Mathematics*, Dover, New York, 1986.

12   The precise implications of Gödel's work are the subject of an impressive monograph by Torkel Franzen, *Godel's Theorem: An Incomplete Guide to Its Use and Abuse*, A. K. Peters, Wellesley, Massachusetts, 2005. See especially Chapter 4, 'Incompleteness Everywhere', pp. 77-95.

13   Hermann, *To Think Like God: Pythagoras and Parmenides*, p. 101.

14   Richard McKeon (ed.), *The Basic Works of Aristotle*, Random House, New York, 1941, p. 898; Aristotle, *Metaphysics* 1080b19.

15   Ibid, 986a 2-12. See also the beginning of Book II of Aristotle's treatise *De Caelo (On the Heavens)* 293a19-26, McKeon, p. 428: 'It remains to speak of the earth, of its position, of the question whether it is at rest or in motion, and of its shape. As to its position, there is some difference of opinion. Most people – all, in fact who regard the whole of heaven as finite – say it lies at the centre. But the Italian philosophers known as Pythagoreans take the contrary view. At the centre, they say, is fire, and the earth is one of the stars, creating night and day in its circular motion about the centre. They further construct another earth in opposition to ours to which they give the name counter-earth. In all this they are not seeking for theories and causes to account for observed facts, but rather forcing their observation and trying to accommodate them to certain theories and opinions of their own.'

16   Karl Popper, *The Open Society and Its Enemies*, 5th edn., one volume hardback, Routledge, London, 2002. The book was first published by Routledge and Kegan Paul, London, 1945, and went through many new editions and reprintings over the following six decades.

17   Charles Freeman, *The Closing of the Western Mind: The Rise of Faith and the Fall of Reason*, William Heinemann, London, 2002, reflected on the opposite process occurring in late classical antiquity with the triumph of Christianity over the sceptical and empirical philosophies of the Greeks. He cites the Church Fathers on the importance of faith and the triviality or even danger of rational inquiry. We must, declared St John Chrysostom, 'restrain our own reasoning, and empty or mind of secular learning, in order to provide a mind swept clear for the reception of divine words.' Basil of Caesarea, similarly: 'Let us Christians prefer the simplicity of our faith to the demonstrations of human reason ... For to spend much time on research about the essence of things would not serve the edification of the church.' Lactantius, another of the towering figures of the early church, declared: 'What purpose does knowledge serve – for as to knowledge of causes, what blessing is there for me if I should know where the Nile rises, or whatever else under the heavens the "scientists" rave about?' (p. 322.)

18   Ray Monk, *Bertrand Russell 1872-1921: The Spirit of Solitude*, Jonathan Cape, London, 1996; and *Bertrand Russell 1921-1970: The Ghost of Madness*, Jonathan Cape, London, 2000.

19   This was Russell's own characterisation of his philosophical development, in his *History of Western Philosophy*, George Allen & Unwin, London, 1946. In a summation of Russell's philosophy for the Royal Society, Ray Monk observes that, having discovered Euclidean geometry through his older brother Frank's tutoring, as a boy of eleven, Russell conceived 'the hope that *all* knowledge could be like that: "I liked to think of the applications of mathematics to the physical world, and I hoped that in time there would be a mathematics of

human behaviour as precise as the mathematics of machines. I hoped this because I liked demonstrations, and at most times this motive outweighed the desire, which I also felt, to believe in free will." ... This conception of philosophy is exemplified by the figure of Pythagoras, around whom Russell built a kind of myth that reveals much about his attitude to the subject. Pythagoreanism, on Russell's understanding, was a reformed version of Orphism, which was, in turn, a reformed version of the worship of Dionysus. Central to all three was the exaltation of ecstasy, but, in the cult of Pythagoras, this ecstasy is to be achieved, not by Bacchanalian revelries, but by the exercise of the intellect. The highest life, on this view, is that devoted to "passionate sympathetic contemplation". It is with such passages in mind that one should understand Russell's characterisation of his own philosophical development as a "retreat from Pythagoras". Interestingly, in 1943, Kurt Gödel, who still took a Platonist view of mathematics, attempted to suggest a reversal of Russell's retreat from Pythagoras, but Russell did not engage him in the debate. Monk, *Bertrand Russell 1921-1970: The Ghost of Madness*, p. 269.

20   'Though Wittgenstein never lost his admiration for Russell's early work in logic,' Monk wrote, 'he vehemently disapproved of Russell's popular writing of the inter-war period. "Russell's books should be bound in two colours", he once said, "those dealing in mathematical logic in red – and all students of philosophy should read them; those dealing with ethics and politics in blue – and no-one should be allowed to read them".' Perhaps his most famous book is his *History of Western Philosophy* (1946) which was, in Monk's words, 'greeted with almost universal disdain by the academic philosophers who reviewed it. Even C. D. Broad, an ex-pupil and admirer of Russell's ... could not bring himself to overlook the book's outrageous and cavalier superficialities and simplifications.' Yet 'despite its many flaws (or perhaps to some extent because of them), the book became a runaway best seller and placed Russell's finances on a secure footing for the rest of his life.' *The Ghost of Madness*, pp. 278-9.

21   Monk, *Bertrand Russell 1872-1921: The Spirit of Solitude*, p. 315.

22   The philosophy of Spinoza clearly appealed to Russell from quite an early age, but he never made of it or saw in it what Antonio Damasio has recently discovered: the vital co-existence of feeling and emotion with reason in the brain. See his *Looking for Spinoza: Joy, Sorrow and the Feeling Brain*, Harcourt, 2003.

23   Monk, *The Spirit of Solitude*, pp. 316-7.

24   Ibid., pp. 436-7.

25   His first marriage to Alys Pearsall was all but celibate, childless and ended in divorce; his second marriage to Dora Black produced two children, but ended in alienation, divorce and a complete refusal of Russell to communicate with Dora; his third wife, Patricia Spencer, withdrew from him, attempted suicide and, as Monk expresses it, 'For the third time in his life, Russell found himself living in a hollow shell of a marriage, in which love, passion and affection were replaced by a fragile and brittle courtesy' (*The Ghost of Madness*, p. 260) – and which ended, also, in a bitter divorce.

26   The poem is quite a long one and concludes, '... The winged words on which my soul would pierce/Into the height of love's rare Universe/Are chains of lead around its flight of fire. /I pant, I sink, I tremble, I expire.' For the complete text, see Shelley, *Everyman's Library Pocket Poets*, Alfred Knopf, New York, 1993, pp. 167-189.

27   When Morrell died, in April 1938 after a long illness, Russell wrote to her husband, Philip, 'The news is a terrible blow. A great part of my life ... is gone dead with her. I do not know anything consoling to say.' Monk, *The Ghost of Madness*, p. 212.

28   Monk, *The Spirit of Solitude*, p. 403.

29   'In a series of speeches and journalistic articles, beginning in September 1945, Russell argued passionately that, in order to preserve peace, America had to act firmly and immediately to impose its will on the rest of the world and, in particular, on the Soviet Union. From the very beginning, these articles had a bellicosity that contrasted markedly with the pacifist views he had expressed in the 1930s.' Monk, *The Ghost of Madness,* p. 298.

30   Ibid., p. 177.

31   Monk writes of the 'self-delusion to which Russell was prone whenever he wrote on social, political or historical subjects.' He believed his books, such as *Power* (1938) or *Freedom and Organization* (1934) to be of very great importance, when in fact they were lightweight pieces of reflection, lacking in serious scholarly weight or conceptual originality. Ibid., p. 212.

32   'There is increasing scientific evidence that reason and emotion need to live side by side in the healthy mind. It appears that some degree of irrationality acts as healthy corrective to the aridity of narrowly logical thought,' writes Charles Freeman towards the end of his defence of the classical tradition, *The Closing of the Western Mind*, p. 327.

JS 269, Now I remember Jerusalem
Jörg Schmeisser, etching, 1980.

Part Three – Contemporary Concerns

Boyd ... defined strategy as 'a mental tapestry of changing intentions for harmonizing and focusing our efforts as a basis for realizing some aim or purpose in an unfolding and often unforeseen world of many bewildering events and many contending interests.' Its aim was 'to improve our ability to shape and adapt to unfolding circumstances, so that we (as individuals, or as groups or as a culture or as a nation state) can survive on our own terms.'
- Grant T. Hammond (2001)[1]

I believe that we face the task of developing cooperative practices that will enable us to undertake a series of low intensity conflicts. Failing this, we will face an international environment of increasingly violent anarchy and, possibly, a cataclysmic war in the early decades of the twenty first century.'
- Philip Bobbitt (2002)[2]

We discovered real defects within the United States government ... We found national security institutions built to fight and win the cold war, yet poorly designed to combat the stateless and shadowy enemy of al Qaeda and Islamist terrorism. We found a lack of counterterrorism capabilities across our government – from our border protection, to our military and covert action capacity, to the communications capability of our emergency responders. We found a disturbing lack of unity of effort in the way government shares, analyzes and acts on information.
- Thomas H. Kean and Lee H. Hamilton (2006)[3]

1. On Strategy in the Twenty-first Century World

Writing at the very end of the twentieth century, Ernest R. May, one of the doyens of strategic studies in the United States, remarked that the history of Germany's 'strange victory' over France in 1940, of which he had just written an account, 'is particularly well worth recalling now.' In the wake of the Cold War, he observed, 'the United States and the other seemingly victorious Western democracies exhibit many of the same characteristics that France and Britain did in 1938-40 – arrogance, a strong disinclination to risk life in battle, heavy reliance on technology as a substitute, and governmental procedures poorly designed to anticipating or coping with ingenious challenges from the comparatively weak.'[4]

Philip Bobbitt was, at that time, in the advanced stages of writing a monumental study of war and peace, *The Shield of Achilles: War, Peace and the Course of History.* It was sent to press before the attacks of 11 September 2001 dramatically vindicated Ernest May's misgivings. His own argument, looking back over hundreds of years of Western history, was that the dangers that could be seen arising in the uncertain era after the Cold War years were every bit as great as those just faced down. They were all the more so for being of a different nature to the dangers of the era that had just passed, not least because they would be characterised by unprecedented uncertainty.

Bobbitt argued that the new dangers would compel profound reassessments of the nature of 'national' security itself and, with it, the civil laws, force structures and rules of engagement that buttress such security. Indeed, he suggested that unless such reassessments were made and reforms undertaken in anticipation of what could happen at any time, we could face a catastrophic breakdown in global order. The destruction of the World Trade Center, so richly symbolic of the broader aims of the perpetrators, took place before the book had come off the press. Bobbitt, therefore, added a postscript, in which he warned that what had just happened was more a symptom of what lies in store than an aberrant event. The great imperative in its wake was less to act within our means, than to think outside the square.

Perhaps the oldest and most enduring classic of strategic thought is Sun Tzu's *The Art of War.* Few passages from that ancient book are so commonly cited as that in which the sage remarked: 'one who knows the enemy and knows himself will not be endangered in a hundred engagements.

One who does not know the enemy but knows himself will sometimes be victorious, sometimes meet with defeat. One who knows neither the enemy nor himself will invariably be defeated in every engagement.'[5] But what does it mean to know oneself when that 'self' is a vast and complex state? And what does it mean to know your enemy, when you have just defeated a series of mighty enemies and appear to be confronted only by bandits, terrorists and stateless fanatics?

Bobbitt's answer is that we need to delve into the depths of our history if we are to understand ourselves and see from whence our enemies are arising. Who we are consists not simply in what we call ourselves and what language we speak, but how we think, what infrastructure we depend upon to sustain ourselves, what laws and institutions govern the ways in which we act, or constrain our capacity to deal with or even foresee certain kinds of dangers. All this changes over time and can overtake our collective comprehension of who we actually are, making us more vulnerable in Sun Tzu's sense, especially to new, shrewd and ruthless enemies. Bobbitt's endeavour was to deepen our perspective on the challenges we face by analysing the patterns of change that have taken place over the past five hundred years and the strategic consequences those changes have brought.

The Shield of Achilles was greeted by those best placed to assess it as a masterpiece for our times. William Shawcross greeted it as an awe-inspiring survey of the roots of strategic thinking, international law and the constitutional structure of states over the past five hundred years. Historian of war Michael Howard declared that it will surely rank as one of the most important works on international relations published during the last fifty years. He remarked in a foreword to the book that it is a remarkable and perhaps unique book. 'There have been many studies of the development of warfare, even more of the history of international relations, while those on international and constitutional law are literally innumerable. But I know of none that has dealt with all three of these together, analyzed their interaction throughout European history, and used that analysis to describe the world in which we live and the manner in which it is likely to develop.'[6]

There are three central premises to the argument Bobbitt advances in *The Shield of Achilles*, and from them he draws four striking inferences. The first premise is that modern history is best understood as a series of epochal wars that have shaped both state constitutions and the international society of states. The second is that strategy, law (both constitutional and international) and history (as a study) are inextricably intertwined, since they shape one another. Properly speaking, none can be understood without close reference to the others. The third is that the state is not withering away in a globalising world, but *nation* states are turning into what he called *market* states, which have a different constitutional and strategic logic than nation states.

Bobbitt's first premise is grounded in a view of history that goes back via Hobbes and Machiavelli to Thucydides. Indeed, he opens his argument with the claim that Thucydides was the first to write the history of an epochal war when he realised, in 413 BCE or shortly thereafter, that the struggle between Athens and Sparta was not merely a series of wars but a prolonged, fundamental conflict which would shape the whole future of the Greek world. 'So it is with all epochal wars,' Bobbitt asserts, ' – the Hundred Years' War, the Thirty Years' War, the Punic Wars – and so it will be seen of the war of the twentieth century.'[7] 'Epochal wars,' he writes, 'put the constitutional basis of the participants in play and do not truly end until the underlying constitutional questions are resolved.'[8]

From this premise he derives his first inference: 'We should regard the conflicts now commonly called the First World War, the Second World War, and the Korean and Vietnam Wars, as well as the Bolshevik Revolution, the Spanish Civil War and the Cold War as a *single war*, because all were fought over a single set of constitutional issues that were strategically unresolved until the end of the Cold War and the Peace of Paris in 1990.'[9] That single war, which he dubbed *the Long War*, was fought to determine which form of constitution – liberal parliamentary, fascist or communist – would replace the imperial states of Europe that had emerged after the epochal war of the Napoleonic period and had dominated the world between the Congress of Vienna and August 1914. This competition was itself triggered, he argues, by the instability of two imperial states – Germany and Russia – which mutated into the fascist and communist forms that the liberal democracies then had to master in order to survive.

Bobbitt sees this process as rigorously analogous to earlier epochal struggles which shaped the emergence and fate of princely states, kingly states, territorial states and nation states between the sixteenth and nineteenth centuries. Just as earlier epochal wars were resolved by major international settlements – Westphalia, Utrecht and Vienna – so the Long War was resolved by the 1990 Peace of Paris. This settlement, setting the seal on the victory of the liberal parliamentary nation state over fascism and communism, encouraged Francis Fukuyama to declare the 'end of history'. Bobbitt offers a more challenging and realistic appraisal of what had actually happened and a far more complex prognosis as to what possible futures we now face. At the heart of his prognosis is the claim that, having resolved the great constitutional issue of the twentieth century that divided them, nation states of the early twenty-first century had become 'uncertain as to how to configure, much less deploy, their armed forces.' The uncertainty had arisen because the traditional answers depended 'on certain assumptions about the relationship between the State and its objectives that the end of this long conflict has cast in doubt.'[10]

This brings us to Bobbitt's second premise: that strategy, law (both constitutional and international) and history (as a study) are inextricably intertwined since they reciprocally shape one another. It is most concisely stated as follows: 'The State exists by virtue of its purposes, and among these are a drive for survival and freedom of action, which is strategy; for authority and legitimacy, which is law; for identity, which is history.'[11] Law cannot come into being until the state secures a monopoly on the legitimate use of violence. Strategy cannot be formulated unless law prevails, for in its absence there is only civil war or banditry. 'Yet the legitimacy necessary for law and for strategy derives from history, the understanding of past practices that characterizes a particular society.'[12]

The most important thing about this premise is the inference he draws from it (his second): that the key technologies produced by the Long War – weapons of mass destruction, information technology and global communications – are undermining the very possibility of nation states in the twentieth century sense. Just as surely as cannon and muskets in an earlier era undermined the possibility of principalities and feudal baronies, these technologies, he argues, began to undermine the nation state at just the point when its liberal parliamentary form had triumphed over its darker rivals for primacy. They render it increasingly difficult for the nation state to maintain the kind of sovereignty by which it was defined. They render the defence and governance of territories and populations increasingly problematic. They do so by creating threats which transcend borders or territorial conflict, by creating economies that transcend any national base, and by making possible histories which the nation state cannot master through the control of information.

This brings us to Bobbitt's third premise: that the state is not withering away, but nation states are turning into market states, which have a different constitutional and strategic logic than nation states. They have their roots in the liberal parliamentary nation state, but are being forced to evolve under the above pressures and there are decidedly different ways in which things might turn out. One thing which will differentiate them will be the capacity they exhibit to reshape their military and security forces for what Paul Bracken, writing in 1993, called 'an entirely new operational environment, taking account of revolutionary changes in military technology and the possible appearance of entirely new kinds of competitors.'[13]

What all variants of the market state would discover to their cost, according to Bobbitt, is that the permeability of borders due to the uncontrollability of capital flows and of information, combined with the social consequences of these developments, the novel dangers posed by weapons of mass destruction in many more hands and forms than during the Cold War, environmental and viral hazards, the difficulties in managing consensus or marshalling resources for strategic purposes and the unrelenting nature of economic competition require new ways of thinking about what they themselves are and how they must cooperate.

These considerations gave Bobbitt his third and fourth inferences. The third inference is that the old strategic paradigm of threat, deterrence and retaliation must be replaced by a new one based on vulnerability, pre-emption and resilience. This will require fundamental rethinking of strategic doctrine, force structures and international law. Such rethinking has barely begun, Bobbitt observes, but must accelerate or be overtaken by events in possibly catastrophic ways. His fourth inference is that international institutions, already, by 2001, in large measure discredited, will be necessary, but must be transformed or reinvented if they are to play the constructive role required of them. Stated broadly, these third and fourth inferences may seem quite common fare. Certainly, variations of them have been in circulation for some time. What Bobbitt has put together, however, is a powerful synthesis with historical and conceptual roots that give it considerably more leverage than other more superficial reflections along similar lines. Taken together with his other premises and inferences, this world view offers quite a powerful set of lenses through which to re-examine both recent developments and future prospects.

Bobbitt called his postscript, the only part of his book written after 9/11, 'The Indian Summer'. The phrase, he writes, 'usually evokes a pleasant sensation of warm autumn weather that gives us a second chance to do what winter will make impossible'. Its origin, however, 'is more menacing. The early American settlers were often forced to take shelter in stockades to protect themselves from attacks by tribes of Native Americans. These tribes went into winter quarters once autumn came, allowing the settlers to return to their farms. If there was a break in the approaching winter – a few days, or weeks of warm summery climate – then the tribal attacks would be resumed, and the defenceless settlers became their prey. Once again, the settlers were forced to band together or to become victims, attacked one by one.'[14]

He goes on to argue that the first decade or so of this century should be seen as such an Indian summer and that we would do well to look to our defences. If the 2001 attacks 'inspire us now to deal realistically and creatively' with the emerging dangers of the twenty-first century, then the sacrifice of thousands on 9/11 could yet be turned to our common advantage. But if we *disregard* the implications of those attacks, he warns, we could find ourselves confronted by 'a world-rending cataclysm' as global institutions fracture, states lapse into turmoil, weapons of mass destruction proliferate and are used, and civil law is warped by fear into new authoritarian forms. For, he concludes, 'we are entering a fearful time, a time that will call on all our resources, moral as well as intellectual and material.'

Why did he apprehend such a cataclysm? First, because, in his own words, 'War is not a pathology that, with proper hygiene and treatment, can be wholly prevented. War is a natural condition of the State, which was organized in order to be an effective instrument of violence on behalf of society ... On September 11 2001, the nascent community of market states came to this knowledge as every society that preceded it has: through violence.'[15] Second, because the nature of the war that hit home that day is something existing laws and strategic doctrines are not equipped to deal with. Third, because the cascading consequences of not being so equipped could trigger crises far worse than most people can readily imagine.

Immediately after 9/11 and for some time thereafter there was a vigorous debate about whether it should be responded to as a crime or an act of war. Bobbitt was not in doubt that it was an act of war. The problem was that it was not an act of war by a nation state and therefore put customary usage of the laws and machinery of war out of their reckoning. Al Qaeda, he argues, was a virtual state, not a territorial one, not a nation state, 'which means that our classical strategies of deterrence based on retaliation will have to be rethought.' They cannot be effective in these circumstances, because 'what threatens the states of the world now is too easy to disguise and too hard to locate.'

We are, Bobbitt urges, on the cusp of a new epoch of war, the nature of which will confound those who think of war merely along the lines given by twentieth century experience. The liberalism that emerged triumphant from the twentieth century will have to reshape itself to cope with what is coming and most of its citizenry are unprepared for what this will entail. Their very concepts of security are outmoded and confused. Most fundamentally, 'National security will cease to be defined in terms of borders alone, because both the links among societies as well as the attacks on them exist in psychological and infrastructural dimensions, not on an invaded plain marked by the seizure and holding of territory.'

'In such a world,' Bobbitt remarks, 'we must move our thinking from threat based strategies that rely on knowing precisely who our enemy is and where he lives, to *vulnerability* based strategies that try to make our infrastructure more slippery, more redundant, more versatile, more difficult to attack ... There will be no final victory in such a war. Rather, victory will consist in having the resources and the ingenuity to avoid defeat. So long, however, as states rely on a deterrence and retaliation model for their strategic paradigms – that is, a model that requires a threat-based analysis – they will inevitably neglect those steps, including enhanced intelligence collection, pre-emption, the development of defensive systems (including sensors), vaccinations, the pre-positioning of medical supplies and advanced methods of deception that provide the basis for operating within a different paradigm, one that relies on a vulnerability analysis.'

Against such a background, the strategic policy debate in Australia over the past decade looks different from the way it looks if viewed solely from a provincial Australian perspective. Many of those wedded to the view that Australia should not involve itself more than marginally in conflicts outside its immediate region have tended to dismiss even preliminary attempts to rethink the broader strategic context in which Australia is situated, willy nilly. Bobbitt's formidable thesis was and remains a useful stimulus to more serious debate. Even those more inclined to reassess the country's strategic outlook would do well to study *The Shield of Achilles* however, for, while some tinkering with the force structure took place between 1999 and 2007, it would appear to have taken place on a somewhat improvised basis, rather being rooted in systematic thinking.

Australia might well be seen as a nation state evolving into a market state for the reasons identified by Bobbitt. As much as any such state, though with our own peculiar variations on the common themes, it is becoming more and more implicated in world order and global infrastructure security challenges, less and less likely to be threatened by conventional territorial invasion. Yet a preoccupation with conventional continental defence and denial of the so-called sea-air gap to a notional major adversary continues to consume the lion's share of the defence budget. Over the decade of the Howard government this began to change in response to the crisis in East Timor in 1999, the shock of 9/11 in the United States, the Bali bombings of 2002, and the political meltdown in the Solomons in 2003.

Even as our actual military and security commitments more and more came to resemble those that Bobbitt's world view would anticipate, the defenders of an older way of conceiving the strategic challenges facing the country insisted that nothing fundamental had changed and that we must still be prepared to fight conventional nation state wars. It is not necessary to postulate that such wars will not occur in order to see that, as a matter of practical reality, they have become extremely unlikely to occur in ways that directly threaten Australian sovereignty in the conventional sense of the term.

Australia, by the first decade of the twenty-first century, was faced with growing problems in maintaining its armed forces because of budgetary constraints compared with the expensive nature of contemporary advanced platforms, the lure of the marketplace on service personnel, the contradiction between a force structure configured for continental defence in depth and the realities of constant overseas deployments requiring more and different capabilities than have been developed under late Cold War conditions. All of this was true in the 1990s. The strategic environment that was thrown into stark relief by 9/11 accentuated these problems. Yet there were many who were reluctant to see that there was any fundamentally new kind of danger to consider.

Bobbitt's treatise, coming in the wake of the spectacular Islamist assault on the United States, was a call for thinking long and hard about what our vulnerabilities are, instead of what threats we face; about how resilient we can make ourselves, rather than about how we can deter or retaliate against some notional conventional aggressor; about how we can best contribute to the security of international economic and informational infrastructure, participate in pre-empting emerging dangers and build new alliances against unconventional dangers, rather than about whether we can blast an imagined conventional enemy out of the sea-air gap between our northern shores and the islands of Indonesia.

These were and are fairly radical thoughts. But thinking is what the strategic environment of the early twenty-first century calls for. The tinkering that was done between 1999 and 2007 was reasonably intelligent. Only a few well-placed strategic thinkers, though, appear to have understood that what was occurring was an actual paradigm shift. Indeed, those defending the old strategic doctrine hardly seem to have perceived that doctrine as a paradigm at all, but rather took it to be simply an unchanging reality. Even among those leading the incremental change, however, there was scant evidence of really systematic thinking. This lack of a really thorough strategic review had become so apparent by the last year of the Howard government that the Opposition was able to gain credibility by declaring that they would produce a new Defence White Paper if elected. They were elected and a new White Paper will be written in 2008. The challenge in preparing such a White Paper will be to review the country's strategic outlook and capabilities even more thoroughly than was done in the mid-1980s when the last such major review took place.

If Bobbitt was even approximately correct in his diagnosis of the new strategic environment and his prognosis for the epochal struggle ahead, between or among market states and anarchic forces, we shall be compelled to undertake such a radical rethinking of our strategic policy and force structure in the not too distant future. What will it take to prompt such thinking? We have certainly had a good deal of early warning about what is brewing. But, sunk in old paradigmatic ways of seeing our security, we have lumbered on, tinkering and tarrying, rather than thinking hard and coherently about where we are heading. We have always had the luxury of living in a continent-sized country remote from nation state threats. Like our rich natural resource endowment, that encourages a certain complacency. In the Indian summer of the 2000s, we cannot afford such complacency. For the dangers that are now looming are, in their own way, every bit as great as those of the past century. We must prepare ourselves to deal with them, or run the risk of coming to grief in grimly unfamiliar ways.

1.   *Endnotes*

1   Grant T. Hammond, *The Mind of War: John Boyd and American Security*, Smithsonian Books, Washington, 2001, p. 161.

2   Philip Bobbitt, *The Shield of Achilles: War, Peace and the Course of History*, Alfred A. Knopf, New York, 2002, p. xxiv.

3   Thomas H. Kean and Lee H. Hamilton, *Without Precedent: The Inside Story of the 9/11 Commission*, Alfred A. Knopf, New York, 2006, p. 318.

4   Ernest R. May, *Strange Victory: Hitler's Conquest of France*, Hill and Wang, New York, 2000, pp. 10-11.

5   Ralph D. Sawyer, *The Seven Military Classics of Ancient China*, Westview Press, 1993, p. 162.

6   Bobbitt, *The Shield of Achilles*, foreword by Michael Howard, p. xv.

7   Ibid., p. 21.

8   Ibid., p. xviii.

9   Ibid., p. 24.

10   Ibid., p. 7.

11   Ibid., p. 6.

12   Ibid., p. 7.

13   Paul Bracken, 'The Military After Next', *Washington Quarterly*, 16 (1993) p. 157.

14   Bobbitt, *The Shield of Achilles*, pp. 821-2.

15   Ibid., p. 819.

The birth of Mohammed was fortunately placed in the most degenerate and disorderly period of the Persians, the Romans and the barbarians of Europe: the empires of Trajan, or even of Constantine or Charlemagne, would have repelled the assault of the naked Saracens, and the torrent of fanaticism might have been obscurely lost in the sands of Arabia.
- Edward Gibbon (1776)[1]

Our aim is to change the jaahilii [pagan] system at its very root – this system which is fundamentally at variance with Islam and which, with the help of force and oppression, is keeping us from living the sort of life which is demanded by our Creator.
- Sayyid Qutb (1964)[2]

Now, in the early years of the twenty first century, the nation faces the intertwined menaces of global terrorism and proliferation of weapons of mass destruction. A city can be destroyed by an atomic bomb the size of a melon ... Smallpox virus bio-engineered to make it even more toxic, then aerosolized and sprayed in a major airport might kill millions of people. Our terrorist enemies have the will to do such things and abundant opportunities, because our borders are porous both to enemies and to containers. They will soon have the means as well ... This is not the time to let down our guard.
- Richard Posner (2006)[3]

2. On Challenges Laid Down by the Islamists

This essay was written within a few days of 11 September 2001 and published on 21 September 2001. It was my first attempt to come to terms with the significance of those events. I have left it almost entirely unchanged because it is of some interest to look back on it and consider the extent to which first impressions have been vindicated or otherwise by subsequent developments.

'Bull's-eye!' exclaimed a Cairo taxi-driver as he watched United Airlines Flight 175 spear into the mid-section of the World Trade Center south tower and explode in a fireball. That exclamation, half a world away from New York geographically and a whole world away in feeling, gives us our best clue as to how this war of terror is likely to play out. This is not, as President Bush and many others want to believe, simply a war of good against evil. It is a war of power against resentment. It will have to be thought through and fought accordingly, or the effort to crush terrorism will merely breed it and cause chaos to proliferate like the plague.

In the ten days that have elapsed since the World Trade Center was destroyed and the Pentagon scarred, countless learned and thoughtful people have expressed opinions as to what it all portends, who the culprits are, what should now be done. This is simply my own effort to reckon with the situation. It takes the form of seven theses or propositions. Each is discrete, in that it can be contested on its own. Each is also a counsel of discretion against presumption and self-righteous anger in the wake of the catastrophe.

First thesis: *This is an act of war, not merely a nihilistic crime.* Geoffrey Robertson has argued that the destruction of the World Trade Center 'should be characterised and prosecuted as an international crime, not as a war.' I think his motives in saying this are admirable, but the available evidence suggests to me that the perpetrators of this deed intended it as an act of war. They are not finished with us yet, they have just begun. They are not accessible to a working system of justice and have many ways to elude attempts to bring them to justice. They will have to be fought and we would be foolish to assume that they have not long thought through the probable reaction to this devastating strike against the American heartland.

Second thesis: *This is the first serious blow in a long-planned war by an international coalition of Islamic militants against the West and its allies.* While it is very important not to allow paranoia to take over in the wake of a shock of this kind, there seems to be sound evidence that the planning of this attack was orchestrated within a global organisation. The immediate suspect has been Osama bin Laden, who has very quickly assumed something of the status of the dark criminal genius in *The Usual Suspects*. His repeated denials that he is responsible may simply be lies, but we should take them as pointing indirectly to something intelligence professionals have suspected for some time.

Just over five years ago, Mehdi Chamran, a senior Iranian intelligence figure, brought together a coalition of militant Islamist groups under the rubric Hizballah International. Its purpose was, in the words of Iran's spiritual leader, Ayatollah Ali Khameini, to carry *jihad* into 'all continents and countries'. In mid-June 1996, there seems to have been a summit of Islamic militant groups in Tehran. It is said to have been attended by leading Palestinian, Lebanese, Egyptian, Kurdish and Gulf States activists. An agreement was supposedly reached there to pool financial resources and coordinate training, so as to make possible military interoperability across at least thirty countries.

According to Washington-based sources, Osama bin Laden, Imad Mughaniya of the Lebanese *Hizballah* and Ahmad Salah of the Egyptian *Jihad* were appointed at this summit as a coordinating committee under the chairmanship of Mehdi Chamran to oversee a global strategy of Islamic war against both the infidel Western world and its allies in the Islamic world. According to the same sources, a further meeting took place in July 1996 on the Afghanistan/Pakistan border, involving not only representatives from the aforementioned groups, but also from Algeria, along with 'senior officers from the Iranian and Pakistani intelligence services.'[4]

In short, the probability is that this operation was the work of an international organisation with roots in many countries and branches inside the intelligence services of much of the Islamic world. It was conceived and carried out with consummate professionalism and the meticulous calculation of training, timing, targets and the physics of demolition. Scores of operatives were involved in preparations lasting months if not years and yet the plot did not leak. This was clearly not the work of inspired amateurs. It was, we should suspect, the work of Mehdi Chamran's high command of the Islamic *jihad*. In this light, the comments coming out of Tehran in the immediate wake of the attacks, and not least those of Ali Khamenei, need to be analysed very closely.

Third thesis: *This blow was long premeditated and its perpetrators have thought ahead several moves.* In the aftermath of the catastrophe, a number of people have stated that whoever did this has miscalculated, in that they may have believed they would demoralise and divide the United States and its allies, but they have instead brought them together as almost nothing else could have done. This may be true, but we would be unwise to assume so.

The commitment of the *jihad* coalition to its goals may blind it to the real consequences of its own actions, but we would be prudent to assume that the stirring-up of enormous anger was *exactly* what these people sought to achieve. We would do well to surmise that they sought to provoke a massive American response which would, in turn, enable them to spark upheaval all over the Islamic world and topple pro-Western regimes from North Africa to the Indonesian archipelago.

Fourth thesis: *This is not the clash of civilisations, but a profound challenge to civilisation.* In 1996, even as the Islamic militants convened under Mehdi Chamran, Samuel Huntington's book, *The Clash of Civilisations and the Remaking of World Order*, created something of a sensation.[5] This development will tempt many people to embrace his argument. Henry Kissinger was an enthusiastic reviewer of Huntington's book when it came out, but he is correct in stating on this occasion that 'America and its allies must take care not to present this ... as a clash of civilisations between the West and Islam. The battle is against a radical minority that disgraces the humane aspects Islam has displayed in its great periods.'

The challenge laid down in front of us is to deepen our collective understanding of Islam as a civilisation and to extend the idea of civilisation itself. In 1996, at exactly the time that Mehdi Chamran and his allies were meeting to plot a global *jihad*, I visited Toledo during a stay in Spain and was deeply moved by the realisation that this city, which dates back to pre-Roman times, was at its peak during the centuries of Islamic civilisation. It was a city where Muslims, Jews and Christians lived in peace and mutual toleration.[6] It was not Islam, but Catholicism that brought intolerance and religious terror to Spain. This is the sort of thing of which we need to remind ourselves even as we watch people invoking Christian righteousness and just war.

In the early 1990s there was much talk of a new world order. The only form that new world order took was the military primacy of the United States and the economic liberalism known as the Washington consensus. The September 11 attack was manifestly directed at both these things. Simply responding in rage will not get to the root of the problem. We need to get beyond dismissing the enemies of the United States and of global economic liberalism as irrational. Until we do so, even militarily effective action against them will serve only to deepen the feelings that have led to this pass and will fan the flames of the *jihad*.

Fifth thesis: *This is a spectacular case of what Chalmers Johnson has called 'blowback'.*[7] Johnson argues that American arrogance and imperial presumption are generating ever more resentment around the world. Americans, he warns, should prepare themselves for a century of revolt and retribution, what he called 'blowback'. I can well imagine that he is currently giving many talks enlarging on this theme. Certainly it is something which Americans in particular and all those around the world who do *not* resent American primacy would do well to think much harder about than they have up until now.

In a recent thoughtful reflection, the veteran American social historian Studs Terkel commented, 'There is an article in the current edition of *Harper's* magazine[8] in which we are compared to the Romans. But they did not pretend to do good, they were just conquerors. We are always doing good, we are always innocent. But we are always the ones looked on badly. Why?' Alas, he is wrong on both counts. The Romans did, indeed, 'pretend to do good' as they built their empire. Anyone who thinks otherwise is simply ignorant of Roman history. And the United States is not as innocent as many of its citizens would like to believe.

The Romans of the second and first centuries BCE, like the Athenians three centuries earlier, debated the rise of empire and were concerned with the very issues that Americans have been in the past hundred years – since the Spanish-American War. They were not mere conquerors, but saw themselves as making the world safe for the Roman way of life, extending order and civilisation into the barbarian world and replacing Asiatic decadence with Roman virtue. The parallels with the American sense of manifest destiny are actually quite striking and instructive.

The American defeats of Germany and Japan and then of Soviet Communism have strong parallels with the expansion of Roman power. Those who wish to respond that the United States has been uniquely distinguished by the magnanimity of its victories would do well to remember that Rome saw itself, also, as terrible in war but magnanimous in victory. It was Virgil who wrote that it was the destiny of Rome 'To hold dominion and seal peace with custom; Spare those who yield, the defiant utterly subdue.'[9] This is very much the spirit of the *Pax Americana*. And 'the defiant' are all those who see fit to rebel against the imperial peace put in place over half a century and maintained with scores of military bases right around the globe.

If this latest revolt against the *Pax Americana* is to be understood, it needs to be put in the deep context of the impact of the *Pax Americana* (and the *Pax Britannica* before it) on the Islamic world. It is a quaint and self-indulgent illusion to think that that impact has been a matter of Americans 'always doing good' and 'always being innocent'. Those who believe this have just been given what ought to be a very rude awakening. The Anglo-American imperial goal in the Islamic world has been one of dominating access to oil and playing the many tribes and factions of the Islamic world against one another.

The uprising in Iran in 1977-79 should have led to deep learning, but it led only to more of the same. Those wishing to learn now should begin with James Bill's lucid and deeply informed book *The Eagle and the Lion*.[10] They should read with special attention the final chapter, The Politics of Foreign Policy Failure: A System of Reinforcing Errors. To broaden the understanding beyond the pivotal case of Iran, they might do well to read Paul Bracken's provocative little book, *Fire In the East*, published in 1999.[11] The message is this: Do not go about your work thinking that 'we' always do good and 'we' are always innocent. We, in fact, the citizens of the dominant Western civilisation, have established that dominance through a will to power every bit as relentless as that of the Romans. And there will be blowback.

Sixth thesis: *The attempt to root out and destroy the perpetrators could catalyse a wider disorder.* US Secretary of State, Colin Powell, has declared that the United States will root out and destroy international terrorism. Senator John McCain has vowed 'We are coming. God may have mercy on you. We will not.' President Bush has declared that those who attacked the United States have always found that 'they have chosen their own destruction.' The risk is that drastic measures will bring in their train more destabilisation, more resentment, more chaos and a swelling of the ranks of the *jihad*. Given the social conditions in wide parts of Africa, the Middle East and Central Asia, the use of US military forces in various parts of these regions could trigger a landslide into what Robert Kaplan has called 'the coming anarchy'.

Afghanistan is under the greatest threat because it harbours Osama bin Laden. It is already in a horrifying state of political and economic degeneration, caused chiefly by the ruthless war by proxy between the Soviet Union and the United States in the 1980s. Pakistan, which has been the ally of the Taliban from its inception, is being pressed very hard by the United States to provide base support for American military action against the Taliban, which in turn threatens war against Pakistan. Pakistan is far gone in economic degeneration and brewing social disorder itself and could erupt into chaotic conflict internally and with India. Given that both are now nuclear powers, dire scenarios could ensue.

Consider, further, that – if my third thesis is correct – the Hizballah International has anticipated something along these very lines. If would-be reformers in Iran are trying to moderate the more extreme elements of the Islamic blowback, Mehdi Chamran and his allies seek to drive things in the opposite direction. Their strategy is to draw the United States into ill-considered, multiple and unsustainable confrontations. Che Guevara, in the 1960s, vowed to catalyse 'one, two, many Vietnams.' These people may well intend to produce 'one, two, many Somalias'. Will the US send its forces into Iran? Into a disintegrating Pakistan? Into Iraq? Into Sudan? Where will lines be drawn? And when Henry Kissinger calls for military measures directed at the source, we would do well to remember his secret and utterly counter-productive bombing of Cambodia thirty years ago and what that catalysed.

Seventh thesis: *This war has begun with a deeply symbolic assault on America. It will be won, if it is won, by an even more powerful symbolic response and not by arms alone.* By using civilian airliners in suicide assaults on the World Trade Center and the Pentagon, the perpetrators have signified the deep roots of their motivation and the implacable nature of their intentions. It is conceivable that the combined efforts of the United States and its many allies will find and destroy at least some of the perpetrators. It is even arguable that some such effort is imperative, lest they become too emboldened by this daring and stunning success. However, if the deeper question of what is ultimately at stake is not addressed with searing honesty, then the military and police effort will generate only frustration, anger and despair on a global scale.

What is ultimately at stake is the very idea of a humane global civilisation. What is called for is a paradigm shift in our common appreciation of what such a world order requires of us individually and collectively. It must be a world order in which Western secularism, Christian, Muslim and Jewish monotheism, Confucian humanism, Hindu polytheism, Buddhist compassion and the countless other variations on the human search for meaning and dignity can find common cause. As human beings, not as true believers, we have a common future in the stewardship of this stunningly beautiful and utterly isolated oasis of life, this earth.

For as long as we do not see this, we will fail to comprehend how a Cairo taxi-driver could shout 'Bull's-eye!' at the sight of the World Trade Center being hit by a civilian airliner. If we do see it, then we shall be able to transcend our anger at what has just happened. We might then feel able to reach out to the disinherited and pathological people who greeted it with joy. Their joy, after all, is not so foreign to us as we tell ourselves in our current grief and rage. It was exactly such joy that the ancient Israelis expressed on the downfall of Babylon, or that many Americans felt at the fire-bombing of Tokyo in 1945. So, let this catastrophe become not merely the catalyst for crying havoc and letting loose the dogs of war, but rather a defining moment in the twenty-first century enlargement of universal humanism.

2.   *Endnotes*

1   Edward Gibbon, *The Decline and Fall of the Roman Empire*, Chapter LI, Everyman's Library, Dutton, New York, 1969, vol. 5, p. 297.

2   Sayyid Qutb, *Milestones* (1964), 1978 edn, Holy Koran Publishing House, Beirut, p. 34, cited in Ahmed Bouzid, 'Man, Society and Knowledge in the Islamist Discourse of Sayyid Qutb', PhD dissertation, Virginia Polytechnic Institute and State University, Blacksburg, Virginia, 1998, p. 110.

3   Richard A. Posner, *Not a Suicide Pact: The Constitution in a Time of National Emergency*, Oxford University Press, 2006, pp. 2-3.

4   Michael Griffin, *Reaping the Whirlwind: The Taliban Movement in Afghanistan*, Pluto Press, London, 2001, pp. 135-6, citing Yossef Bodansky from *Strategic Policy* vol. XXIV, no. 8, 3 August 1996.

5   Samuel P. Huntington, *The Clash of Civilisations and the Remaking of World Order*, Simon & Schuster, New York, 1996.

6   Maria Rosa Menocal, *The Ornament of the World: How Muslims, Jews and Christians Created a Culture of Tolerance in Medieval Spain*, Little, Brown and Co., Boston, New York and London, 2002.

7   Chalmers Johnson, *Blowback: The Costs and Consequences of American Empire*, Metropolitan Books, Henry Holt & Co., New York, 2000.

8   Lewis Lapham, 'The American Rome: On the Theory of Virtuous Empire', *Harper's*, August 2001, pp. 31-8.

9   Virgil, *The Aeneid*, Book VI, lines 852-3.

10   James A. Bill, *The Eagle and the Lion: The Tragedy of American-Iranian Relations*, Yale University Press, 1988.

11   Paul Bracken, *Fire in the East: The Rise of Asian Military Power and the Second Nuclear Age*, Perennial, HarperCollins, 1999.

In those days, a government denial was all it took to make a reality disappear. LBJ had come to be widely perceived as a liar, yet the press and public still had no real awareness of just how much they had been lied to, even by LBJ, let alone by his predecessors and now by his successor.
- Daniel Ellsberg (2002)[1]

Daniel Ellsberg is the most dangerous man in America. He must be stopped at all costs ... [O]nce we've broken the war in Vietnam, then we can say this son-of-a-bitch nearly blew it. Then we have, then we're in strong shape – then no-one will give a damn about war crimes.
- Henry Kissinger (1971)[2]

Ordinary public information was not worth considering at all; it was hardly even real information. It became valuable only when it was created by the government and legitimized by classification. The priesthood maintained its position by seeming to know things that ordinary people did not, or could not, or should not know ... The information was important because they had created it or had access to it; they were important because of the information they possessed.
- Peter Schrag (1974)[3]

3. On Official Secrecy and Mendacity

This essay was written in 2002 as the debate over Iraq was coming to a head. As with the 2001 essay on the significance of 9/11, so in this case, I have left the essay almost entirely unaltered because it shows both the extent and the limitations of my understanding of some things to do with the debate on Iraq. This is more clearly apparent in the essay 'On the Decision to Overthrow Saddam Hussein'.

Daniel Ellsberg leaked more top secret documents into the public domain than anyone else before Vasili Mitrokhin brought the KGB's secret archives West.[4] Mitrokhin was hailed as a hero for revealing the secrets of the KGB. Ellsberg revealed the Vietnam War secrets of the Pentagon and the White House in the *Pentagon Papers*, as they became known. This precipitated both the end of Congressional support for the war and the downfall of President Nixon.

Was Ellsberg a hero? Your judgement very likely depends on what you think of the Vietnam War and Richard Nixon. Before pronouncing any judgement, however, read Ellsberg's *Secrets: A Memoir of Vietnam and the Pentagon Papers*. It tells the tale of a highly intelligent, deeply informed insider of uncommon integrity who learned disturbing truths, wrestled with his conscience and dramatically altered his sense of where his moral duty lay in keeping government secrets.

This is important not only for the Vietnam War and Nixon. It is vitally important to the whole ideal of open and democratic government. It is also vital to the possibility of humanity avoiding in the twenty-first century the kind of calamities that scarred the twentieth century. Indeed, it is important right now, in the current confrontation between the United States and Iraq, as the 71-year-old Ellsberg has been vigorously and publicly arguing in recent months.[5] But I'll come to that last.

There is no better way into the significance of Ellsberg's story than the way he begins his memoir – with his account of his inside knowledge of the Gulf of Tonkin incident, which was used by President Johnson to win Congressional approval for the escalation of American military intervention in Vietnam in August 1964. But before looking there, you need to know who Ellsberg was, in 1964.

Born in 1931 in Chicago, Ellsberg was the grandchild, on both sides, of Jews who had emigrated from Russia in the 1880s and settled in America. He was ten years old when the Japanese attacked Pearl Harbor. His father, a structural engineer, spent the ensuing war years designing factories to build bombers. Ellsberg tells us that nothing he was aware of during those years seemed so 'incomprehensibly evil as the deliberate bombing of women and children' – by the Nazis, in Warsaw, Rotterdam and London. He didn't know of the Holocaust until after the war.

He also didn't know that the Allies, too, bombed women and children. As he puts it: 'We didn't see films of what was happening to people on the ground under our bombers, or in the firestorms in Hamburg, Dresden, or Tokyo. We believed what we were told: that our daylight precision bombing with the Norden bombsight was aimed only at war factories and military targets, though some civilians were inevitably hit by accident ... Official secrecy and lies concealed a deliberate British campaign of terror bombing targeted directly on [the civilian] population, in which the United States joined in the later years of bombing Germany and throughout the bombing of Japan.'[6]

Educated (on a Pepsi-Cola Co. scholarship) at Harvard University between 1949 and 1952, as an economist, Ellsberg considered himself a Cold War liberal democrat. He admired Franklin Roosevelt and Harry Truman. He had wanted to specialise in labour economics but became fascinated by 'the field of decision theory, the abstract analysis of decision making under uncertainty' and it was here that he was to make his mark. Being a talented student, he won a Woodrow Wilson fellowship for graduate study at Cambridge University in 1953. He then took military training in the Marine Corps – because 'the corps didn't bomb cities ... it fought soldiers, not civilians'; returning to Harvard in 1957 to do a PhD in decision theory.

'My three years in the Marines,' he relates, 'had left me with a respect for the military, an interest in strategy, and a greater readiness to apply intellectual concepts to military problems than I would have felt otherwise.' Since his theoretical interests had clear application to problems of military strategy, he sought and won an invitation from the Economics Department of the Rand Corporation in Santa Monica, California, 'to spend the summer of 1958 there as a consultant during my graduate study at Harvard.'[7] This rapidly led to bigger and better things.

In 1959, aged only twenty eight, he was invited to give a series of lectures at Harvard, in the Lowell Lectures series, under the rubric 'The Art of Coercion'. Two of the lectures were given to Henry Kissinger's seminar on international relations. These were entitled 'The Theory and Practice of Blackmail' and 'The Political Uses of Madness'. They were an analysis of Hitler's coercive diplomacy against Austria and Czechoslovakia in 1937-38. More than a decade later, Kissinger was to say that these lectures had provided him with a conceptual model for how to deal with Vietnam.[8]

From the summer of 1959, he became a permanent employee at Rand and chose to specialise in an issue which he had come to believe was crucial to the avoidance of nuclear war: 'the command

and control of nuclear retaliatory forces by senior military officers and especially by the president.' The enormous sensitivity of the subject led to his being granted extraordinary access, entailing 'knowledge of some of the most highly protected and closely held secrets in our military structure. These included military plans for general nuclear war that were generally inaccessible even to the highest civilian authorities.'[9]

This knowledge included the top secret estimate by the Joint Chiefs of Staff that, in the event of general nuclear war with the Communist bloc, American nuclear weapons were expected to kill five to six hundred million people, most of them in the first few days. He was stunned. 'A hundred Holocausts ... It was the JCS's best estimate of the actual results, in terms of human fatalities, of our setting into motion the existing machinery for implementing the current operational plans of the JCS for general war ... I still remember holding that graph in my hand and looking at it in an office of the White House annex in the Executive Office Building on a spring day in 1961. I was thinking: This piece of paper, what this piece of paper represents, should not exist. It should never in the course of human history have come into existence.'[10]

Given his expertise in this area, he was summoned to Washington in 1962 to work for the Executive Committee (ExComm) of the National Security Council (NSC) during the Cuban Missile Crisis. This was a major learning experience. It left him, he recalls, 'with a vivid sense of how thermonuclear warfare might actually come about in a crisis, not only by the failures of high level control I had begun to foresee – which were exhibited by both sides in this confrontation – but as a result of major miscalculations at the highest levels and of prior commitments made without any adequate sense of where they were likely to lead. Each side had grossly misunderstood the other, wrongly estimated its behaviour, failed to understand actions of the other as responses to ... their own words and actions.'[11]

Out of these hair-raising studies, he was invited by Rand to embark on a highly classified project, examining patterns of high-level American governmental decision-making in Cold War crises, with a view to improving 'the president's understanding and control of his own bureaucracy'. Walt Rostow, chairman of the policy planning staff in the State Department, convened an interagency panel for him, consisting of senior officials just below the highest level from State, Defence, CIA and JCS who undertook to facilitate access to classified papers that would assist his inquiry.[12]

He was fully engaged in this inquiry when, in July 1964, Assistant Secretary of Defense for International Security Affairs, John McNaughton, asked him to become his special assistant and to focus on the problem of Vietnam. The idea was that he would not be writing the history of the Vietnam problem, but seeing it from inside as it unfolded. He requested and got the civil service rank and pay equivalent to that of a major general in the military and access to everything McNaughton himself was seeing on Vietnam or any other matter assigned to him. He was thirty three-years old.[13]

This was the man who experienced the Tonkin Gulf incident from inside – on his first day in the new job as special assistant to McNaughton. He was there as flash cables came in from Captain John Herrick, commander of a two destroyer flotilla - the *USS Maddox* and the *USS Turner Joy* - operating off the North Vietnamese coast in the Gulf of Tonkin. A long series of them suggested that the ships were being fired on by Vietnamese torpedo boats. Then came one stating that all earlier reports may have been erroneous and attributing reports of 'torpedoes' to 'freak weather effects and overeager sonar men'.[14]

Herrick advised caution in assessing and responding to this 'attack', but Ellsberg saw that this was not the disposition of Washington's top decision-makers. Preparations were immediately put on foot for a retaliatory air strike. Robert McNamara, the Secretary of Defense, gave a press conference after midnight in which he stated that US ships on a routine patrol in international waters had been the object of deliberate aggression; that the evidence for the attack was unequivocal; that the aggression was unprovoked; and that, in responding to it, the United States sought no wider war. Even then and certainly within forty-eight hours, Ellsberg testifies, 'I knew that each one of these assurances was false.'[15]

He had long been initiated into the realm of secrecy. This was his initiation into the realm of official lying. 'In my new job I was reading the daily transcripts of [McNamara's and Rusk's] secret testimony [to congressional committees in closed hearings] and at the same time I was learning from cables, reports and discussion in the Pentagon the background that gave the lie to virtually everything told both to the public and, more elaborately, to Congress in secret session.'[16] He believed in the system, though, in the secrecy, the Cold War world view and the presidency, so it took him seven more years to rebel against the lying.

On the evening of 4 August, at an NSC meeting, President Johnson asked John McCone, head of the CIA, 'Do they want war by attacking our ships in the middle of the Gulf of Tonkin?' McCone answered, 'No. The North Vietnamese are reacting defensively to our attack on their off-shore islands. They are responding out of pride and on the basis of defense considerations.' Yet Johnson went before Congress on 7 August, calling for sweeping war powers to respond to North Vietnam's 'unprovoked aggression.' The Gulf of Tonkin resolution was passed 416 to 0 by the House and 88 to 2 by the Senate. LBJ's Vietnam War had begun.[17]

Ellsberg had remarkable access to both secret documents and high-level personnel for the following six years as the war escalated. Like his boss McNaughton, he believed from the outset that LBJ's war was a mistake and that America should disentangle itself from Indochina. Very early on, he learned that there were others at very high levels who believed Johnson was blundering into a disaster. But they did not speak out. In addressing the press or Congress, all of them supported and even boosted Johnson's policy line.[18]

Ellsberg was initially more fascinated than appalled to observe, as he tells us, that 'journalists had no idea, no clue, even the best of them, just how often and how egregiously they were lied to.'[19] The fact is, he writes, that there was 'an apparatus of secrecy, built on effective procedures, practices and career incentives that permitted the president to arrive at and execute a secret foreign policy, to a degree that went far beyond what even relatively informed outsiders, including journalists and Congress, could imagine.'

This leads him to what is certainly one of the most significant judgements in the book. 'It is a commonplace,' he remarks, 'that "you can't keep secrets in Washington", or "in a democracy", that "no matter how sensitive the secret, you're likely to read it the next day in the *New York Times*." These truisms are flatly false. They are in fact cover stories, ways of flattering and misleading journalists and their readers, part of the process of keeping the secrets well ... the fact is that the overwhelming majority of secrets do *not* leak to the American public ... The reality unknown to the public and to most members of Congress and the press is that secrets that would be of the greatest import to many of them can be kept from them reliably for decades by the executive branch, even though they are known to thousands of insiders.'[20]

It was against this system that Ellsberg finally rebelled in 1971. But getting to that point was, as he has described it elsewhere, 'quite simply the most frustrating, disappointing, disillusioning period of my life.'[21] As a brilliant and successful American, a committed liberal Democrat and anti-Communist, an insider and a high-level consultant, he took a lot of disillusioning. It began in McNaughton's office. It deepened when he spent two years in Vietnam attached to Edward Lansdale's staff. It turned critical when he worked for Robert McNamara in 1967-68 on a secret study of US decision-making on Vietnam since 1945. It was that study that he leaked to the *New York Times* in 1971. His open rebellion was triggered, more than anything else, by Henry Kissinger's lies, in his capacity as Richard Nixon's national security adviser in 1969-70.

Ellsberg left the Pentagon and went with Lansdale to Vietnam in August 1965, in order, as he put it, to watch 'the cable traffic from the transmitting end of the channel.' What he discovered, as he wrote to LBJ's Head of Pacification, Robert Komer, was that there were problems at that end of pervasive ignorance and 'bureaucratic practices that poison this flow of information.' In June 1966 he wrote to Robert McNamara that official reporting was 'grossly inadequate to the job of educating high-level decision-makers to the nature of the essential problems here.' Above all, he saw that lying and self-deception, 'pressures for optimistically false reporting at every level' were endemic. Learning was absolutely vital, but was systematically blocked by what he dubbed 'anti-learning mechanisms.'[22]

What he had still to acknowledge was the extent to which the lying went all the way to the top. In October 1966 he was on a flight back from Vietnam with his former boss McNaughton, McNamara and Komer. He recalls how McNamara called him to the rear of the plane and said, in Komer's presence, 'Dan, you're the one who can settle this. Komer here is saying that we've made a lot of progress in pacification. I say that things are *worse* than they were a year ago. What do you say?' Ellsberg concurred. When the plane landed, however, McNamara went straight into a press conference and stated, 'Gentlemen, I've just come back from Vietnam, and I'm glad to be able to tell you that we're showing great progress in every dimension of our effort. I'm very encouraged by everything I've seen and heard on my trip.'[23]

When he left Vietnam, in early 1967, Ellsberg decided to return to Rand rather than to the Pentagon. The freedom that he had to choose speaks to the high regard in which he was held. The reality was that he 'wanted to be free again to tell what I knew and what I believed about our Vietnam policy to officials across the board in government agencies without having to worry if I was contradicting the position of a boss or a department.' What he still believed was that insider consulting through Rand would be 'the perfect institutional base' for talking not to the public, but to 'people with clearances, people with responsibility for making or advising on national security policy.'[24]

In the middle of 1967, McNamara asked McNaughton to prepare a historical study of Vietnam decision-making. McNaughton assigned the task to his deputy, Morton Halperin, who, in turn assigned it to his own deputy, Leslie Gelb, a former graduate student of Henry Kissinger's at Harvard. Ellsberg was enlisted because Halperin and Gelb 'wanted people with analytical skills, an ability to see patterns and propose lessons learned.' With his trademark energy and acumen, he went to work. What he learned stunned him.

By his own account, the hypothesis he brought to this work, for all his earlier experience of lying in the system, was the one publicly aired by journalist David Halberstam and historian Arthur Schlesinger Jr in 1966: the quagmire model. This model held that escalation in Vietnam at every step, since the early 1950s, had occurred because of optimistically false reporting and ill-founded judgements by White House advisers that the United States could have its way in Vietnam. Ellsberg discovered, looking first at the Kennedy years, that 'this assumption was mistaken. Every one of these crucial decisions was secretly associated with realistic internal *pessimism*, deliberately concealed from the public.'

Ellsberg's vocational commitment to insider consulting began to fray as he discovered, in 1967, in the official Pentagon records, that in every one of the major Vietnam decisions, by Truman, Eisenhower, Kennedy and Johnson, 'the president's choice was *not* founded upon optimistic reporting or on assurances of the success of his chosen course.' On the contrary, 'escalation ... was always immediately preceded and accompanied by a breakthrough of gloomy realism, including an internal consensus that the new commitment the president was choosing would probably be inadequate for success.'[25]

'I had never questioned,' Ellsberg reflects concerning his hard learning in 1967, 'the assumption of many students of presidential power that secrecy is vital to preserve a president's range of options. But I now saw how the system of secrecy and lying could give him options he would be better off without, or it could dangerously prejudice his choice.'[26] When, therefore, in March 1968, Johnson chose not to run for re-election, Ellsberg saw some hope in bringing the Pentagon study and its lessons to the attention first of Robert Kennedy, whom he hoped to see elected, and then Henry Kissinger, national security adviser to Richard Nixon, who was elected.

Robert Kennedy may well have listened to what Ellsberg had to say. Indeed, he did listen, when Ellsberg spoke to him in late 1967. When he was assassinated in June 1968, however, that option was shut down. Henry Kissinger's response was a different matter. Kissinger visited Rand on 8 November 1968, three days after Nixon had won the election, and told a Rand audience 'I have learned more from Dan Ellsberg than from any other person in Vietnam.' That learning had been chiefly, from 1965 onward, about how to ask the right questions in Vietnam. It had led Kissinger to publicly state, in 1967-68, that the United States had no rational alternative but to get out of Vietnam, while trying to arrange a 'decent interval' between its exit and a Communist takeover.[27]

Kissinger asked Henry Rowen, president of Rand, for a study of options in Vietnam. Ellsberg was chosen to head the project. The paper was completed by Christmas Eve and taken by Ellsberg to New York to show to Kissinger. They discussed it on and off for several days and Kissinger listened closely to Ellsberg's advice as to how the incoming president might overcome bureaucratic obstacles to divergent and well-informed advice reaching the top. Then he gave Kissinger a very specific piece of advice about the dangers of entering the secret world, which he was about to do.

'Henry, you're about to receive a whole slew of special clearances, maybe fifteen or twenty of them that are higher than top secret ... First you'll be exhilarated by some of this new information ... But second ... you'll feel like a fool for having studied, written, talked about these subjects ... for years without having known of the existence of all this information ... what amounts to whole libraries of hidden information, which is much more closely held than mere top secret data ... The danger is you'll become something like a moron. You'll become incapable of learning from most people in the world, no matter how much experience they may have in their particular areas that may be much greater than yours.'[28]

What Ellsberg did not know in December 1968 was that Nixon, despite his campaign promise about peace with honour, had no intention of ending the war in Vietnam – and that Kissinger was complicit in what was, from the outset, a whole new round of lying to the American Congress and people. Morton Halperin called him in August 1969 to tell him, from inside the administration, 'Nixon's staying in; he's not getting out.'[29] This was clearly contrary to public hopes and perceptions at that time, but as Ellsberg remarks, those who believed Nixon's public rhetoric had not read the

Pentagon study. His own work on that study had led him to see that the president was part of the problem and had burned out of him the desire to be 'in any sense a "president's man"'. Still, he conceived the hope that 'Nixon could be induced to think again', through Kissinger.[30]

He went to see Kissinger at San Clemente in August 1970 with the idea of encouraging him to read the Pentagon study. 'In effect, I had the idea of leaking information *into* the White House about what was actually visible from the outside ... I wanted Kissinger to worry that the trend of his policy was foreseeable, so that it might seem less viable to him.'[31] This time he found Kissinger unresponsive and indisposed to learn, in just the manner that Ellsberg had warned him of a mere twenty months earlier.

Did he have a copy of the study in the White House? Yes. Had he read it? Answer, 'No, should I?' Ellsberg: 'Absolutely.' Kissinger: 'But do we really have anything to learn from this study?' Ellsberg's heart sank. 'I thought: My God! He's in the same state of mind as the rest of them all along. They each thought that history started with his administration and that they had nothing to learn from earlier ones. Yet, in fact, each administration, including this one, repeated the same patterns in decision-making and pretty much the same (hopeless) policy as its predecessors without even knowing it. *That* was what there was to learn from the study, and Kissinger obviously needed it.'[32]

'I was suddenly depressed, but I went on to answer, "Well, I certainly do think so. It's twenty years of history and there's a great deal to be learned from it." He said, "But after all, we make decisions very differently now." My depression deepened. I said: "Cambodia didn't look all that different." ... He said, "You must understand, Cambodia was undertaken for very complicated reasons." I said, "Henry, there hasn't been a rotten decision in this area for twenty years that wasn't undertaken for very complicated reasons. And they were usually the same sort of complicated reasons."'[33]

Ellsberg's disillusionment was complete. He could now see that 'This wasn't a policy, or a pattern of decision-making, that was going to be changed from the inside, from "speaking truth to power" as a consultant. Cambridge professors come to power couldn't learn from the failure of a former close colleague any more than Republicans could learn from Democrats or the Americans from the French.'[34] He had decided even before then, however, that he would no longer agree to serve a system that 'lies automatically, at every level from bottom to top.'

'It occurred to me that what I had in my safe at Rand was seven thousand pages of documentary evidence of lying, by four presidents and their administrations over twenty-three years, to conceal plans and actions of mass murder.' Why mass murder? Because the French had, in 1946 through 1954, attempted to overthrow the independent state of Vietnam and reimpose colonialism on it; America had taken over from the French and America, not North Vietnam, had violated the Geneva Agreement of 1954, which explicitly denied Vietnam was two states and called for a plebiscite in 1956 to reunify the northern and southern zones. As he expressed it to himself in belated shock, the United States was not fighting on the wrong side. It *was* the wrong side.[35]

Out of this conviction came his decision to leak the Pentagon study to the *New York Times*. Publication began on 13 June 1971. The immediate reaction of Nixon and Kissinger was fury and an attempt to have a legal injunction placed on publication of the study, quickly dubbed *The Pentagon Papers*.[36] Nixon became obsessed with Ellsberg, fearing a conspiracy at the NSC to undermine his secret war policy, and created a Special Investigative Unit at the White House to attack Ellsberg. It was this unit, dubbed the 'Plumbers', who were arrested at Watergate on the night of 17 June 1972. Nixon's efforts to prevent the Watergate investigation from finding out about the Plumbers' efforts to silence or defame Ellsberg were what led to his downfall.[37]

The Nixon administration's efforts to gaol Ellsberg foundered on the fact that Ellsberg had violated no law in leaking seven thousand pages of top secret documents to the press. Indeed, it was the executive branch which was acting unconstitutionally in endeavouring to suppress their publication, in violation of the First Amendment. Nixon's Plumbers had, in any case, violated the criminal law repeatedly in their efforts to counter release of the *Pentagon Papers* by burglarising the office of Ellsberg's psychiatrist, Lewis Fielding, forging documents, illegally tapping phones and arranging for Cuban thugs to beat Ellsberg up in Washington DC.[38]

On 10 May 1973, the House of Representatives voted to cut off all funding for US combat operations in Indochina, including bombing. Nixon vetoed the measure. It was, as Larry Berman has shown, only the Watergate prosecution which prevented Nixon from resuming bombing in Cambodia, Laos and North Vietnam.[39] Preparation of the Pentagon study had not brought about a policy shift. Publication of it had clearly shown there should be one. What is sobering, as Ellsberg reflects at the end of his book, is that even this did not end Nixon's war. Ironically, it was Nixon's own efforts to stop the mouth of the best informed and most conscientious objector to the war that ended his war by ending his presidency.

What, then, are we to learn from all this in present circumstances? That an American president is again embarking on an illegal war and keeping crucial facts and strategic advice secret from the American public and the rest of us? Paradoxically, I think that is probably not true. Without any doubt, the really egregious liar in this case is Saddam Hussein.[40] In any case, the debate over the advisability of going to war with Iraq has been vigorous and relatively open. That the US is about to get into another Vietnam? I think not. Iraq is not Vietnam, it does not have great power backing and Saddam Hussein lacks Ho Chi Minh's moral, ideological or even nationalist credibility.

That the US has learned nothing since the debacle in Vietnam? No, for it has learned a great deal and will not fight in Iraq the way it fought the Viet Cong. Moreover, the Iraqis will not fight the US as the Viet Cong did – tenaciously and spiritedly. They are likely, for the most part, to want to see Saddam and his thugs, including his two sons, strung up or shot like Mussolini or Ceausescu – as indeed they should be.

What, then? This, I think. Decades of questionable US (and Western European and Russian) policy lie behind the looming war in the Persian Gulf. Too much of this history remains secret. If a study like the one McNamara organised in 1967 was now to be done, it would likely reveal valuable lessons for policy-makers and also for the world public. The policy-makers are almost certainly as largely ignorant of it as Henry Kissinger was of the history of Vietnam policy in 1970, something of which their critics tend to be unaware.[41]

If they are serious about transforming Iraq after Saddam Hussein is dead, then they will do well to read deeply into their own secret history. They will do even better to make it public and accessible, so that informed commentary can add to their understanding of the roots of present conflicts and policy dilemmas. Such is the ongoing addiction to secrecy in high places (whether in Washington, London, Paris or Moscow) however, that other Ellsbergs may be required before this will happen.

3.   *Endnotes*

1   Daniel Ellsberg, *Secrets: A Memoir of Vietnam and the Pentagon Papers*, Viking, New York, 2002, p. 331.

2   Ibid., p. 341. The remarks were reportedly overheard by Charles Colson, special counsel to the President, as told to Seymour Hersh.

3   Peter Schrag, *Test of Loyalty: Daniel Ellsberg and the Rituals of Secret Government*, Simon and Schuster, New York, 1974, pp. 252-3.

4   Mitrokhin had worked for thirty years in the archives of the KGB's foreign intelligence service when he was exfiltrated by MI-6 in 1992. What he brought with him covered KGB operations from the 1920s to the 1980s. It was described by the FBI as 'the most complete and extensive intelligence ever received from any source.' See Christopher Andrew and Vasili Mitrokhin, *The Mitrokhin Archive: The KGB in Europe and the West*, Allen Lane, Penguin, 1999, and *The Mitrokhin Archive II: The KGB and the World*, Allen Lane, Penguin, 2005.

5   See the archive of his media interviews and public speeches at www.ellsberg.net.

6   German bombing of Britain is estimated to have killed 75,000 people. Allied bombing of Germany is estimated to have killed 400,000 people. American bombing of Japan killed 500,000 people, of whom perhaps 180,000 were killed by the atomic bombs and 100,000 in the fire-bombing of Tokyo. Truly, as Winston Churchill phrased it, those who sowed the wind, reaped the whirlwind.

7   Ellsberg, *Secrets: A Memoir of Vietnam and the Pentagon Papers*, p. 30.

8   Ibid., p. 344. The occasion was a meeting at San Clemente in August 1970, at which Ellsberg challenged Kissinger, not for the first time, regarding the bombing of Cambodia and North Vietnam. Kissinger's statement, superficially flattering, Ellsberg relates, 'made the hair on the back of my neck stand up.' Hitler had used techniques like that because he was 'crazy, madly aggressive and reckless. But after a certain point it brought the world down around him ... For someone to imitate him in this respect was to cultivate madness and court disaster.'

9   Ibid., p 32.

10   Ibid., pp 58-9.

11   Ibid., pp. 33-4.

12   Ibid., p. 34.

13   Ibid., pp. 35-6. He records that when he came into his new Pentagon office on the first morning he was met by two six-foot-high stacks of classified documents. He'd asked for everything and there it was. One day's secret 'take' on Vietnam. The next day there were two more six-foot-high stacks. He reduced this impossible load by limiting his request to top secret documents, a few specific sorts of regular reporting and messages classified Eyes Only, NoDis,

ExDis or LimDis caveats controlling 'who knew – and shouldn't know – in an elaborate hierarchy of responsibility and secrecy.' pp. 37-9.

14   Ibid., pp. 7-10.

15   Ibid., pp. 10-12.

16   Ibid., p. 13.

17   Ibid., p 16. The resolution read: 'Congress approves and supports the determination of the President, as Commander in Chief, to take all necessary measures to repel any armed attack against the forces of the United States and to prevent further aggression ... The United States is ... prepared, as the President determines, to take all necessary steps, including the use of armed force, to assist any member or protocol state of the Southeast Asia Collective Defense Treaty requesting assistance in defence of its freedom.'

18   McNaughton kept a file, out of bounds even to Ellsberg which, he discovered, contained top secret critiques of the Pentagon's war strategies – both those of the Joint Chiefs and McNamara – by such luminaries as McGeorge Bundy, George Ball, Hubert Humphrey, Mike Mansfield, Richard Russell and Clark Clifford. Ellsberg 'first saw these critiques when they were published seventeen years later'. pp. 81-3. Ellsberg remarks 'These exhortations to withdraw were coming from men who were both charter cold warriors and ... as sensitive to Democratic domestic politics as Johnson himself. The fact that the president enjoyed access to advice like this from men like these was the longest and best kept secret of the Johnson Vietnam era.' p. 83.

19   Ibid., p. 41.

20   Ibid., p. 43.

21   Daniel Ellsberg, *Papers On The War*, Simon & Schuster, New York, 1972, p. 17.

22   Ibid., p. 18. This is carried over into *Secrets*, thirty years later, at pp. 185-6.

23   *Secrets*, pp. 141-2. The reality was that McNamara was so disillusioned with the course of the war that, in late 1966, he submitted a top secret sensitive memo to LBJ calling for a negotiated exit strategy. Circulated to the Joint Chiefs, this memo provoked what Ellsberg describes as 'a storm of protest that marked the beginning of the end of his influence with the president and of his tenure[as Secretary of Defense].' Ibid., p. 182.

24   Ibid., p. 181.

25   Ibid., p. 189.

26   Ibid., p. 205. 'I realized something crucial [in March 1968]', Ellsberg writes, 'that the president's ability to escalate, his entire strategy throughout the war, had depended on secrecy and lying and thus on his ability to deter unauthorized disclosures – truth telling – by officials.' (p. 204)

27   Ibid., pp 227-9.

28   Ibid., pp. 238-9.

29   Ibid., p. 257.

30   Ibid., p. 261.

31   Ibid., p. 343.

32   Ibid., p. 347.

33   Ibid., p. 348.

34   Ibid., p. 349.

35   Ibid., pp. 255-6.

36   For a detailed contemporary study of the publication controversy, see Sanford Ungar, *The Papers and the Papers*, Dutton, New York, 1972.

37   See Ellsberg's chapter 31, 'The Road to Watergate', in, *Secrets*, pp. 422-443. It was the tape of Nixon's instruction to Haldeman and Ehrlichman on 23 June 1972 to get CIA officials to induce the FBI to stop its investigation short of Howard Hunt and Gordon Liddy so that they could not be questioned about the Ellsberg case which, when finally extracted from Nixon in August 1974, was seen as the 'smoking gun' that led to his resignation.

38   For a detailed account of all this, see Peter Schrag, *Test of Loyalty: Daniel Ellsberg and the Rituals of Secret Government*, Simon & Schuster, New York, 1974.

39   Larry Berman, *No Peace, No Honor: Nixon, Kissinger and Betrayal in Vietnam*, Free Press, New York, 2001. See Ellsberg, *Secrets*, pp. 452-4.

40   Kevin Woods, James Lacey and Williamson Murray, Special Report - 'Saddam's Delusions: The View From the Inside', *Foreign Affairs*, May/June 2006, pp. 2-27.

41   David Fromkin, *A Peace to End all Peace: The Fall of the Ottoman Empire and he Creation of the Modern Middle East*, Phoenix Press, London, 2000, is a good guide to the deep background. Daniel Yergin, *The Prize: The Epic Quest for Oil, Money and Power*, Simon and Schuster, 1991, covers both the deep background and much of the more recent history.

... according to the researchers who discovered the oldest known stone artefacts, created in Ethiopia two and a half million years ago, their makers knew the basic principles of knapping. The work of these extremely ancient hominids indicates 'a clear understanding of conchoidal fracture mechanics ...'

\- Marek Kohn (1999)[1]

Churchill's great history of World War II has been cleaned of every single reference to Allied communications intelligence except one (and that based on the American Pearl Harbour investigation), although Britain thought it vital enough to assign 30,000 people to the work. The intelligence history of World War II has never been written. All this gives a distorted view of why things happened.

\- David Kahn (1967)[2]

At the very end of months and months of long and careful debriefing, Alibekov was invited to write a study paper for the CIA of all the information in his possession on the entire Soviet BW programme. This long paper was considered so potentially dangerous should it ever fall into the wrong hands that it was given the highest US security classification that exists, and even Alibekov was not permitted to keep a copy or ever see it again.

\- Tom Mangold and Jeff Goldberg (1999)[3]

4. On Technology and Moral Accountability

*A*dolf Hitler's racial censuses, slave labour programs, genocide of the Jews and apocalyptic war-making were all facilitated by something called Hollerith machines. The Third Reich used thousands of them. They were manufactured by a company called Deutsche Hollerith Maschinen Gesellschaft (German Hollerith Machine Company), or Dehomag, for short. Dehomag was owned and controlled from New York by IBM. IBM was dominated by one man: Thomas J. Watson.

Between 1933 and 1940, IBM courted the Nazi Party. It also lobbied the defence ministries of the European continent to catch onto what can only be described as a foretaste of the present-day Revolution in Military Affairs: the use of Hollerith machines for data analysis and logistics coordination. The Wehrmacht was its greatest customer.

Thomas Watson micro-managed this massive sales campaign from New York. Then he masterminded a scheme to keep IBM's profits coming in from Nazi Germany right through the Second World War and the Holocaust. As France fell to the Nazis, Watson realised that his long closeness to the Nazi regime, for which Hitler had personally decorated him in 1937, had become a public relations liability. He therefore ostentatiously returned the decoration, piously denouncing Hitler's aggression. The outraged Nazis wanted to seize Dehomag. But they couldn't and they didn't. Watson's IBM had them by the proverbials.

In his 2001 book *IBM and the Holocaust*, Edwin Black describes the dilemma that confronted Hitler's regime in June 1940. 'Since the birth of the Third Reich, Germany had automated virtually its entire economy, as well as most government operations and Nazi party activities, using a single technology: Hollerith. Elaborate data operations were in full swing everywhere in Germany and its conquered lands. The country suddenly discovered its own vulnerable over-dependence on IBM machinery.'[4]

Wasn't it possible to simply Nazify Dehomag and produce the machines without regard to IBM? Doubtless, but it would have taken years and the war was under way. As Black puts it, 'Holleriths could not function without IBM's unique paper. Watson controlled the paper ... Holleriths could not function without cards. Watson controlled the cards ... Hollerith systems could not function without machines or spare parts. Watson controlled the machines and the spare parts ... Even if the Reich confiscated every IBM printing plant in Nazi-dominated Europe and seized every machine, within months the cards and spare parts would run out.'

Even the all-conquering Third Reich could not out-manoeuvre one of the world's first true multinational corporations. Dehomag was not a simple, national operation that the Nazis could commandeer. It was part of a global conglomerate, relying on financial, technical and material support from IBM headquarters in New York. Watson quite simply held all the cards. And Watson was not about to quit the game.

It got much better for Watson and IBM when America also entered the war. The market for their goods in America boomed. As Emerson Pugh has shown, IBM produced vast quantities of weapons and military equipment for America.[5] It became as vital to Anglo-American intelligence and logistics operations as it was for the Nazis. Special mobile Holleriths were developed for the Pacific War. IBM military units were formed to deploy and maintain the IBM technology in the war theatres.

When the war was over, Dehomag was absorbed back wholly into IBM and profits from the war years were protected from reparations exactions. No IBM executives were charged with war crimes of any description. German bankers, industrialists and military leaders were so charged, but not the information technology men. Indeed, Watson offered IBM's services – Holleriths – in the handling of the voluminous records for the Nuremberg war crimes trials – *pro bono*.

Edwin Black first became interested in the role of Holleriths in the Holocaust in 1993. He visited the new Holocaust Museum in Washington, D.C., and found that among the first exhibits was a Hollerith machine. His parents were both Holocaust survivors. From that time, he says, 'I was haunted by a question whose answer has long eluded historians. The Germans always had the lists of Jewish names ... But how did the Nazis get the lists? For decades no-one has known. Few have asked.' The answer was the Hollerith machine.

The Hollerith machine was a device using punch cards to sort and record data. Invented in the 1880s by a German American, Herman Hollerith, it was first put to dramatically effective use in the US national census of 1890. In 1910, a German engineer named Wilhelm Heidinger licensed Holleriths for use in Germany and created Dehomag. Thomas Watson, born in 1874, was a salesman of a particularly unscrupulous kind. Initially working for John Patterson's National Cash Register Company, in 1913 he was hired as manager of the Computing Tabulating Recording Company, which had taken over the Hollerith technology from its inventor. By 1923 he was the maestro at CTR and it was at his instigation that it was renamed International Business Machines. It was also in that year that, taking advantage of Germany's hyperinflation, he presented Willy Heidinger with an offer he couldn't refuse: face bankruptcy or sell out a dominant interest to IBM.

Even before the rise of the Nazi regime, Dehomag was one of IBM's great milch cows, providing more than half of its total overseas income. The Nazis, however, fascinated Thomas Watson. He himself ran IBM as a charismatic totalitarian and was dubbed by *Fortune* magazine 'The Leader'. In Mussolini and in Hitler, Watson recognised himself. Even better, though, he saw, in Hitler's drive to create a controlled society of relentless regimentation and surveillance, a market Frederick Winslow Taylor-made for IBM. It could make every step in the process more efficient. It could even make new steps possible.

Watson never committed the indiscretion of openly avowing Nazi sympathies. He gave liberally to charities and universities and the arts. He preached liberal free trade and peace. He cultivated friendship with America's democratic elite and sextupled IBM's domestic revenues through working with the Roosevelt New Deal on Social Security – using Hollerith machines. Yet even as he did all this he assiduously courted the lords of fascism and custom-designed for them the solutions to their problems – including the final solution to what Hitler saw as his biggest problem.

As soon as Hitler came to power, IBM/Dehomag offered a solution to his desire for a thorough racial census. Friedrich Zahn, one of Germany's leading statisticians and a committed Nazi, declared that statistics would provide the Party with 'the road map to switch from words to deeds.' In early 1934, Willy Heidinger, at Dehomag, described precisely how punch card technology would enable Hitler to eradicate all the unhealthy cells in the German cultural body.

'We are very much like the physician, in that we dissect cell by cell the German cultural body. We report every individual characteristic ... on a little card. These ... cards are sorted at a rate of 25,000 per hour ... with the help of our tabulating machine,' Heidinger told a Nazi gathering, with Watson's representative alongside him. His speech was promptly copied to Watson. Did The Leader cable Heidinger that he was appalled and that IBM could have nothing to do with such an agenda? Not at all. He promptly congratulated him on both the content and implications of his address. He also dramatically increased IBM investment in the Hollerith plant in Germany.

Neither mass demonstrations in New York and elsewhere calling for American companies to boycott Nazi Germany, nor the evidence of Nazi racism and blatant aggression deterred Thomas Watson from doing all he could to expand operations in Germany. Yet on IBM Day at the World's Fair in New York, 13 May 1940, he told 30,000 people that war was bad and that right-thinking people should avoid any involvement with it. They knew nothing of his investments in Germany, since these were made through Dehomag.

Watson's key man in Geneva, Jurriaan Schotte, had just cabled him a long memo headed 'Our Dealings With War Ministries in Europe'. It detailed how IBM had worked tirelessly, between *Kristallnacht* (10 November 1938) and the invasion of France, to persuade the general staffs of twenty-four European countries of the military applications of Hollerith machines and punch card technology. They were, finally, buying big.

On 20 January 1942, leading Nazis assembled at Am Grossen Wannsee 56-58 to plan the final solution to the extermination of Europe's Jews. The names Reinhard Heydrich, Heinrich Himmler, Adolf Eichmann are well known as the perpetrators of this extermination. Less well known are Roderich Plate and Richard Korherr, who put together the statistical databases from which the planners of genocide worked.

Richard Korherr ran something called the Statistical Scientific Institute in Berlin. It was a Hollerith complex. Using IBM technology, designed and manufactured by Dehomag under contracts monitored from Geneva and overseen in New York by Jurriaan Schotte, Korherr was, as Black expresses it, 'keeper of the state's most incriminating genocidal secrets.' His Institute statistically coordinated genocidal operations right across Europe, involving the rounding-up, transportation and extermination of millions of human beings.

IBM would claim that it did not know the Nazis were doing such things with its technology. The denial reeks of the rankest corporate hypocrisy. What Black's history makes abundantly clear is that Thomas Watson and his corporate team put profit ahead of every possible and every available ethical or political consideration from 1933 right through until 1945 – and beyond.

Watson could have declined to do business with the Nazis in 1933, when mass movements in the United States demanded that American corporations do just that. He could have pulled up stumps after the *Anschluss*, or after *Kristallnacht*, or after the invasion of Poland, but there is no evidence that he even considered doing anything of the kind. Once the United States was in the war, his trade with the Nazis was illegal and treasonable, but even then he kept it going, using every device of concealment and clandestine control. The conscious and sustained hypocrisy of the man is breathtaking, even in the perspective of more recent feats of international arms merchants and bankers.

Amid all the darkness of the story he has to tell, Edwin Black shows us at least one shining hero. Rene Carmille, Comptroller General of the French Army, promised the Nazis in 1940 that his Holleriths could 'deliver the Jews of France'. Instead, he secretly used them to build a database of 800,000 former French soldiers who could be quickly mobilised to fight for liberation.

Carmille had worked for French counterintelligence since 1911 and was one of the top members of the resistance's Marco Polo network. He delivered no Jews to the Nazis. Arrested in February 1944, he was tortured by Klaus Barbie, but did not crack. He died at Dachau on 25 January 1945. His moral courage puts Thomas Watson and his entire IBM corporate staff utterly to shame.

Richard Evans has remarked that the outcome of the David Irving trial 'vindicated our capacity to know what happened after the survivors are no longer around to tell the tale. It showed that we *can* know without reasonable doubt, even if explaining and understanding will always be a matter of debate.'⁶ Edwin Black's determined investigation of the deep complicity of IBM in the murderous ravages of the Hitler regime is a grim contribution to such knowing.

In her remarkable interviews with the children and grandchildren of Nazis, former Japanese soldiers and survivors of the bombing of Hiroshima, Erna Paris has shown that the truth of history is not only difficult to ascertain, but is extremely difficult for many people to face.⁷ Edwin Black's history shows us a truth which took a lot of unearthing and must be very difficult for IBM to face. It is not only very disquieting, but cuts right to the heart of the great modern debates about free trade and corporate accountability, technology and civilised humanism.

Information technology is the very symbol of twenty-first century civilisation. It is also the technology which, more than any other, symbolises what Martin Heidegger called the 'enframing' of human beings: their reduction to statistical units in vast labour and marketing processes, to the status of 'utter availability and sheer manipulability'.⁸ IBM under Thomas Watson embodied and enacted the massive use of information for social manipulation and control. It was, as Edwin Black remarks, 'gripped by a special, amoral corporate mantra: if it *can* be done, it *should* be done.' The direct consequence was Hitler's automated genocide and the *blitz* in his *krieg*.

Our way of being human in the information age is at stake here. As Hubert Dreyfus has put it, 'we embody an understanding of being that no one has [explicitly] in mind. We have an ontology without knowing it.'⁹ Under the long shadows cast by IBM, the ontology we increasingly embody without knowing it is quite possibly just that apprehended by Martin Heidegger: reduction to statistical units seen as without moral significance and subordinated to the quantitative calculations of planners. We also risk, in our own right in such a world, becoming beings that fundamentally cease to *think*.

Thinking, Heidegger argued, was not merely a matter of *calculation*, but 'a transformed way of being in the world.' Ironically, Heidegger's *own* thinking collapsed into totalitarianism after he had written *Being and Time* in the late 1920s. He became an enthusiastic Nazi in 1933, in the belief that Hitler would lead the German people in a 'collective breakout from the [Platonic] cave'.¹⁰ Even his greatest insights have, ever since, been overshadowed by this extraordinary abdication of intellectual and moral responsibility.

Nonetheless, Heidegger had a powerful point, just as Watson had a powerful technology. Our thinking *must* encompass the social implications of our technologies and enable us to transform our way of being in the world. If it does not, then it falls away from being thinking at all. Nowhere was this more so than at IBM in the 1930s, where Thomas Watson had had five words engraved on the steps of the IBM School in New York: READ, LISTEN, DISCUSS, OBSERVE, THINK. The word 'THINK' was everywhere at IBM. What Watson meant by it and what Heidegger meant by it were poles apart conceptually, but both, at that very time, were embracing Hitler. What *thinking* were they really doing?

Rene Carmille lived out the better part of what Watson's naive students may have taken him to mean and what generations of existentialists have taken Heidegger to have meant. In 1943, already aware that the Gestapo were on his trail, he addressed the graduating class of the Ecole Polytechnique in Paris. He told them to remember their dignity and their rational freedom, both as human beings and as French citizens.

'No power in the world can stop you from remembering', Carmille told the students, 'that you are the heirs of Cartesian thought, of the mysticism and mathematics of Pascal, of the clarity of the writers of the sixteenth century ... All this is written in your soul, and no-one can control your soul, because your soul belongs only to God.' Think about Carmille's words. For there are technologies at work in the twenty-first century world which make Thomas Watson's Hollerith business look like child's play – and there has never been a greater need for real thinking.

4. *Endnotes*

1   Marek Kohn, *As We Know It: Coming To Terms With An Evolved Mind*, Granta Books, London, 1999, p. 58.

2   David Kahn, *The Codebreakers: The Story of Secret Writing*, Scribner (1967), 2nd edn, 1996, p. xi., preface to the original edition.

3   Tom Mangold and Jeff Goldberg, *Plague Wars: A True Story of Biological Warfare*, Pan Books, 1999, p. 183.

4   Edwin Black, *IBM and the Holocaust: The Strategic Alliance Between Nazi Germany and America's Most Powerful Corporation*, Little, Brown & Co., New York, 2001.

5   Emerson Pugh, *Building IBM: Shaping an Industry and its Technology*, MIT Press, Cambridge, MA, 1995.

6   Richard J. Evans, *Lying About Hitler: History, Holocaust and the David Irving Trial*, Basic Books, New York, 2001.

7   Erna Paris, *Long Shadows: Truth, Lies and History*, Bloomsbury, London, 2001.

8   David Farrell Krell (ed.), *Martin Heidegger: Basic Writings*, Harper, San Francisco, 1993.

9   Hubert L. Dreyfus, *Being in the World: A Commentary on Heidegger's Being and Time, Division I*, Cambridge, MIT Press, 1991.

10 Rudiger Safranski, *Martin Heidegger: Between Good and Evil*, Harvard University Press, 1998.

We have made a thing, a most terrible weapon that has altered abruptly and profoundly the nature of the world ... a thing that by all the standards of the world we grew up in is an evil thing. And by so doing ... we have raised again the question of whether science is good for man ... Atomic weapons, even with what we know today can be cheap ... atomic armament will not break the economic back of any people that want it. The pattern of use of atomic weapons was set at Hiroshima ... against an essentially defeated enemy ... it is a weapon for aggressors, and the elements of surprise and terror are as intrinsic to it as are the fissionable nuclei.

- J. Robert Oppenheimer (1945)[1]

Keep strong, if possible. In any case, keep cool. Have unlimited patience. Never corner an opponent and always assist him to save face. Put yourself in his shoes – so as to see things through his eyes. Avoid self-righteousness like the devil – nothing is so self-blinding.

- John F. Kennedy (1960)[2]

... proliferation to new countries cannot be considered in isolation from possession of nuclear weapons by the old, Cold War powers, where the motherlode of nuclear danger still resides ... the inescapable question is whether the possessors of nuclear arms will insist on keeping them indefinitely or will be prepared to accept their abolition.

- Jonathan Schell (2003)[3]

5. On Strategic Calculation in the Era of WMD

The Cuban Missile Crisis of October 1962 was the most dramatic and dangerous crisis of the nuclear age – so far. For thirteen days, the United States and the Soviet Union teetered on the brink of World War III because of mutual fear and suspicion about the fate of Cuba and the installation there of Russian nuclear missiles. When they stepped back from the brink, the world heaved a collective sigh of relief. Yet they had so very nearly toppled in. 'We were in luck,' remarked John Kenneth Galbraith, then American Ambassador in India, 'but success in a lottery is no argument for lotteries.' Declassified information now shows that the crisis was, in the words of Robert McNamara, US Secretary of Defense at the time, 'more dangerous than is generally recognised even today – and that its lessons have yet to be learned.' Since we live more than ever in the era of ballistic missiles and debates about how to defend ourselves against them, it is time that learning was done.

Roger Donaldson's splendid 2001 film about the Cuban Missile Crisis, *Thirteen Days* should be required viewing for everyone. It is an immensely impressive effort to condense the drama and significance of those climactic days of the Cold War into just over two hours. It is not a documentary and is not wholly accurate. The number of characters is culled to make the drama theatrically manageable. Lines historically spoken by one person are sometimes given to another for the same reason. Kenny O'Donnell, a key staffer of President John F. Kennedy, is given a leading role, played by Kevin Costner, that he did not have in reality. Yet the film is morally, intellectually and historically serious in ways that Oliver Stone's films had made one despair of ever seeing from Hollywood again. See it. See it more than once. It is a rare cinematic invitation to being a citizen, rather than a mere consumer, a politically thoughtful adult, rather than a mere entertainment junkie.

The film opens, as Stanley Kubrick's classic *Doctor Strangelove* ended more than forty years ago, with an awesome sequence simulating nuclear war. Donaldson's opening, however, is even more powerful than Kubrick's ending. This time, there is no Vera Lynn singing 'Wish Me Luck as You Wave Me Good-bye'. There is simply stunning footage of flaming missile launches and gigantic nuclear explosions. It is comparable, in disturbing ways, to the strongly realistic opening war scene in the 2000 hit film *Gladiator*, in which the Roman legions fire a storm of burning missiles at barbaric warriors in the German forests. Romans archers (commonly Syrians or Scythians), however, fired arrows through up to 400 metres of gloomy air. Donaldson simulates intercontinental ballistic missiles fired through space to the other side of the world, each able to kill – indeed to obliterate – more human beings than there were in the entire Roman armed forces when the Roman Empire was at its peak. Such has been the remarkable 'progress' in the human genius for war-making over the past eighteen centuries.

The inferno of the last great explosion gives way to a dark sky and the underbelly of an American U-2 spy plane on a photo-reconnaissance mission over Cuba. The CIA and the Pentagon have suspected for some months that the Soviet Union could be seeking to implant offensive weapons in revolutionary Cuba. The Soviet Union has been reassuring top American officials that it has no intention of doing any such thing. The U-2 is looking for the hard evidence – and finds it. On the morning of 16 October, President Kennedy is informed of the finding. With this the drama begins. The *possibility* of the kind of nuclear holocaust we have just witnessed in simulation feeds the fear and strategic anxiety with which the American high command reacts to that evidence. President Kennedy, to whom his predecessor Dwight Eisenhower had referred less than two years earlier as 'Little Boy Blue', is faced with a supreme challenge to his leadership abilities. Donaldson's film dramatises how he faced that challenge and won a reputation for statesmanship.

The casting for the film is one of its many strengths. Little-known actors Bruce Greenwood and Steven Culp play John and Robert Kennedy with breathtaking finesse. The Kennedy brothers are so famous and their personalities so spiked with myth and controversy, that taking on either role must have seemed an almost terrifying challenge. Yet Greenwood and Culp conjure the very body language, tones of voice and intimate fraternal partnership that characterised JFK and his 'brother protector' Robert.[4] They portray them, however, as knights in shining armour. What Seymour Hersh called 'the dark side of Camelot' is nowhere in evidence.

No dramatic impersonation can ever be complete, of course. Those inclined to quibble about the Kennedys, however, might do well to use John Julius Norwich's excellent study *Shakespeare's Kings* (1999) as a foil. Shakespeare was not always accurate – notoriously so in the case of his famous *Richard III*, for example – but he was a supreme dramatist and his history plays are as vivid an education in the politics of power as one can find. Donaldson and his script writer, David Self, are not at the level of Shakespeare, but they have given us a drama of great power and integrity. And Bruce Greenwood has given us a very fine JFK enactment.

Kevin Costner's Kenny O'Donnell was an intriguing dramatic gambit. The film's promoters stated that the screenplay was based on ninety-odd hours of interviews with O'Donnell before he died, which suggested that this was his version of events. That seems rather doubtful, given what the available historical record reveals – especially the White House tapes of the missile crisis debates. What Donaldson's team did was craft a workable drama out of a few major sources, the O'Donnell interviews, Robert Kennedy's own, famous memoir, *Thirteen Days* and transcripts of the White House tapes, deliberately *enhancing* the role of O'Donnell. Certainly, none of the standard histories records O'Donnell as playing nearly so significant a role as he does in the film.[5] What Donaldson and Self did was use poetic licence to simplify Washington power politics and allow the ordinary viewer to feel implicated in the drama, through identification with the little-known figure of O'Donnell. That said, it should be understood that O'Donnell was no mere hanger-on in the Kennedy camp – far from it. He had roomed with Bobby Kennedy at Harvard University in 1946 and captained the football team in which the undersized Bobby Kennedy played. The film alludes to this background, but the uninformed will miss the depth of its significance. When John Kennedy was first running for political office in 1952, it was O'Donnell who steered the campaign away from disaster. He got Bobby Kennedy to step between the rank and file of the Democratic Party and old Joe Kennedy, the family patriarch, whose arrogance and interference threatened to derail his son's fledgling career. He remained a key member of the Kennedy brothers' so-called 'Irish Mafia' from that point forward.[6] In the film, he is simply given this old role writ large. It is a combination of his actual historical presence with a sort of 'Greek chorus' function, mediating between the tragic actors and the audience.

I think the device works, though purists may cavil. O'Donnell, alone among the characters in the film, is shown in his family setting and talking with his wife, Helen, about the human implications of what is unfolding on the world stage. This is one of our best clues to the fact, as I read the film, that O'Donnell is meant to 'humanise' the drama, whereas the other characters are somehow caught up in the abstractions and preoccupations of power politics. He moves between the two worlds and carries us with him. He is also made to speak candidly to the Kennedy brothers at critical moments, in ways congruent not only with the historical record but with the moral tenor of the drama. This is what a 'Greek chorus' was designed to do in the ancient world. In this case, of course, the 'tragedy' was averted. The figure of O'Donnell helps the audience to focus on how this is made possible and on how desperately close-run a thing it was.

Powerful though it is, the film actually understates the gravity of the danger that confronted the United States in October 1962. This seems to be because Donaldson, Self and their team relied overwhelmingly on *Thirteen Days*, the O'Donnell interviews and *The Kennedy Tapes* (White House recordings of the crisis), overlooking the later analytical literature on the subject. Their sources are good. Their drama is very good. But the truth is both more complex and more dramatic than they allowed or seem to have imagined. Paradoxically, this is one of the reasons the film is such a good *introduction* to the subject. It vividly presents the heart of the matter, making it easier for those who are stirred by it to then see where further detail enlarges the story.

In the film, emphasis falls on Soviet deceit about the installation of the missiles and the fear of the Russians gaining (and exercising) a first-strike nuclear option against the United States. The audience is given only hints of the deeper background to the crisis. There is no real indication, for example, that the Soviet move was prompted in considerable measure by American efforts to topple Fidel Castro and was intended to deter an invasion of Cuba. There is nothing but the most oblique suggestion that the United States was still considering a first-strike nuclear option against the Soviet Union – before the Russians could develop an effective deterrent force. Yet these were the historical realities.

The film also focuses on the tension between the Kennedy brothers and the Joint Chiefs of Staff over whether to use force against Cuba and the Soviet missile bases at the risk of general war. This is largely accurate, but the film-makers seem to have been unaware that, throughout the whole crisis, the Joint Chiefs of Staff believed they could carry out a successful invasion of Cuba because they and the CIA had *radically underestimated* the size of Soviet nuclear and conventional forces in Cuba. The hidden realities, with which the Pentagon's forces would have collided had they launched the planned invasion, were that both the number of Soviet troops in Cuba and the number of Soviet nuclear weapons there were *four times as large* as estimated. There were not 10,000 but 43,000 troops and not forty but one hundred and sixty-two nuclear warheads there. And Khrushchev had authorised the Soviet commander on the spot to use them if the American invasion began.

As Robert McNamara remarked long afterwards, the option urged by the Joint Chiefs of Staff would have resulted in 'utter disaster'. Given these sombre facts, it is all the more stunning to know that, when Khrushchev backed down and the crisis abated, when President Kennedy declared that this was not a time for gloating or for talk of 'victory', the Joint Chiefs were adamant that 'We have been had.' They – and Curtis LeMay in particular – wanted to bomb and even invade Cuba and overthrow Castro and believed that an *opportunity* to do so had been forfeited.

LeMay, the head of the Strategic Air Command, notorious for his 'bomb them back to the Stone Age' rhetoric in both Korea and Vietnam, blurted out, 'We lost! We ought to go in and make a strike on Monday anyway.' As one leading historian of the matter has commented, 'the smell of smouldering hawk feathers hung in the air' on the morning that Kennedy and Khrushchev stepped back from the edge of the precipice. LeMay is the leading hawk in the film, as he was in real life, but neither he nor any of the Chiefs was given such scandalous lines in *Thirteen Days*. Truth, it seems, is not only stranger but at times more lurid than 'fiction'.

The film gives an interesting and economical version of the crucial meeting between Robert Kennedy and the Soviet Ambassador, Anatoly Dobrynin, on the night of 27 October. It omits altogether the meeting between the two the previous night, which is as good an illustration as one can get of the rigorous economies entailed in compressing real history into any form of theatre. The meeting that it does show follows Robert Kennedy's own account quite closely. It adds an interesting line spoken by Dobrynin, however, which is a fascinating condensation of what Dobrynin actually said and felt at the time.

The Soviet Ambassador in the film says to Robert Kennedy that he believes the Kennedy brothers are good men, but that there are some military people in the United States who 'want war'. 'I assure you', he says, 'that there are other good men. Let us hope that men of good will can counter the terrible power of this thing that has been put in motion.' The real Dobrynin cabled Moscow after the meeting, saying Kennedy had told him, exhausted and near tears, that there was overwhelming pressure from the military on the Kennedy brothers and they did not know how much longer they could resist it.

When this cable reached Moscow, according to Oleg Troyanovsky who was Khrushchev's special assistant for international affairs at the time, it created a 'state of alarm in the Presidium of the Central Committee'. They feared that the Kennedys would 'soon lose control of the situation' to the 'many hotheads in Washington who wanted an invasion.' The decision was: step back from the brink and deny these hotheads their opportunity.[7] That's what happened. Not a moment too soon. Dramatic as it is, the film does not capture the full dimensions of this chilling nuclear shadow-play

In reflecting, in the 1990s on the lessons of 1962, Robert McNamara realised that 'the decisions of each of the three nations before, during and after the crisis, had been distorted by misinformation, miscalculation and misjudgement.' It will not do, he argued, to merely explain the pressures that led national leaders to do what they did at that time. We must, if we are to avoid major disasters in future, inquire profoundly into the sources and logic of such misjudgements. We must create leaderships better able to 'resist pressures or "forces" of this sort, to understand more fully than others the range of options and implications of choosing such options.'

We must, in other words, develop a code of *cognitive responsibility* – an ethic which demands that pivotal perceptions and passionate beliefs be rigorously cross-examined as a fundamental part of the intelligence and policy process. When McNamara remarked in 1999 that the 'lessons' of the Cuban Missile Crisis have still not been learned, this was the real burden of his claim. The greatest virtue of Roger Donaldson's film is that it *showed* this to a mass audience who, overwhelmingly, will have had no acquaintance with the serious historical literature on the subject. A film is not an adequate substitute for a good set of books, but it is an invaluable stimulant to reading them and, at a minimum, a means of becoming vividly aware of realities that otherwise remain locked away on library shelves. *Thirteen Days* is a particularly fine example of what film can accomplish and is a film for our time, when so many strategic dilemmas confront us and so much cool, clear thinking is called for.

5.   *Endnotes*

1   J. Robert Oppenheimer, 'Atomic Weapons and the Crisis in Science', an address to the American Philosophical Society, Philadelphia, 16 November 1945. Cited in Kai Bird and Martin J. Sherwin, *American Prometheus: The Triumph and Tragedy of J. Robert Oppenheimer*, Alfred A. Knopf, New York, 2005, pp. 323-4.

2   John F. Kennedy, quoting and reviewing Basil Liddell Hart's *Deterrent or Defense*, *The Saturday Review*, 3 September 1960. Quoted by Arthur Schlesinger Jr., in foreword to Robert F. Kennedy, *Thirteen Days*, W. W. Norton, New York and London, 1999, p. 11.

3   Jonathan Schell, *The Unfinished Twentieth Century: The Crisis of Weapons of Mass Destruction*, Verso, 2003, p. xiv.

4   James W. Hilty, *Robert Kennedy: Brother Protector*, Temple University Press, Philadelphia, 1997. See also, David Talbot, *Brothers: The Hidden History of the Kennedy Years*, Free Press, New York, 2007.

5   James Blight et al., *Cuba on the Brink: Castro, the Missile Crisis and the Soviet Collapse*, Pantheon, New York, 1993; Ernest R. May and Philip D. Zelikow, *The Kennedy Tapes: Inside the White House During the Cuban Missile Crisis*, Harvard University Press, 1997; Aleksandr Fursenko and Timothy Naftali, *One Hell of a Gamble: The Secret History of the Cuban Missile Crisis*, John Murray, London, 1997; Lawrence Freedman, *Kennedy's Wars: Berlin, Cuba, Laos and Vietnam*, Oxford University Press, 2000.

6   On the deeper Kennedy family history, see Ronald Kessler, *The Sins of the Father: Joseph P. Kennedy and the Dynasty He Founded*, Warner Books, New York, 1996; Doris Kearns Goodwin, *The Fitzgeralds and the Kennedys*, Pan Books, 1987; and Rose Fitzgerald Kennedy, *Times to Remember*, Pan Books, London and Sydney, 1974. The role of Kenny O'Donnell in the Kennedy family's venture into politics is well attested by Rose Kennedy at pp. 163, 345, 350 and 393-4.

7   For a detailed and thoroughly informed account of all this from inside the Soviet leadership, see William Taubman, *Khrushchev: The Man and His Era*, Free Press, New York, 2003, pp. 529-583.

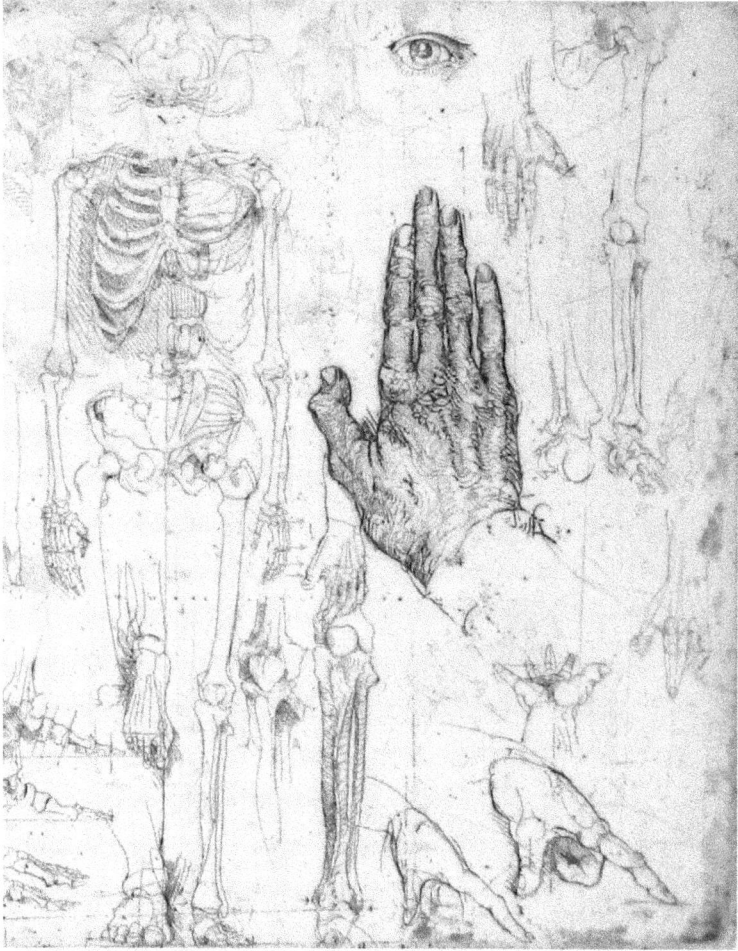

JS 110, *Sketches Also After Dürer* (detail)
Jörg Schmeisser, etching, 1974.

It is difficult to estimate the ability of Iraq to recover its delivery capability in regards to weapons of mass destruction. Accurately assessing Iraq's capacity to rebuild its nuclear weapons program is equally problematic. It is clear, however, that Iraq possesses significant potential in these areas and that it is almost certain to apply itself to developing both weapons of mass destruction and a means to deliver them.

- Anthony Cordesman (1997)[1]

[Richard] Butler felt that [Kofi] Annan had perhaps forgotten how long a spoon you need when you sup with the devil. 'Saddam is the devil; his track record on keeping agreements is appalling. And he's a very unsentimental person. The instant he thinks it's in his interest, he will drop Kofi down the crocodile hole. It would be a tragedy if such a good man and his gifted initiatives for peace should then get trashed in the dustbin of history because he didn't realize the realities of dealing with the devil.'

- William Shawcross (2000)[2]

The US military commanders who battled their way to Baghdad and endured the long hot summer of 2003 believe that there was a window of opportunity in the early weeks and months of the invasion, which was allowed to close. Though some degree of opposition was unavoidable, the virulent insurgency that emerged was not inevitable, but was aided by military and political blunders in Washington.

- Michael Gordon and Bernard Trainor (2006)[3]

6. On the Decision to Overthrow Saddam Hussein

This essay was written on the very eve of the Iraq War and published the day the war commenced. Like the essay about 9/11, it has been left unchanged, save for the occasional cosmetic adjustment, in the belief that it is of more interest as a piece written with the prospect of war than as one based on the wisdom of hindsight. The failure to discover WMD in Iraq and the chaos and violence there between late 2003 and mid-2007 vindicated many of those who opposed the decision to go to war. As of early 2009, the surge strategy has restored the situation and hopefully laid a secure foundation for exit.

Saddam Hussein is long overdue to be terminated with extreme prejudice. He is one of the most atrocious tyrants of the lamentably many who have plagued humanity in the past hundred years.[4] He is utterly brutal, faithless and incorrigible. A great many of the people of Iraq long for his demise, but are unable to effect it without American military intervention. But beliefs such as these can be serious *obstacles* in the way of thinking dispassionately about Washington's recourse to war against him. That the decision has been made to go to war does not eliminate the need to confront the many *objections* to doing so.

Of the countless outpourings on the subject in 2002-3, an especially significant comment is one by William Shawcross, in mid-February 2003. The mass demonstrations against the very idea of war were under way. Shawcross wrote of them, 'Have people gone completely and utterly mad?... They say that they are supporting the United Nations and marching against war. But that is absurd. How come they don't realize that they are undermining the United Nations, and making war more likely?'[5]

While the popular opinion was that Bush and Blair had gone mad, Shawcross believes that popular opinion itself had done so. This is significant because of Shawcross's *credibility* in making such a judgement. After all, he made his reputation, in the late 1970s with a withering critique of the American invasion of Cambodia between 1969 and 1972.[6] He has since had an active and abiding concern with the challenges of United Nations collective security measures, peacekeeping operations, human rights advocacy and protection of refugees. His most recent book is a scrupulous and deeply informed reflection on these issues.[7]

He concluded his recent article with the ringing words, 'Be bold, be honorable, be radical – support Blair and Bush, not Saddam. If the UN is to survive, its resolutions must be enforced. Saddam must not be comforted – he must be disarmed.' As Australian human rights specialist Robert Horvath wrote, also in February, 'As long as peace activists aspire to defend the Iraqi people from everyone but its executioners and torturers, their cause will be nothing but a distasteful reminder of the West's long complicity in [Saddam]'s tyranny.'[8] Yet anti-war pacifists insist that any support for war is a sign of hopeless naiveté about American imperial ambitions, lack of moral sense or testosterone poisoning in male brains.

That's how arguments tend to go.[9] In the early 1980s, Shawcross had a no-holds barred exchange with Henry Kissinger's aide Peter Rodman concerning his book *Sideshow: Kissinger, Nixon and the Destruction of Cambodia*.[10] It shows, even now, how tenaciously people cling to their opinions, using evidence and rhetoric like stones in a slingshot against their opponents, rather than asking where careful examination of the evidence might enable both sides to learn.

Rodman wrote, 'Close scrutiny of the materials shows that the evidentiary basis of the book is so seriously flawed as to discredit his whole enterprise ... His vaunted research turns out to be slipshod, distorted by bias, and in some cases bordering on the fraudulent. It is a compendium of errors, sleight of hand, and egregious selectivity; he has suppressed entirely a mountain of evidence in his possession that contradicted his principal points.'[11] Shawcross responded with equal vigour: 'Rodman's article ... like Kissinger's memoirs,[12] does not even dent the argument of my book ... the feebleness of its criticisms, and its low intellectual quality, far from devaluing *Sideshow*, vindicated it ... Even as a hatchet job, it failed completely.'

Two highly articulate individuals stood unwavering in their opinions, after each surveying a mountain of evidence and critically scrutinising the other's case. How can this be? Alas, it is all too common. What, at the end of the day, were they arguing about? Many points of detail, but one fundamental proposition: that the secret bombing of Cambodia, initiated by Nixon and Kissinger in early 1969, was *responsible* for the atrocities committed by the Khmer Rouge after April 1975.

Arguably, Shawcross demonstrated that the US invasion was illegal, ill-advised and terribly destructive and that it destabilised Cambodia. But Rodman was correct that American bombing ended twenty months before the Khmer Rouge came to power and that 'by no stretch of moral logic' can their killing of something like two million of their own people be blamed on those who had done all they could think of to *prevent* them coming to power at all.[13]

This is a useful foil to the current debate about war against Saddam Hussein, both because similar passionate antagonisms have arisen and because, in this case, Shawcross believes we should *support* Washington in going to war. Yet there are many who passionately believe that 'by no stretch of moral logic' can Washington's war be justified.[14]

Once people disagree on a very basic, not to say sweeping, moral or political perception of this kind, there is a very strong tendency for them to lapse into the twin traps of what cognitive scientists call 'confirmation bias' and 'belief preservation'. The former consists in looking only for evidence that will support an opinion, while disregarding or devaluing that which does not. The latter means clinging to a belief even when considerable evidence which contradicts it is presented.

The great need therefore, if you believe the present war is justified, that, as Shawcross urges, the pressure should be on Saddam Hussein, not on Bush and Blair, is to consider less the grounds for *supporting* this judgement, than all the *objections* to it. This requires disentangling the various objections in order to be able to consider each on its merits. For the biggest source of frustration in complex arguments is the way claims and counter-claims jump all over the place, making rational discourse extremely difficult.

What, then, *are* the various objections that have been advanced against the idea of disarming Saddam Hussein by force? There are at least twelve. Most have some merit. The weight accorded different ones varies from one anti-war protestor to another, but all warrant consideration. A rational assessment of the case should involve weighing each in turn, giving it its due, and then seeing if there is any telling rebuttal to it, the objective being to see how the *balance of considerations* looks when all reasonable claims have been weighed in the scales. This is *not*, of course, what we naturally or intuitively do, nor is it easy.

Here are the twelve objections: first, that without the sanction of the UN Security Council, the use of force to disarm, much less overthrow Saddam violates international law. Second, that there is no hard evidence that Saddam Hussein is guilty as charged – that he actually has weapons of mass destruction, or links with al Qaeda. Third, that even if he has them, pre-emptive war would still be impermissible. Fourth, that Washington and London are rushing into war without exhausting the alternatives.

Fifth, that war will kill too many innocent Iraqis and destabilise Iraq. Sixth, that the United States is really only after Iraq's oil. Seventh, that it is hypocritical of the United States and Britain to overthrow Saddam Hussein for having weapons of mass destruction or for human rights abuses. Eighth, that this war will anger Arabs and Muslims generally. Ninth, that even if Saddam has, or might acquire, weapons of mass destruction, he could have been deterred from using them, so there is no call to fight a war to disarm him.

Tenth, that if he has such weapons, then attacking him with the intent to overthrow and even kill him will guarantee that he will use them. Eleventh, that to invade Iraq without the sanction of the UN Security Council will undermine the moral authority of the United Nations and with it the whole architecture of global collective security. Twelfth, that it is not in Australia's national interest to participate in a Washington-led war in the Persian Gulf.

It will be seen readily enough that these twelve objections fall into roughly three more general categories: invading Iraq would be illegal (objections 1 to 4), immoral (objections 5 to 7) and stupid (objections 8 to 12). A first approximation of the case against this war would, then, look something like the following: the United States, Britain and Australia should not have invaded Iraq to overthrow Saddam Hussein because doing so was illegal, immoral and stupid.

It is illegal because it violated international law; because there was no hard evidence that Saddam Hussein was guilty as charged; because even if Saddam *was* guilty as charged, pre-emptive war remains impermissible; and because Washington and London rushed into war without exhausting the alternatives.

It is immoral because the war will kill too many innocent Iraqis and destabilise Iraq; because the United States is really only after Iraq's oil; and because it is hypocritical of the United States and Britain to overthrow Saddam Hussein for having weapons of mass destruction or for human rights abuses.

It is stupid because it will anger Arabs and Muslims generally; because, even if Saddam has, or might acquire, weapons of mass destruction, he could be deterred from using them; because, if he has such weapons, attacking him will *guarantee* that he will use them; because invading Iraq, without the sanction of the UN Security Council, will undermine the moral authority of the United Nations and with it the whole architecture of global collective security; and because it is not in Australia's national interest to participate in a Washington-led war in the Persian Gulf.

Just setting out even a basic schema like this shows why it is so difficult to generate rational agreement on the subject. Those whose gut instinct (what I would call their primary judgement) is that there should *not* be a war are likely to see this formidable array of objections to it as conclusive, but are unlikely to have thought through most of them, much less to have done so with an eye to possible rebuttals. Conversely, those whose primary judgement is that a war to overthrow Saddam is *justified* are unlikely even to have realised there are so many grounds for objection, much less to have carefully considered each of them.

When an argument begins, people on both sides almost inevitably violate all principles of good argument in seeking to justify their primary judgements. They jump up and down the levels of generality. They omit all manner of considerations. They fall into confirmation bias and belief preservation, rather than seeking out objections to their primary judgement or its notional grounds. Small wonder, then, that they commonly end up accusing one another of bad faith, irrationality, ignorance and moral incompetence. That's what Shawcross and Rodman both did.

As individuals and even as groups, we have tremendous difficulty in coping with the *cognitive burden* of handling complex arguments under pressure. By *default* and without, for the most part, any conscious decision to be dishonest or slipshod, we do many of the things Shawcross and Rodman accused one another of doing. Why? Because our hominid brains are wired for cutting to the chase, not for patiently disentangling complex arguments, formulating and testing hypotheses, exploring counterfactual propositions and analysing the probabilities involved in various risk assessments. At the intuitive level, unaided by specialist tools, we are just not very good at these things.

Clearly, there are many areas of uncertainty and some imponderable risks in embarking on the war. Expert international lawyers have argued both for and against the legality of war being resorted to.[15] The moral case is complicated by uncertainties regarding how many innocent Iraqis will be killed or otherwise harmed by a war, as well as by questions concerning the *consistency* of making war on Iraq while not doing so in the case of North Korea, which has flagrantly violated its non-proliferation commitments and almost certainly has already built a small nuclear arsenal.

There are also serious prudential grounds on which war might be considered ill-advised. The most substantial of these seems to be the treatise by William Nordhaus of Yale University, *War With Iraq: Costs, Consequences and Alternatives*, which warns that a long, widened or unsuccessful war could inflict very grave costs on the US and global economies. Moreover, while the swift defeat of Iraq might serve as a deterrent to North Korea or Iran against persisting with its weapons of mass destruction program, it might as readily quicken their determination to bring such programs to fruition in order to deter US military action against them.

How, then, can the case for war be made in face of these formidable objections? What follows cannot claim to be exhaustive, given the complexity of the matter. It is simply intended to suggest the lines along which the above objections have been or could be rebutted. When the objections and rebuttals are set side by side in an at least roughly structured way, it becomes easier for those debating the matter to disagree rationally, by pinpointing where evidence or new claims can make a difference. And there will be new evidence, whatever the fortunes of war. We should all, therefore, stand ready to adjust and even radically modify our primary judgements in the light of what emerges.

The *legal* justification for war pivots on the claim that UN Resolutions 687 of 1991 and 1441 of 2002 mandate the use of force to compel Saddam Hussein to disarm.[16] The Anglo-American powers seek to enforce the law, not violate it. France and its partners on the Security Council passed Resolution 1441 last year, but are now backing away from it when it is mocked. Saddam is guilty as charged, in that he is in material breach of Resolutions 687 and 1441. Hans Blix, though he still believes more inspections would be useful, has declared that Saddam lacks all credibility when he denies having any weapons of mass destruction.[17]

Iraq's links to al Qaeda are far closer than any government has thus far attempted to argue publicly. On close examination, the much dismissed meeting between Mohammed Atta, the 9/11 field commander, and Iraqi intelligence officer Samir al-Ani in Prague in April 2001, *did* take place.[18] It has further been credibly established that two of the 9/11 suicide pilots, Marwan al-Shehhi and Ziad Jarrah, met with Iraqi intelligence officers in the Gulf States in the months leading up to 9/11;[19] that senior Iraqi intelligence officer Farouk Hijazi went to Afghanistan in 1998 and met with Osama bin Laden; that there are literally scores of reliable CIA reports of other dealings between the two parties going back to 1992; and that Iraq ran a training camp at Salman Pak in the late 1990s, at which terrorists were trained in how to hijack passenger airliners using only knives.[20]

Far more fundamentally, an inquiry which has received altogether too little attention suggests that the masterminds of the biggest terrorist operations against the United States since the Gulf War, including *both* attacks on the World Trade Center, are very possibly Iraqi intelligence officers under deep cover. They are a group of terrorists who all turn out to claim birth in Kuwait and family ties to one another starting with Ramzi Ahmed Yousef and the recently apprehended Khalid Sheikh Mohammed. The surmise is that they are not who they claim to be, but an Iraqi intelligence ring using as cover the identities of a group of Kuwaitis whose lives were ended and whose records were interfered with by Iraqi intelligence during the occupation of Kuwait in 1990-91.[21]

Pre-emptive war against Iraq is warranted because Saddam Hussein is incorrigible, criminal and better eliminated now than allowed to hatch any further mischief.[22] There is a serious debate over the *strategic prudence* of this step. That is not, however, a legal question. Nor has there been a rush to war. It has taken twelve years to get to this point, during which time Saddam Hussein has resisted every diplomatic and coercive measure directed at him. The cost to his people and to the rest of the world has been exorbitant. It is simply untrue to say that alternatives to war have not been exhausted. They had, in fact, been exhausted by the end of 1998 and only Western restraint averted war at that time and until now.[23]

War is *morally* justified because Saddam cannot be trusted to keep any agreement, because he has used chemical weapons against his own people as well as against Iran and because he has one of the worst human rights records of any ruler in the world at present – certainly the worst in Iraq's twentieth century history. It is difficult to know how many innocent Iraqis will be killed in the war, but it is not difficult to know how many innocent Iraqis have died as a result of Saddam's tyranny and defiance of UN sanctions. The number is around one million, at a conservative estimate.

The most damning account of his slaughter of his own people can be found in the brilliant and anguished work of an Iraqi exile, Kanan Makiya, *Cruelty and Silence*. In 1988 alone, in Kurdistan alone, well over 100,000 people were slaughtered in mass executions.[24] A swift Anglo-American decapitation of Saddam's regime would be unlikely to kill more than a fraction of that number. Savage and prolonged urban warfare, it must be admitted, could be a different proposition.

Yet, if the moral case for war depends on the fate of innocent Iraqis, their own opinion concerning the morality of war should count for something. Many Iraqi exiles believe, as they put it in Sydney last week, that 'War is bad, but Saddam is worse.' Iraqi Shiites in exile told journalist Sarah Miles last week 'they will celebrate the day the US topples Iraqi President Saddam Hussein ... they say Mr. Hussein is loathed by most Iraqis and should be overthrown.'[25] Iraqi exiles in America have had telephone conversations with relations inside the country in which the latter have expressed their fervent hope and belief that the Americans are coming, that it is *serious* this time and that, at last, the tyrant will be toppled.

One Iraqi in the United States, Ramzi Jiddou, declared, 'I am a pacifist, but it will take a war to remove Saddam Hussein and, of course, I'm for such a war.'[26] Julius Strauss was, in his own words, 'intensely skeptical of the wisdom of Washington's insistence on deposing Saddam', until he saw at first hand the appalling realities of Saddam's rule in Iraq. He then concluded, 'As the drums of war beat ever louder, I am still unsure of the strategic wisdom' of the war, 'but of the moral rectitude of such a course there can be no doubt.'[27]

Bernard Kouchner, socialist and founder of *Medecins Sans Frontières* has declared, 'The removal of [Saddam] is the primary concern. What is worse than war is leaving in place a dictator who massacres his people. I wish people would actually listen to the ones most threatened by all this, the Iraqi people who are subjugated by this dictatorship.'[28] Whose moral integrity would you trust more, Bernard Kouchner's or that of French President Jacques Chirac, who has been close to Saddam for many years?

The weakest moral argument against war seems to me to be the assertion that the Americans are only going to war in order to gain control of Iraq's oil. Oil is, of course, vital to the whole global economy, not to a few Americans alone. But if oil was the only motive for war, one would have to assume that the United States has been frustrated by a lack of access to oil. It has not been. Indeed, it has restricted Saddam's right to sell oil and has resisted French and Russian efforts over the years to get these restrictions lifted. What a war will do is put Iraq's oil wealth back into the hands of Iraqis other than Saddam and his brutal thugs. Intelligently used, it could still make Iraq a model of prosperity and modernisation in the Arab world.

What of the argument from hypocrisy? It is true that, in the 1980s, the Western powers turned a blind eye to Saddam's egregious human rights abuses and lent him active support in his war of aggression against Iran. Yet it seems perverse on the part of the anti-war movement to claim that the Anglo-Americans were wrong to turn a blind eye then and are equally wrong to condemn Saddam now. Which is it to be? Is Saddam to be indulged or condemned? Condemned, surely, both then and now. It is not Bush and Blair who are morally at fault at this point, but the once and forever Saddam crony Jacques Chirac.

Is the war, then, a *stupid* idea? If the worst fears expressed by many well-informed people – of savage urban warfare, the destabilisation of Iraq and wider upheavals in the Middle East – come to pass, the war could turn out to have been a strategic error.[29] If Saddam, having denied that he has any weapons of mass destruction, unleashes chemical and biological weapons *in extremis*, would it be rational to conclude that the war was justified, but tragically costly, or would it look as though we should have settled for attempting deterrence and containment?

This is the most intractable objection to the primary judgement that war is justified. It was the thrust of John Mearsheimer and Stephen Walt's widely read article late last year that Saddam can be deterred, provided that is the unambiguous intention of Western strategists.[30] They were responding to Kenneth Pollack's argument that 'For more than thirty years, Saddam's pattern has been to coldly miscalculate the odds, with disastrous results for Iraq and its neighbours.'[31]

Stalin said, in the late 1940s, 'The West thinks I'm the next Hitler, but I'm not Hitler. I know when to stop.' Since Saddam admires Stalin, perhaps we could count on him knowing when to stop? Suppose, however, this megalomaniac was let off the hook this time, as in 1991 and 1998? He would quite rightly conclude that no future threat of force by the West, or any part thereof, would be acted upon. He would therefore proceed to rebuild his weapons of mass destruction programs. He would, in all probability, eventually hand on power to his sinister son Qusay, if not to the even viler Uday. Perhaps he is containable, but the danger is likely to increase over time, not diminish.

Has going to war undermined the United Nations and thus global collective security? Fear and resentment of US power could trigger various forms of backlash around the world. It is France, Germany and Russia, however, that undermined the United Nations by refusing to enforce Resolution 1441, just as they violated the sanctions in the 1990s. They, not Washington and London, are responsible for undermining collective security.

Paris, Berlin and Moscow are showing that their threats are idle and their words empty. Washington and London are saying to an international criminal: enough is enough. That and the principle of enforcing United Nations covenants on weapons of mass destruction against recalcitrant tyrants are why it is in Australia's national interest to fight alongside the Americans and the British. We should heed William Shawcross's clarion call: 'Be bold, be honorable, be radical' – and overthrow the tyrant of Iraq,[32] not the will of the Anglosphere to stand up to him.[33]

6.   *Endnotes*

1   Anthony Cordesman, *Iraq: Sanctions and Beyond*, Westview Press, 1997, p. 290.

2   William Shawcross, *Deliver Us From Evil: Warlords and Peacekeepers in a World of Endless Conflict*, Bloomsbury, London, 2000, pp. 247-8.

3   Michael Gordon and Bernard Trainor, *Cobra II: The Inside Story of the Invasion and Occupation of Iraq*, Atlantic Books, London, 2006, p. 506.

4   Paul Monk, 'Bearing Witness against Tyrants', *AFR Review*, 21 February 2003, pp. 10-11.

5   William Shawcross, 'Support for Bush and Blair is good for the UN, bad for Saddam', *Sunday Age*, 16 February 2003. The reason, of course, that he thought such demonstrations made war more likely is that they showed Saddam that the West was divided and uncertain about standing up to him and thus encouraged him to believe what he wanted to believe, which is that the threat of force would never come down to war and that he could, therefore, temporise and survive. The classic study of the problem of democracies which are averse to war, bargaining with ruthless dictators is Telford Taylor's monumental *Munich: The Price of Peace*, Random House, New York, 1979.

6   William Shawcross, *Sideshow: Kissinger, Nixon, and the Destruction of Cambodia*, rev. edn, Cooper Square Press, New York, 1987. The book was originally published in 1979.

7   Shawcross, *Deliver Us From Evil*.

8   Robert Horvath, 'Long March to Ignorance', *The Australian*, 18 February 2003, p. 11.

9   A classic passage from Thucydides, which everyone should commit to heart as a school student, is his reflection on the effect of war and revolution on the minds of human beings, as shown by the revolution at Corcyra. Robert B. Strassler (ed.), *The Landmark Thucydides: A Comprehensive Guide to the Peloponnesian War*, III: 70-84, pp. 194-201.

10   These exchanges did not, of course, appear in the original (1979) edition of the book, but in the 1987 edition, as an appendix.

11   Peter Rodman, *Sideswipe: Kissinger, Shawcross and the Responsibility for Cambodia*, in Shawcross (1979) p. 416.

12   Henry Kissinger, *The White House Years*, Hodder & Stoughton, Sydney, 1979. Shawcross pointed out in 1987 that Kissinger's account of the bombing of Cambodia in his memoirs was deceitful and incomplete.

13   Shawcross, *Sideshow*, pp. 14 and 430-2. Rodman used the figure of three million Cambodians killed, but the actual number seems to have been about 1.7 million. This is the figure cited by Geoffrey Robertson in *Crimes Against Humanity: The Struggle For Global Justice*, Allen Lane, Penguin, 1999, p. 259. The best overall account of Pol Pot's rise to power and devastation of his own country is David Chandler's *Brother Number One*, Allen & Unwin, Sydney, 1993. By Chandler's account, writing before Ben Kiernan had conducted a reasonably thorough investigation in Cambodia itself, a minimum of 100,000 Cambodians were tortured to death or executed by the Khmer Rouge, while at least one million died of starvation, overwork or disease as a direct consequence of KR tyranny (p. 4).

14   On of the most forthright expressions of this view is that by former US President Jimmy Carter: 'Just War – or a Just War?' *New York Times*, 9 March 2003.

15   Darin Bartram, Suri Ratnapala, Greg Rose, David Flint, Stephen Hall et al., 'The Case for a Legal Attack', *The Australian*, 8 March 2003, p. 11. They were responding to a letter by another large group of international lawyers in *The Age* and *The Sydney Morning Herald*, 26 February 2003, who argued that war would be illegal without a further, specific UNSC resolution mandating it.

16   Janet Albrechtsen, 'Academics Fail Court Examination: With International Lawyers on Their Side, No Wonder the World's Thugs Aren't Afraid of a Global Criminal Court', *The Australian*, 5 March 2003, made this point with her accustomed bluntness.

17   'All Eyes on the Inspector', *Time*, March 3, 2003, pp. 26-8. Former UNSCOM Chief Weapons Inspector Scott Ritter played an odd and inconsistent role in public debate on this subject after he resigned from UNSCOM in August 1998. In the immediate aftermath of Saddam's expulsion of UNSCOM in late 1998, he catalogued the weapons of mass destruction the Iraqi dictator still had, despite seven years of inspections and numerous UN resolutions. Scott Ritter, *Endgame: Solving the Iraq Problem Once and For All*, Simon & Schuster, New York, 1999, Appendix, pp. 217-224.
Richard Spertzel, former head of the biological weapons inspection team for UNSCOM between 1994 and Saddam's expulsion of UNSCOM in October 1998, is particularly relevant here, as expressed in January 2003: 'It should be recalled that in early 1995 Iraq was denying that it ever had a BW program, in spite of the accumulating evidence by UNSCOM to the contrary. Iraq then also made a great display for the news media to assure the world that its Al Hakam complex was only for animal-feed production when in reality that was only a cover story for its largest BW-agent production facility. Does the world forget so easily? Or is it only that the world body does not care what Iraq possesses?' Richard Spertzel, 'No Smoking Gun Farce Revealed', *National Review Online*, 13 January, 2003.

18   Various people, not least on the editorial staff of the *New York Times*, deny that this meeting took place, but as Edward Jay Epstein pointed out, the Czech government repeatedly contradicted assertions by the *New York Times* that the meeting never took place. It did take place. As Epstein remarks, Atta and the Iraqi intelligence officer, Ahmad Khalil Ibrahim Samir al-Ani, might have discussed something other than 9/11, or al-Ani might have rebuffed overtures for help by Atta, but that the meeting *occurred* was confirmed by those in the best position to know – Czech intelligence and the highest officials of the Czech government.

19   David Rose, 'Saddam and Al Qaeda', *The Evening Standard*, London, 9 December 2002. 'My own doubts emerged more than a year ago, when a very senior CIA man told me that, contrary to the line his own colleagues were assiduously disseminating, there was evidence of an Iraq-Al Qaida link. He confirmed a story I had been told by members of the anti-Saddam Iraqi National Congress – that two of the hijackers, Marwan Al-Shehhi and Ziad Jarrah,

had met Mukhabarat officers in the months before 9/11 in the United Arab Emirates. This, he said, was part of a pattern of contact between Iraq and Al Qaida which went back years.' This has been confirmed by CIA Director George Tenet himself. See 'Iraq and al Qaeda: Who's Campaigning to Deny the Links?' *Wall Street Journal* Review and Outlook, 25 October 2002.

20   David Rose, 'An Inconvenient Iraqi', *Vanity Fair*, January 2003, pp. 73-5 and 132-4.

21   Laurie Mylroie first put the foundation of the case together in the wake of the trial and conviction of Ramzi Yousef for the 1993 World Trade Center bombing. See her article, 'The World Trade Center Bomb: Who is Ramzi Yousef? And Why It Matters', *The National Interest*, Winter 1995/96, pp. 3-15. She followed with this with a book, published in 2000, then revised and reissued in the wake of the 2001 destruction of the World Trade Center, *The War Against America: Saddam Hussein and the World Trade Center Attacks*, Regan Books, HarperCollins, 2001 – a book greeted by Vincent Cannistraro, former head of CIA counterterrorism operations as 'One of the most brilliant pieces of research and scholarship I have ever read.' Mylroie's argument was never publicly confuted, but appears to have subsided. What precisely the flaw in it was remains unclear as of this writing.

22   President Bush made this explicit in his final address on the eve of war, as reprinted in *The Australian*, 19 March 2003, p. 4: 'Instead of drifting along towards tragedy, we will set a course towards safety. Before the day of horror can come, before it is too late to act, this danger will be removed.'

23   William Shawcross provides a lucid account of the exhaustion process in the two Iraq chapters (nine and twelve) of *Deliver Us From Evil* (2000). The following passage perhaps conveys the sense of the case: 'By this stage (late 1998) ... many of the US principals were arguing for war. William Cohen, in particular, felt that, unlike the Grand Old Duke of York, he could not go on marching his men up and down the hill indefinitely. Also aggressive was the British Prime Minister Tony Blair. He wanted war, unless the guarantees from Saddam could at least be presented as cast iron.' (p. 297).

24   Kanan Makiya, *Cruelty and Silence: War, Tyranny, Uprising and the Arab World*, W. W. Norton & Co., New York and London, 1993, p. 168.

25   Sarah Miles in Damascus, 'Bring It On: Shiite Exiles Await War and Freedom', *The Age*, 16 March 2003, p. 12.

26   Stephen F. Hayes, 'The Horrors of Peace: Saddam's Victims Tell Their Stories', *The Weekly Standard*, 10 March 2003.

27   Julius Strauss, 'Why This Monster Must Go', *The Age*, 3 March 2003.

28   Sophie Masson, 'To arms, oh citizens, form up in rebellious ranks: Not all leading French figures oppose a war to disarm Saddam Hussein', *The Australian*, 18 March 2003, p. 11.

29   The war on terrorism that was declared after 9/11 is redolent of the sweeping command given to Gnaeus Pompeius (Pompey) by the Roman Republic to crush the plague of pirates in the Mediterranean under the Lex Gabinia of 67 BCE. Robin Seager, *Pompey the Great*, Blackwell, London, 2002, pp. 43-7.

Note: This passage has been left unchanged since it was written in March 2003. Plainly, this scenario is what came to pass, in 2003-2006. The surge and counter-insurgency strategy, led by General David Petraeus in 2007, began to bring the situation under tolerable control, but it remained, as of the beginning of 2008, an open question whether this level of control, could be sustained, if and as US troops were withdrawn.

30   John J. Mearsheimer and Stephen Walt, 'An Unnecessary War', *Foreign Policy*, January/February 2003.

31   Kenneth M. Pollack, *The Threatening Storm: The Case For Invading Iraq*, Random House, New York, 2002, p. 422.

32   William Shawcross, 'Justice Demands Invasion of Iraq', *The Australian*, 20 March 2003, p. 13, underscores this call. 'The French would like us to believe we are rushing into war. Nonsense. For twelve years Hussein has defied UN demands that he hand over and destroy his biological, chemical and nuclear weapons programs. They have been "12 years of humiliation for the UN", in the words of Britain's Foreign Minister Jack Straw. Twelve years in which the international community has failed to enforce its own laws against one of the vilest dictators on earth.'

33   The idea of the Anglosphere is the coinage of American internet entrepreneur James Bennett. See Ramesh Ponnuru, 'Anglosphere of Influence', *The Australian*, 17 March 2003, p. 11.

Democracy? Freedom? What do these words mean? I don't want any part of them.
- The Shah of Iran (1974)[1]

... the United States never really knew the trees, the bark or the soil that composed the Iranian forest-politics. Those experienced analysts who had some of this expertise were routinely ignored. Long before the revolution, [Earnest] Oney [who worked on Iran for the agency from 1951 until January 1979] proposed an in-depth study of the religious leaders of Iran. His bureaucratic superiors vetoed the idea, dismissing it as 'sociology'.
- James Bill (1988)[2]

For the United States, trying to change the regime in Tehran is not just a lost cause, it would be a mistake. Whenever we have tried, we have ended up worse off than when we started.
- Kenneth Pollack (2004)[3]

7. On Constraining Islamist Iran

This essay, written in 2005, remains topical, not least with the publication in late 2007 of the controversial US National Intelligence Estimate (NIE) stating that Iran had suspended its nuclear weapons program in late 2003. Once again, however, the essay has been altered only cosmetically in the belief that the subject can always be updated, but that reflections of this nature gain in value with the passage of time because they allow us to see what has changed, instead of constant current intelligence lulling us into a belief that we can grasp things at any given point in time.

Will Iran be next?' asked James Fallows in December 2004.[4] Next to be invaded by American forces, because it is on the verge of completing nuclear weapons. Iran probably will not be invaded because of the daunting costs that could entail, but that means it may well complete nuclear weapons, which raises significant problems for both the non-proliferation regime and the balance of military power in the Middle East. Regime change in Iran would be a much better way to solve these problems.

The idea that the theocratic regime in Tehran might implode is not fanciful. The regime is deeply unpopular among its own people. Ayatollah Khomeini's dream, of creating an Islamic state inspired as much by Plato's *Republic* as by the Koran, has failed badly. Had it not been for its brutal political thuggery, the regime of the mullahs could well have ended some years ago. The momentum for change generated by the overthrow of Saddam Hussein in April 2003, and the holding of national elections in Iraq in January 2005, could bring things to the tipping point.

This is not, of course, a matter of certainty, but there is pretty good empirical evidence that a majority of Iranians are deeply disillusioned with radical Islam and want to see political reform in their country. Add to this the way the tide seems to have turned in much of the region and you have the makings of what could sweep the mullahs from power and bring a representative, multi-party democracy to Iran. This would not only bring considerable benefits to Iran itself, but would have profound implications for the whole struggle for 'hearts and minds' in the Islamic world. It would also open up the possibility of Iran abandoning its nuclear weapons program of its own volition – as South Africa did more than a decade ago.

The reflexive response to the idea of regime change in Iran might be scepticism, but it is worth remembering that, throughout 1989, as one Eastern European Communist regime after another toppled from power, conservatives and sceptics insisted that each was an exception and that the others would endure. Poland was supposedly unique. Poland, Hungary and Czechoslovakia had Catholic backgrounds. East Germany would not be let go by the Soviet Union. Bulgaria and Romania were tougher and more backward than the rest. They were the last to go – then, in 1991, the Soviet Union itself imploded.

That something like this could be brought about in the Middle East was the great hope of those who encouraged the controversial invasion of Iraq in 2003. There is a long way still to go, but there is, at last, a sense of momentum. Charles Kurzman's study of the genesis of Iran's 1979 revolution suggests that this momentum could be the decisive catalyst.[5] It shows that revolutions in general have a certain unpredictable dynamic about them. They overturn regimes by upending expectations – as in the case of Eastern Europe in 1989.

Kurzman's case is that all the standard social scientific explanations for the downfall of the Shah and the rise of Ayatollah Khomeini turn out to be falsified by intractable factual anomalies. According to the available evidence, the revolution of 1979 'shouldn't have happened when it did, or at all.'[6] Happen it did, however, like a social *tsunami*. Indeed, it was a far more popular revolution, in terms of mass participation, than the classic revolution of 1789 in France or the political earth tremor that finally brought down Soviet Communism.[7]

Kurzman's 'anti-explanation' for this is that various sets of circumstances can generate outbursts of mass protest which take unpredictable turns because they occur amid confusion, disrupt people's expectations and overturn 'one of the dominant premises of contemporary social science, the stability of preferences.'[8] This premise gives way under circumstances of acute social disturbance. From this he draws a conclusion which brings to mind the spirit of the 1960s:

> If we want to change the world – and who doesn't? – then we are marching boldly toward a situation of confusion, the moment when old patterns begin to be disrupted and new ones take their place. For change as significant as a revolution, we cannot know in advance who will cling to the old ways and who will embrace the new. All that remains is to pursue the goal for its own sake, because we consider it the right thing to do. All we can do is try to make the unthinkable thinkable. That is what Khomeini did. Whether or not we agree with his goals, we can learn from his pursuit of them.9

L'imagination au pouvoir, as they declared in Paris in May 1968. All power to the imagination! But imagination is not enough. How power is shaped and exercised is absolutely crucial. Kurzman's thinking here runs the risk of opening the door to the kind of disaster that occurred again and again in modern revolutions: 'idealists' taking advantage of confusion to seize power and then imposing fascist, communist or theocratic tyrannies, as happened in Iran. Those hoping for moderate and liberating change get swept along by the tide of revolution, then subjected to a new regime that will brook no opposition.

This has been the great flaw of all the radical revolutions since that in France in 1789. They failed abysmally in the task of creating constitutions of liberty and ended up, in short order, magnifying rather than overcoming tyranny. The twentieth century ended with this lesson being, at last, more or less absorbed. The East European revolutions of 1989 were democratic ones, not 'radical' ones – American rather than French revolutions.[10] So it is now in the Middle East – or so it has begun. The vital task is to keep things developing on these lines. Iraq is a work in progress, the Palestinian Authority has made the merest of beginnings, Egypt and Syria are beginning to feel the pressure. Iran would be the greatest breakthrough.

Why? For three reasons. First, because Iran has, since 1979, been the heartland of the radical Islamist cause and a rallying point for all Muslim resentment of the West in general and the United States of America in particular. It has been the single most persistent sponsor of Islamist terrorism for a generation and a regime change could put an end to this era. Second, because the theocratic regime in Iran has failed its people and it is time to demonstrate that there is a better way to bring Muslim cultures into alignment with the modern world. Showing this in Iran would be especially potent precisely because radical Islam has been tried there and has failed.

The third reason, of course, is that it may be that only regime change can end Iran's quest for nuclear weapons. If war is not a viable option, as most informed observers seem to believe, and if the existing regime will not change course short of war, then it follows that the downfall of that regime is the necessary condition for ending the nuclear weapons program. What is clear is that the social preconditions for a regime change exist in Iran and the momentum for change in the Middle East could provide what we might call the 'Kurzman Shift' in regime stability.

But why should Iran be harassed over its nuclear program? The question has two levels: what should be done about the wider set of circumstances in which Iran has chosen to build nuclear weapons; and what can be done in those circumstances to contain or remove the danger that such weapons might represent in the hands of Iran? The key consideration at the first level is the decision by the nuclear powers, the United States chief among them, to retain their own nuclear arsenals despite the end of the Cold War. The key consideration at the second level is how to handle Iran, if it cannot be dissuaded from completing nuclear weapons in the next few years.

In 1982, in *The Fate of the Earth*, Jonathan Schell deplored the vast nuclear arsenals in existence as a gigantic trap into which our whole species had stumbled.[11] Twenty years passed without the trap being sprung. The Cold War even ended with the two nuclear superpowers agreeing to greatly scale down their nuclear arsenals. Nonetheless, they both retained substantial nuclear arsenals. In 1998, in *The Gift of Time*, Schell argued that this could lead to a second round of nuclear arms races and a breakdown in the non-proliferation regime. Even as his new book went to press, India and Pakistan exploded nuclear weapons.[12]

Then came September 11, 2001. 'If there proves to be a silver lining to the terrible events of September 11,' wrote Jessica Mathews, President of the Carnegie Endowment for International Peace, in 2002, 'it may be that they restored proliferation of weapons of mass destruction to its rightful place at the top of the global security agenda and added a sense of urgency to controlling their continuing spread.'[13] It did exactly that, of course. What it did not do is add a sense of urgency to the *abolition* of such weapons. On the contrary, it seemed to reinforce a belief among the guardians of existing nuclear arsenals that these were necessary and safe; only such weapons in the hands of rogue states or terrorist groups were a problem.

Schell, as tireless an abolitionist as any of those who stood up against slavery in the nineteenth century, believes this is a compound delusion. It is bad enough for the United States to have held onto its nuclear arsenal for the purposes of 'deterrence' after the end of the Cold War but 'after September 11, an even more radical departure from deterrence was announced – the doctrine of pre-emptive war. Its aim was to stop – or head off – nuclear proliferation by military force ... But proliferation to new countries cannot be considered in isolation from possession of nuclear weapons by the old Cold War powers ... the inescapable, underlying question is whether the possessors of nuclear arms will insist on keeping them indefinitely or will be prepared to accept their abolition.'[14]

Schell's argument is weakened somewhat by the consideration that, overwhelmingly, the world's states have voluntarily chosen not to acquire nuclear arms, despite their retention by the major nuclear powers. Those who have acquired them cannot, therefore, claim to have done so on the basis of a universal imperative. Moreover, those who have done so fall into two categories: the states which have never acceded to the Non-Proliferation Treaty (NPT) because they believed both that it was inequitable and that they might need nuclear weapons; and those who have acceded to the NPT but have then used it to cloak their quest for nuclear weapons in deceit.

Israel, India and Pakistan belong in the first category.[15] Iraq belonged, while Iran and North Korea still belong in the second category. That all three had or have repressive regimes, anti-status quo agendas and track records of conducting or sponsoring terrorist acts has added to a sense that they constitute a particularly troubling case of nuclear weapons proliferation. It was not, after all, a desire to use nuclear weapons, but a fear that Hitler would get them first and use them indiscriminately which prompted Franklin Roosevelt to set up the Manhattan Project early in the Second World War. That atomic bombs were then used by the United States against Japan has been a matter of moral controversy ever since and their use has not been repeated.[16]

Given these considerations, it is possible to acknowledge the first level problem without dismissing the second level one. There can be no defence on moral grounds, even given Schell's argument, for Iran's being permitted to both belong to the NPT and build nuclear weapons. Had it never joined the NPT, there would still be concern about its going nuclear, but given that, like Iraq under Saddam and North Korea, it has been cheating on commitments it freely undertook under the NPT, there is, at least, a clear case for international pressure against it to reverse course and abide by its commitments.

The question is, how can it be induced to do so? It is this situation and its deep historical background that Kenneth Pollack set himself to explore in a book published in late 2004: *The Persian Puzzle: The Conflict Between Iran and America*.[17] He spent a good deal of his book discussing the history of Iranian relations with America on the basis that 'Anyone who cannot master that history cannot understand how to move beyond it.'[18] He is correct in saying this, as he is in stating bluntly that 'Americans are serial amnesiacs; as a nation, we forget what we have done almost immediately after doing it.'[19]

Yet he proceeded to argue that, basically, America may have made the odd mistake in regard to Iran, but has almost always had good intentions and has erred more through benign neglect than wilful purpose. The great exception is the decision, in 1953, famous in the annals of CIA covert operations, to engineer the overthrow of Mohammed Mossadeq and put Mohammed Reza Pahlavi back in control of Iran.[20] Even in this case, however, he argued that the problem was more with Mossadeq than with the powers that be in Washington.[21]

James Bill, in 1988, was more trenchant and insightful in his account of US relations with Iran than Pollack in 2004, which is disconcerting given how extensively Pollack draws on Bill in putting together his history of the matter up to 1988. He doesn't appear to have reckoned with Bill's argument that the American policy-making and intelligence system, as it applied to Iran between the early 1950s and the late 1980s, exhibited profound and enduring systemic flaws. Consequently, he doesn't even ask whether those flaws have persisted. The problems with Iraq over the past three years might be seen as suggesting that they have. Yet his prescription for dealing with Iran depends on that policy-making and intelligence system functioning with a high degree of efficacy.

Before considering Pollack's prescription, therefore, it is worth considering at least a few of James Bill's insights of almost two decades ago. He remarked that Iran policy had been dominated by unexamined premises. The most notable one he called 'the Pahlavi invincibility premise' – that the Shah was a rock on which leftist or Islamic dissidents would break like water.[22] This premise, he argued, took hold in Washington after the death of President Kennedy – whom the Shah detested – and lasted right down to the debacle of 1978-79. When, in November 1978, Ambassador Sullivan cabled Washington that the time had come to consider 'some options which we have never before considered relevant', he headed the cable 'thinking about the unthinkable.' President Carter wanted to sack him, as someone who could not be trusted with working to save the Shah's regime.[23]

Bill noted, further, that altogether insufficient attention was paid to developing foreign service and intelligence officers with the language skills and country experience to understand Iran outside the diplomatic cocktail circuit.[24] He went on to remark that there was an obsession with secrecy and a conceit that secret cables somehow had more value than what could be found in open sources, if only one looked. This often generated the most extraordinary ignorance – covered up by the classification of government reports.[25]

There was a great need, Bill argued, for American foreign policy to 'increase its emphasis on long-range analysis.'[26] The upheaval in Iran in 1979 had caught the United States unprepared because for years it had neglected to do such analysis. Unexamined assumptions, irresponsibly reinforced by superficial and self-satisfied scanning of the Shah's arms purchases from America and his rhetoric about Iranian modernisation, blinded official Washington to what was brewing. Lack of an intelligence cadre deeply familiar with the country vitiated political reporting and, when sound reporting did come in, there was an impulse to dismiss it and even to shoot the messenger. Truly, as Stansfield Turner remarked, after heading the CIA, 'Analysis, especially political analysis, is the Achilles heel of intelligence.'[27]

Bill's overall conclusion, in 1988, is stunning when one considers the shock of 2001 and the current calls for reform of the intelligence and policy-making system in the United States. The various problems he had highlighted, he declared, interlocked in 'a system that highly resists reform' and, consequently, produces 'a highly resilient system of errors.'[28] The record of recent years cannot inspire confidence that this system has been reformed or that its resilience in the making of errors has lessened. As it now sets about dealing with Iran, this is of the greatest importance, because there is plenty of scope for error and its consequences could be particularly unpleasant.

If Pollack's history is largely derivative and self-satisfied, his concluding reflection, 'Toward a New Iran Policy' was at least thoughtful and systematic. His major premise was that invasion is not an option the United States should choose. This is chiefly because the grounds adduced in 2002 for invading Iraq do not obtain in the case of Iran. Its regime is oppressive but not genocidal. Its leadership is strategically rational, not compulsively aggressive.[29] It is unlikely to supply terrorists with WMD.[30] It is four times as large and three times as populous as Iraq, with a terrain that would present formidable obstacles to an invader. Even a limited military option intended to eliminate Iran's nuclear facilities would be both problematic to carry out and possibly counter-productive in its consequences.

To military caution, Pollack adds political pessimism. He agrees that 'there is considerable evidence – both anecdotal and quantitative – to show that most Iranians are unhappy with the regime. However, it is a giant leap from that to suggesting that they are on the brink of revolution ... Most of the evidence indicates that Iranians are sick of revolutions and don't want another one.'[31] He has read Kurzman, but he appears to draw precisely the opposite conclusion to Kurzman himself: since the 1979 upheaval was unpredictable, such upheavals are inherently improbable and current evidence about popular preferences or moods is a reliable indicator of how things will remain.

Pollack quite reasonably concludes, then, that 'Iran is a very hard problem,'[32] because none of the 'obvious' solutions work. Since it is not admissible to simply throw up one's hands, he proposes a complex policy approach which he calls 'Triple Track'. These are the three tracks. First, hold open to Iran the prospect of a 'grand bargain' in which, if it plays no games and tells no lies, it will receive considerable benefits. These will be delivered over a protracted period and on a reversible basis to ensure that it keeps the bargain. Second, while offering real 'carrots', wave real 'sticks', in the form of international sanctions. As part of this track, tighten the NPT regime to punish both suppliers and purchasers of illicit technologies and to prohibit 'even civilian nuclear activities that could be related to weapons acquisition.'[33]

Third, prepare a new containment regime as a fallback position. This would require reconfiguring American military forces in the Persian Gulf region; laying down clear 'red lines' to deter Iranian aggression in the area; making unambiguously clear that any use by Iran of its nuclear arsenal would bring down the most dire consequences on its head; massively augmenting intelligence gathering on Iran; and consistently advocating democratisation, the rule of law, human rights and religious tolerance in Iran. In short, settle in for a state of siege on Cold War lines, but with a substantially smaller and also perhaps less tractable adversary.

If Pollack is correct and no Kurzman Shift is likely to occur in Iran, at least for some considerable time, then something like Triple Track may be necessary. But, by his own account, the latter would be very difficult to sustain, especially the second track, because of what he openly and repeatedly describes as 'the perfidy of our allies' – the Europeans and the Japanese – to say nothing of the Russians and the Chinese. Imposing sanctions on Iran proved impossible, for this reason, throughout the 1990s. Tightening the NPT is likely to run into similar problems, especially in the near term which is when it might, in principle, serve some useful purpose.

Problems in achieving multilateral action to reform the NPT will, in any case, have at their root the moral problem highlighted by Schell – that the big powers insist on keeping their nuclear weapons and practising what India's Jaswant Singh, seven years ago, dubbed 'nuclear apartheid'. Meanwhile, any attempt at Triple Track will have to be conducted by an intelligence and policy-making system that shows every sign of still suffering from the flaws James Bill pointed out in 1988. We could, therefore, be in for a rough ride on Iran. If there is hope for a constructive solution, though, it could well be that some form of containment and some degree of reform of the NPT, coupled with further political movement in the Middle East, will trigger a Kurzman Shift in Iran. And if that happens, paradoxically, it may be that we shall have to acknowledge the invasion of Iraq as having opened up the possibility for transformation.

7.   *Endnotes*

1   Quoted in Frances FitzGerald, 'Giving the Shah What He Wants', *Harper's*, November 1974, p. 82.

2   James A. Bill, *The Eagle and the Lion: The Tragedy of American – Iranian Relations*, Yale University Press, Newhaven and London, 1988, p. 417.

3   Kenneth M. Pollack, *The Persian Puzzle: The Conflict Between Iran and America*, Random House, New York, 2004, p. 389.

4   James Fallows, 'Will Iran Be Next?' *The Atlantic Monthly*, vol. 294, issue 5, December 2004, pp. 99-109.

5   Charles Kurzman, *The Unthinkable Revolution in Iran*, Harvard University Press, Cambridge and London, 2004.

6   Ibid., preface p. viii.

7   Kurzman estimates that 10% of the Iranian population participated in the mass demonstrations and general strike that brought down the Shah, whereas fewer than 2% participated in the dramatic events in France in 1789 and fewer than 1% in the public mobilisations of 1991 that brought down Soviet Communism. Ibid., pp. vii-viii.

8   Ibid., introduction, p. 9.

9   Ibid., preface, p. ix.

10   It was Hannah Arendt who most forcefully articulated this distinction at the height of the Cold War, writing, just after the Cuban Missile Crisis, 'It was the French and not the American Revolution that set the world on fire, and it was consequently from the course of the French Revolution, and not from the course of events in America or from the acts of the Founding Fathers, that our present use of the word "revolution" received its connotations and overtones everywhere, the United States not excluded. ... It is odd indeed to see that twentieth century American even more than European learned opinion is often inclined to interpret the American Revolution in the light of the French Revolution, or to criticize it because it so obviously did not conform to lessons learned from the latter. The sad truth of the matter is that the French Revolution, which ended in disaster, has made world history, while the American Revolution, so triumphantly successful, has remained an event of little more than local importance.' *On Revolution* (1963), Penguin, 1982, pp. 55-6.

11   Jonathan Schell, *The Fate of the Earth*, Picador, Jonathan Cape, 1982. pp. 183-4.

12   Jonathan Schell, *The Gift of Time: The Case for Abolishing Nuclear Weapons Now*, Granta Books, London, 1998, p. ix.

13   Foreword to Joseph Cirincione with Jon B. Wolfstahl and Miriam Rajkumar, *Deadly Arsenals: Tracking Weapons of Mass Destruction*, Carnegie Endowment for International Peace, Brookings Institution Press, Washington D.C., 2002, p. vii.

14   Jonathan Schell, *The Unfinished Twentieth Century: The Crisis of Weapons of Mass Destruction*, Verso, London and New York, 2003, introduction, pp. xiii-xiv.

15   On Israel, see Frank Barnaby, *The Invisible Bomb: The Nuclear Arms Race in the Middle East*, I.B. Tauris & Co., 1989, pp. 1-74; and Avner Cohen, *Israel and the Bomb*, Columbia University Press, New York, 1998. On India and Pakistan, see M. V. Ramana and A. H. Nayyar, 'India, Pakistan and the Bomb', *Scientific American*, December 2001, pp. 60-71.

16   The standard account is Gar Alperovitz, *The Decision to Use the Atomic Bomb*, Vintage Books, New York, 1996, pp. 847. For an account of the decisive shift in Soviet thinking under Gorbachev, which had its roots many years earlier, under Khrushchev, see Michael McGwire, *Perestroika and Soviet National Security*, Brookings Institution Press, Washington D.C., 1991, pp. 179-184.

17   In 2002 he produced a book which made the case for invading Iraq. After WMD were not found there, he reflected at some length on how the intelligence estimates had gone wrong. See Kenneth M. Pollack, *The Threatening Storm: The Case For Invading Iraq*, Random House, New York, 2002; and Kenneth M. Pollack, 'Spies, Lies and Weapons: What Went Wrong?', *The Atlantic Monthly*, vol. 293, issue 1, January-February 2004, pp. 79-92.

18   Pollack, *The Threatening Storm*, p. xx.

19   Ibid., p. xxi.

20   Stephen Kinzer, *All the Shah's Men: An American Coup and the Roots of Middle East Terror*, John Wiley & Sons Inc., 2003.

21   Pollack, *The Persian Puzzle*, chapter 3, 'The Ugly Americans', pp. 40-71.

22   Bill, *The Eagle and the Lion*, pp. 440-1.

23   Kurzman, *The Ultimate Revolution in Iran*, p. 2.

24   Bill, *The Eagle and the Lion*, p. 443.

25   'Startling examples of official US ignorance of Iran abound. In 1974, the best-informed American political officer in Tehran had never heard of Ali Shariati, the Paris-trained intellectual whose speeches and writings in Iran provided much of the inspiration for the revolution. What is worse, in 1977 the foreign service officer with the greatest experience in Iran and the one most knowledgeable about internal affairs there also admitted that he had never heard of Shariati. And these were among the best diplomats that the United States posted to Iran.' Ibid., p. 437.

26   Ibid., p. 445.

27   Stansfield Turner, *Secrecy and Democracy*, Houghton Mifflin, Boston, 1985, p. 271.

28   Bill, *The Eagle and the Lion*, pp. 446-7.

29   Pollack, *The Persian Puzzle*, pp. 384-5.

30   Ibid., pp. 419-20.

31   Ibid., p. 387.

32   Ibid., p. 400.

33   Ibid pp. 410-11.

You can't hide from the truth forever.
- Shona (Zimbabwean) proverb

We're taking the African people by the scruff of the neck and saying 'Come with us into the twentieth century.' But they'll be glad they came.
- Garfield Todd (c. 1955)[1]

... there is strong evidence that economic and political freedoms help to reinforce one another, rather than being hostile to one another (as they are sometimes taken to be).
- Amartya Sen[2]

8. On Zimbabwe: Rhodes to Ruin

*Y*ou have inherited a jewel in Africa,' Tanzania's Julius Nyerere remarked to Robert Mugabe in 1980, 'don't tarnish it.'[3] After many years of guerrilla war, Mugabe had wrested control of Rhodesia (Zimbabwe) from its white masters. Hopes were high that he would prove an enlightened statesman, presiding over a genuinely progressive post-colonial regime. Those hopes have been bitterly disappointed. Not only has the jewel of Africa been tarnished, it has been reduced to ruin and famine.[4] This is a disaster for the people of Zimbabwe. It is also an object lesson in what good governance requires, whether one is black or white.

Less than a century before Mugabe became the Prime Minister of Zimbabwe, it had been colonised by the great imperialist Cecil Rhodes at the high tide of British imperialism and the European scramble for Africa.[5] White settlers, including Christian missionaries, encouraged by Rhodes, flocked to 'Rhodesia' and took possession of vast swathes of the best agricultural land, dispossessing the black populations. Mugabe was born, in 1924, under the white regime built on this colonial foundation. He was born, in fact, on a Jesuit mission and educated by Jesuits.[6]

His parents were both trained by Jesuit missionaries: his father, Gabriel, as a carpenter, his mother, Bona, as a catechist and Bible teacher. He himself was mentored by an Irish Jesuit, Father Jerome O'Hea, 'a strong believer in education as the key to emancipation'. O'Hea not only taught the young Mugabe both catechism and Descartes, but tales of the Irish rebellion against the domination of the English.[7] Mugabe was to pay tribute to his Jesuit education and to Father O'Hea personally fifty years later.[8]

A contemporary of Mugabe's, the journalist Lawrence Vambe, has remarked of his education on a Catholic mission in the 1930s, 'Here was young Africa, full of hope and zeal, brought together under the umbrella of this institution, where we were all trying to gain a deeper insight into the white world, as well as break away from our own tribal chains.'[9] Mugabe, despite his education, has never really broken free of such chains. Growing up, he also developed a cold bitterness towards the white Rhodesians. Even their willingness to surrender power to the black majority did not enable him to break free of this bitterness, any more than from his tribal chains.

He gave the appearance of having broken free of both on taking power in 1980, speaking in words worthy of an Abraham Lincoln or Nelson Mandela. He declared to a gathering of world leaders on

17 April 1980 in the old Rhodesian capital, Salisbury, which he was to rename Harare, 'The wrongs of the past must now stand forgiven and forgotten. If ever we look to the past, let us do so for the lesson the past has taught us, namely that oppression and racism are inequalities that must never find scope in our political and social system. It could never be a correct justification that because the whites oppressed us yesterday when they had power, the blacks must oppress them today because they have power. An evil remains an evil, whether practised by white against black or black against white.'

He has so misgoverned Zimbabwe since then, however, that that speech makes him look, in hindsight, less like Abraham Lincoln than like Richard of Gloucester, whose soliloquy, at the beginning of Shakespeare's *Richard III* lays bare his twisted mind. 'Now is the winter of our discontent made glorious summer by this sun of York; and all the clouds that loured upon our house in the deep bosom of the ocean buried', he declares. But he goes on to confess plans to bring about the demise of all who stand in the way of his own power. 'Plots have I laid, inductions dangerous, by drunken prophecies, libels and dreams, to set my brother Clarence and the king in deadly hate the one against the other.'[10]

Yet we know that Shakespeare's brilliant drama was a caricature of the historical Richard III. Far from being the hunchback with a withered arm described by Shakespeare, Richard was actually an attractive and gifted prince, a cultured and intelligent ruler, who earnestly sought the welfare of his subjects. He was also a valiant soldier who was overwhelmed at Bosworth field, in 1485, and subsequently vilified by the house of Tudor. It is worth pausing, therefore, to ensure that we do not judge Robert Mugabe too hastily because of the enmities he has made among the white landowning class in Zimbabwe.

In 1980, as an undergraduate student, I was myself among those who thought Mugabe was a hero, not a villain. Impressed by reports concerning his multiple university degrees, earned while in white prisons, and his magnanimous rhetoric about creating a future free of racism and oppression, I pinned a picture of him to my bedroom wall. As recently as a year ago, I wondered whether the bad reports about him were simply the propaganda of his enemies. Martin Meredith's meticulous new biography demonstrates that, unfortunately, Mugabe was never deserving of trust or confidence and has long since become an irredeemable tyrant.[11]

Mugabe was distrusted by the other black anti colonial leaders in the 1970s, because of his intransigence and his avowed aim of establishing a one party state that would expropriate the farms of the whites and bring Marxism to Zimbabwe. Not only did leading Zimbabwean figures such as Joshua Nkomo and Abel Muzorewa urge the British and Ian Smith's white regime to cut Mugabe out of political negotiations for this reason, but major African leaders such as Samora Machel of Mozambique and Kenneth Kaunda of Zambia (the old Northern Rhodesia to Zimbabwe's Southern Rhodesia) became concerned at that time about Mugabe's express desire to destroy the whites and establish a one party state.

Machel, based on bitter experience in Mozambique only a few years earlier, warned Mugabe in early 1980, 'Don't play make-believe Marxist games ... You will face ruin ... if you force the whites into precipitate flight.'[12] Mugabe seems to have made his 'magnanimous' inauguration address with such counsel on show, but the race hatred and Marxist vitriol were never apparently absent from his mind. He soothed the fears of the whites while eliminating his black rivals for power in the early 1980s. He then dominated the black masses by force, fear and demagoguery, creating a clientele of corrupt black cronies. He never gave up his idea of expropriating the white farmers, however, and has proceeded with this design in recent years in utter disregard of both law and economic rationality. In doing so, he has reduced his own country to the verge of famine.

Ironically, Mugabe's longtime black colleagues in the revolutionary movement were less easily beguiled by him than were the leading figures in the former white regime. Ian Smith wrote in his memoirs regarding his meetings with Mugabe in 1980, that he was 'welcomed most courteously' and saw in Mugabe 'a balanced, civilised Westerner, the antithesis of the communist gangster I had expected.'[13] Yet 'communist gangster' he was and has remained. Smith's former intelligence chief, Ken Flower, was so impressed by Mugabe's sophistication and apparent reasonableness in 1980 that he judged him to be 'someone with a greater capacity and determination to shape the country's destiny for the benefit of all its people than any of his four predecessors.'[14]

It is crucial to the case against Mugabe that he was a man with both the abilities and the means to govern well. Neither Western governments nor the white farmers as a class were hostile to him when he took office. To the contrary, as Martin Meredith points out, there was an abundance of goodwill towards him and a willingness to give him support in the belief that his success would help bring about a peaceful transition to majority rule, also, in South Africa. Well over a billion dollars in foreign aid was offered by the United States and Britain from the outset and numerous 'aid workers and foreign expatriates arrived to help build the new state.'[15]

The apartheid regime in Pretoria, on the other hand, was a ruthless enemy from the beginning. The South Africans resolved to keep Mugabe's Zimbabwe weak and unstable 'to ensure that it presented neither a security threat nor an example of a stable African state.' As Meredith recounts, the South Africans 'recruited into their own defence force some 5,000 former Rhodesian military personnel, including entire special forces units ... and set about establishing a network of agents, informers, spies and saboteurs inside Zimbabwe, finding a large number of serving officers in the army, air force, police and the CIO only too ready to help.'[16] In December 1981, they set off a massive bomb in Harare in an attempt to assassinate Mugabe.[17]

Whether because he had always intended to move in the direction of Leninist dictatorship, or because he was unnerved and angered by South African sabotage and assassination operations, Mugabe soon showed that he was prepared to cut down the rule of law in order to get at his enemies.

Dismissing complaints about the use of torture to obtain confessions from several white saboteurs in 1982, he declared that 'The law of evidence and the criminal procedure we have inherited is a stupid ass. It's one of those principles born out of the stupidity of the procedures of colonial times.'[18] This, not the simple fact of majority rule, was what alarmed those white Zimbabweans who had not prejudged matters by making a pact with the South African apartheid regime. 'Within three years of independence', Meredith informs us, 'about half the white population emigrated.'

It was not the white, but the black population of Zimbabwe, however, which, from the outset, bore the brunt of Robert Mugabe's contempt for the restraints of law and ruthless determination to dominate, rather than serve his country. Mugabe never made any attempt to cultivate his old comrade in arms and revolutionary rival Joshua Nkomo as the leader of a legitimate and 'loyal' opposition, even though that was his natural role. Nkomo's supporters were disproportionately from Matabeleland, Mugabe's from Mashonaland – the two heartlands of the old domain of Cecil Rhodes.

Taking no chances, Mugabe signed a secret agreement with the totalitarian regime of Kim Il-sung in North Korea, in October 1980, his first year in office, for a team of 106 North Korean instructors to train an elite brigade of commandos – 5 Brigade – to be his praetorian guard.[19] It is important to note that this was well before the South Africans showed their hand against him. When 5 Brigade were ready, Mugabe broke violently with Nkomo and used his secret force with utter ruthlessness to physically destroy black opposition to his rule.

 Beginning in January 1983, ninety years after Cecil Rhodes had driven King Lobengula out of Matabeleland, Robert Mugabe unleashed 5 Brigade on the civilian population of Matabeleland. He called this a *Gukurahundi* – a sweeping away of the chaff. Within a fortnight, according to Meredith, at least 2,000 people were killed in mass executions, the violence being 'far worse that anything that had occurred' in the earlier white counter-insurgency war against Mugabe and Nkomo's own rebels. When the Catholic bishops spoke out against his atrocities, he dismissed them as 'sanctimonious prelates' and questioned their 'allegiance and loyalty to Zimbabwe.'[20] 'We do not differentiate who we fight', he declared, 'because we cannot tell who is a dissident and who is not.'

Hundreds of 'dissidents' were abducted and disappeared, never to be seen again, as in Argentina during *la sale guerre* of the mid-1970s. Mugabe was showing his hunch back and withered arm. This was no attractive prince, but a Shakespearean monster. The atrocities of 1983 and 1984 were not an aberration, but the hallmark of Mugabe's political style. Yet he remains in power in 2008. In consequence, the struggle in Zimbabwe has become one not to sustain white privilege and property, but to uphold even the minimum standards of good government and justice against a tyrant who will stop at nothing to preserve his corrupt and ruinous rule.

Mugabe's leading opponent is no white farmer, but the black commoner Morgan Tsvangirai. His most poignant critic is the renowned Zimbabwean musician, Thomas Mapfumo, once a keen supporter of Mugabe, who once praised Mapfumo's stirring folk songs. In 1999, Mapfumo released an album called Chimurenga Explosion, in which he sang of 'the beautiful country that Mugabe has turned to hell.'[21] The slogan of the democratic opposition, which has managed to survive in Zimbabwe against all the odds, is a demand directed at Mugabe: *Chinja maitiro* – 'Change the way you are doing things.'

In his ever more reckless, desperate and incoherent efforts to cling to his power, Mugabe lapses frequently into diatribes against the whites and English laws. In doing so, he betrays his people to a destructive combination of the old 'tribal chains' from which Lawrence Vambe had sought release seventy years ago, and the crudest form of modern tyranny. Such uncouth rhetoric was egregiously exhibited, on 24 November 2000, when Mugabe's goon, Joseph Chinotimba and two hundred so-called 'war veterans', in a 'protest' against legal restraints on the seizure of white farms, broke into the Supreme Court shouting 'Kill the judges!' Thus did Chinotimba play Dick the Butcher to Mugabe's Jack Cade, in a scene straight out of Shakespeare.[22] They seek to shake off 'English laws', but their lawlessness demonstrates precisely why such laws are required.

Yet we should exercise as much restraint in judgement as we can summon. Mugabe was, after all, like so many other African post-colonial rulers, heir to an all but psychologically overwhelming political legacy. He is a man hounded by the furies of Africa's colonial past. His resentment of the white farmers and his awareness of the abuses of colonialism in Africa eat away at him. They have goaded him to his present, wholly destructive attack on the whites.

What we are now witnessing in Zimbabwe is the penultimate act in a political tragedy whose first act was Cecil Rhodes's dispossession of King Lobengula in 1893. Like Orestes, in the classic drama of Aeschylus,[23] Mugabe can, with some reason, claim a 'right' to blood vengeance. He draws on this 'right' in scorning whites or Westerners who decry his abuses while forgetting those of the white past.[24] Like Orestes, however, he thereby ensnares himself in a nightmare blood feud and a cycle of destruction.

In Zimbabwe, as in ancient Athens, only a new rule of law can bring an end to the cycle of revenge. The great insight of Aeschylus was that civilisation requires this right be surrendered under the law. Mugabe's inaugural address and early rhetoric suggested that, perhaps, he had understood this profound point. His tyranny has amply shown that he has not. It is to be hoped that those who displace the octogenarian and stumbling *despotes*, as they are bound soon to do, will both understand this point and act in accordance with it. Then and only then will Zimbabwe begin to recover and become, once more, a young country 'full of hope and zeal' as Lawrence Vambe remembers it being in the 1930s.

In much of Africa, the decades since decolonisation have generally been ones of failed radical government, declining living standards and even genocidal violence. Zimbabwe had a genuine chance to be an exception to this dismal trend and to set an example of hope for the rest of Africa.

Robert Mugabe has run out of all excuses for destroying that chance and that hope. For the sake of his own countrymen and for the sake of Africa, it is time he was removed and replaced – one may hope – with a constitutional regime which will rebuild the institutions inherited by Mugabe and run into the ground over the past twenty eight-years.

8. *Endnotes*

1   Franz Ansprenger, *The Dissolution of the Colonial Empires*, Routledge, London, 1989, p. 185. Todd was Prime Minister of Southern Rhodesia (later Zimbabwe) from 1953 until 1958.

2   Amartya Sen, *Development as Freedom*, Alfred A. Knopf, 1999, preface, p. xii.

3   Ken Flower, *Serving Secretly: An Intelligence Chief on Record. Rhodesia into Zimbabwe 1964-1981*, John Murray, London, 1987, p. 280.

4   David Coltart, 'Zimbabwe's Man-Made Famine', *New York Times*, 7 August 2002, points out that, as a direct result of Mugabe's ruthless actions in driving commercial farmers off their land and utterly disrupting the rural economy, up to six million Zimbabweans will soon face starvation. The estimate comes from the World Food Program. 'What was once the jewel of Africa', he writes, 'has been transformed into a state that increasingly resembles Cambodia under Pol Pot.' Coltart is a member of Zimbabwe's parliament and a leader of the Movement for Democratic Change.

5   Cecil Rhodes obtained mineral concessions in Matabeleland from King Lobengula in 1888, established the British South Africa Company in 1889 and drove King Lobengula out of Matabeleland in 1893. He renamed Matabeleland and Mashonaland 'Rhodesia' in 1895.

6   Martin Meredith, *Our Votes, Our Guns: Robert Mugabe and the Tragedy of Zimbabwe*, Public Affairs, New York, 2002, p. 19.

7   Ibid., pp. 20-21.

8   Ibid., p. 22.

9   Ibid., p. 20.

10   Shakespeare, *Richard III*, act 1, scene 1, lines 1-4 and 32-5. The point is made even more powerfully in Richard Loncrane's 1995 film of *Richard III*, with Ian McKellen in the title role, because there the first part of the famous speech is addressed to the full court, basking in joint celebrations; the second, darker part of it in soliloquy. In act I, scene 3, lines 334-8, Richard declares, again in soliloquy, 'But then I sigh; and with a piece of Scripture, tell them that God bids us do good for evil: and thus I clothe my naked villany with odd old ends stolen forth of Holy Writ; and seem a saint, when most I play the devil.' So does many a ruthless political leader, but Mugabe as much as any.

11   At the end of 1987, he had himself declared executive president, 'combining the roles of head of state, head of government and commander in chief of the armed forces, with powers to dissolve parliament and declare martial law and the right to run for an unlimited number of terms of office.' Meredith, *Our Votes, Our Guns*, p. 79.

12   Ibid., p. 9.

13   Ibid., pp. 41-2.

14   Flower, *Serving Secretly*, prologue p. 3. The four predecessors in question were Garfield Todd (1953-58), Edgar Whitehead (1958-62), Winston Field (1962-64) and Ian Smith (1964-1980).

15   Meredith, *Our Votes, Our Guns*, p. 47.

16   Ibid., p. 51.

17   Ibid., p. 52.

18   Ibid., p. 55.

19   Ibid., p. 62.

20   Ibid., p. 68.

21   Ibid., p. 162.

22   In Shakespeare's *Henry VI Part 2*, act 4, scene 2, lines 72-7, the demotic Jack Cade tells his followers 'there shall be no money; all shall eat and drink on my score and I shall apparel them all in one livery, that they may agree like brothers, and worship me, their lord.' And Dick the Butcher yells out, 'The first thing we do, let's kill all the lawyers.'

23   The best modern English translation is that by Robert Fagles. See his Aeschylus, *The Oresteia*, Penguin, 1979. Don't miss Fagles' luminous 97 page introductory essay, 'The Serpent and the Eagle'. Fagles is an acclaimed translator of the Greek classics, best known for his masterful translation of Homer's *Iliad* and *Odyssey*. He is currently Professor of Comparative Literature at Princeton University.

24   For a searing study of Western atrocities in the Africa of just a century ago, see Adam Hochschild, *King Leopold's Ghost: A Story of Greed, Terror and Heroism in Colonial Africa*, Mariner Books, Houghton Mifflin, 1999. Ludo de Witte's new investigation of the 1961 murder of the Congo's charismatic independence leader Patrice Lumumba at the instigation of Belgian and American secret intelligence services is a vital supplement to the older history: Ludo de Witte, *The Assassination of Lumumba*, Verso, 2001. Lumumba was replaced, of course, by Joseph Mobutu, the CIA's man, whose dismally corrupt and incompetent rule of some thirty years left his vast country a prey to the appalling social disorder and violence which has since torn it apart.

PROSPERO: You do look, my son, in a moved sort, as if you were dismayed: be cheerful, sir. Our revels are now ended ... These, our actors, as I foretold you, were all spirits, and are melted into air, into thin air, and, like the baseless fabric of this vision, the cloud capped towers, the gorgeous palaces, the solemn temples, the great globe itself, yea all which it inherit, shall dissolve and, like this insubstantial pageant faded leave not a rack behind. We are such stuff as dreams are made on; and our little life is rounded with a sleep.
- Shakespeare[1]

... And it shall come to pass in the last days that the mountain of the Lord's house will be established in the top of the mountains and shall be exalted above the hills; and all nations shall flow into it ... And he shall judge among the nations and shall rebuke many people; and they shall beat their swords into ploughshares and their spears into pruning hooks: nation shall not lift up sword against nation, neither shall they learn war any more.
- Isaiah[2]

If you want a lover, I'll do anything you ask me to.
If you want another kind of love, I'll wear a mask for you;
If you want a partner, take my hand,
Or, if you want to strike me down in anger, here I stand.
I'm your man.
- Leonard Cohen[3]

9. On Love, Despair and the Future

This essay was written several years before the release of the documentary film *Leonard Cohen: I'm Your Man* and before I had heard either of Cohen's new CDs of the early part of this decade: Ten New Songs and Dear Heather. While I love many of his new lyrics, especially 'In My Secret Life', 'Here It Is', 'Love Itself', 'By The Rivers Dark' and 'Boogie Street', I have resisted the temptation to rewrite the essay, which was well received on its publication as a tribute to classic Cohen. As with several of the other essays, I have confined myself to a few cosmetic refinements of the piece, including a little updating of the 'tense'.

Leonard Cohen, poet and singer, reached three score and ten years after a half century of dazzling and sombre poetic creation. His *oeuvre*, in verse and song, is rich and should endure. Its ontological horizon is so wide that it swallows up the mass of juvenile and dyspeptic lyrics that the music industry has churned out since the late sixties.

Erotically, Cohen makes Keats seem as though he had freeze-dried testicles and Byron as though he had no heart. Morally, he breathes the spirit of Isaiah over the world after the Holocaust. His unique lyrical style is a wholly idiosyncratic blend of the Psalms and Federico Garcia Lorca, the Chelsea Hotel, Nashville and the Greek islands, Zen Buddhism and the Song of Songs, Franz Rosenzweig and Bob Dylan.

Cohen is not an entertainer of spoiled children, or politically correct white collar workers. He is a master singer of the songs of Zion, by the polluted waters of our post-Christian Babylon. There are few others like him and this makes both his lyrics and his life worth reflecting on.

How does one become a master singer of the songs of Zion? First, be born a Jew. Second, be immersed in poetry, song and the Judaic scriptures from a tender age. Third, explore all this with radical freshness in your youth. Fourth, be a vulnerable sojourner in the world. All these things have been true of Cohen. Born in Montreal in 1934, he was a fourth generation Canadian Jew whose psyche was deeply shaped by Judaism from his earliest years.

He awakened to song through the superb contralto voice of his spirited mother singing European folk songs in Russian and Yiddish in the family home. He was powerfully affected, at the age of eleven, by pictures of the Nazi extermination camps. He was instructed in the scriptures by his maternal grandfather, Rabbi Solomon Klinitsky-Klein, with the Book of Isaiah making an especially deep impression.

Above all, however, he seems to have derived from his Judaism a 'stranger' ethic which withheld him from any orthodox or settled way of life and gave him his haunting lyrics. This was fertilised by his discovery of the songs of Federico Garcia Lorca when he was fifteen. Lorca, he later said, 'taught me to understand the dignity of sorrow through flamenco.'

If there is a single song that could be called Cohen's signature song it is 'The Stranger Song,' recorded on his first album, *Songs of Leonard Cohen* (1968). It expresses a theme which deeply informs his sense of what human life is about. That theme is the burden of our freedom as something we must forever reclaim strenuously against the temptations of 'giving up the Holy Game of Poker.'

Cohen himself has practised Zen Buddhism for many years as an access to freedom from 'attachment,' almost becoming a Zen Cohen, as it were, but his understanding of the burden of freedom is deeply Judaic. The Jew is, from of old, the perennial 'stranger' in the world – leaving Ur, leaving Egypt, exiled to Babylon, dispersed across the face of the world, hunted to death by the *goyim*. At several levels of experience and active metaphor, this is the temper of Cohen's whole body of work.

This stranger ethos runs through his love poetry, his songs of existential fear and despair, and his songs of prophetic darkness. It was surely for this reason that, when his selected poems and songs were published in 1993, he called the book *Stranger Music*. Cohen must have been aware of the ideas of Franz Rosenzweig, since he knew the work of Martin Buber and Gershom Scholem. In any case, Rosenzweig's extraordinary book, *The Star of Redemption*, published in 1921, is probably the ideal theological and philosophical companion to Leonard Cohen's songs.

During the deepening darkness of the Great War, Rosenzweig wrote: 'It is imposed on us that we remain strangers ... strangers even in the very depth of our being.' He went on to articulate a radical defence of the 'high vocation' of Judaism as a witness to the revelation of fundamental ontological truths. His central insight was that Judaic spirituality had, from of old, stood for an ontology of human freedom *against* the idols of the ages, against the very currents of history itself. Its vocation within the Christian civilisation of the West, he argued, was to give witness to this, even as the Christian Church worked *within* the currents of history to convert the pagan world to the God revealed by the Hebrew Bible.[4]

Rosenzweig died four years before Hitler was swept to power in Germany by the Great Depression. The Holocaust then annihilated the Central European Jewish culture and thus its high vocation within Christendom. Leonard Cohen was born, in the safety of Canada, a year after Hitler came to power and the horrifying reality of the Holocaust seems only to have come home to him in 1945. Growing to adulthood in its shadow and the heyday of militant Zionism, he seems nonetheless to have had a poetic affinity with the vision of Rosenzweig. It is an affinity which deepened over time, as Cohen matured and explored his Judaic faith against all other forms of faith or oblivion on offer.

Cohen has been a devotee of the erotic as well as the spiritual all his life. Indeed, he is surely better known for both his loves and his despairs than for his Judaism. He has always been obsessed with the beauty of women and has loved many, but has always moved on, in his restless need for freedom. When he began work on a book called *Death of a Ladies' Man* in the mid-1970s, his long-time publisher Jack McClelland exclaimed, 'Christ, Leonard! *Death of a Ladies' Man*? With a title like that, we don't even need a manuscript!' As his biographer, Ira Nadel, wrote, 'Cohen has sought to witness, touch and experience beauty at close quarters ... [but] when he has obtained beauty he has abandoned it, feeling that it entrapped him.'[5]

He has a reputation, however, for passion, not for seduction. In both his life and his poetry he has been neither a Don Juan nor a Casanova. He has generally remained close to the women in his life and rather than treating them as conquests, has worshipped them – then fled to regain his freedom and creative solitude. Nadel quotes one long-time female friend of Cohen's as saying that he was 'unique and amazing.' He 'really loved women ... He felt that women had a power and a beauty that most did not even know. To be with Leonard was to begin to know your own power as a woman.' It is this richly complex eroticism that gives Cohen's poems of love and departure their profound appeal.

All of Cohen's books of verse and song have been inspired by and dedicated to women he has known, except *Book of Mercy* (1984), which was inspired by his return out of a spiritual desert to the wellsprings of his Judaic faith, in the early 1980s. Marianne Ihlen was his companion on the Greek island of Hydra in the early 1960s, when he left Canada to seek what he called a 'twelfth century lifestyle' in which he could discover himself as a stranger poet. Suzanne Elrod, his most enduring passion, was nineteen when he met her in New York, in 1968. They formed a bohemian partnership that lasted until the late 1970s, and she became the mother of his two children, Adam and Lorca.

The muse of his most mature work was Dominique Isserman in the 1980s. He met her on Hydra in 1982 and she soon became his lover. His album *I'm Your Man* (1988) was dedicated to her with the words, 'All these songs are for you, D.I.' She also directed a video recording of his song, 'Dance Me to the End of Love'. His recording of that song with Perla Batalla and Julie Christensen as supporting vocals and Bobby Furgo on the violin must rank as one of the greatest beauties of his life's work.[6] The lyrics carry us limpidly through 3,000 years of the erotic sublime, from the archaic Biblical world to a dream of violins under summer stars on the cusp of the future:

Dance me to your beauty
With a burning violin
Dance me through the panic
Till I'm safely gathered in
Lift me like an olive branch
And be my homeward dove
Dance me to the end of love.

Let me see your beauty
When the witnesses are gone
Let me feel you moving
Like they do in Babylon
Show me slowly what I only
Know the limits of
Dance me to the end of love.
Dance me to the children
Who are asking to be born
Dance me through the curtains
That our kisses have outworn
Raise a tent of shelter now
Though every thread is torn
Dance me to the end of love.

There is probably no song of Cohen's which better instantiates the judgement of his long-time companion on the road, Jennifer Warnes, that Cohen's love songs are 'the place where God and sex and literature meet ... the songs ... beckon the soul with just the configuration of the lyric.'

Underlying the sensuality there has always been a brooding Judaic spirituality. The two cross over at times, to the point where Cohen can seem to be a devotee of Robert Graves' White Goddess, as in 'Our Lady of Solitude':

And her dress was blue and silver
And her words were few and small
She is the vessel of the whole wide world
Mistress, oh mistress of us all.

In the desolation for which he is so famous, however, we find the resonance of the psalms. 'I will hear a parable, I will speak a dark language with the music of a harp', declares Psalm 49. 'I am afraid and shivering, I am full of horror. And I said, Who will give me wings like a dove?' asks Psalm 55. 'By the stream of Babylon we sat down and wept, when we remembered Zion. We hung up our harps on the willow trees, when the masters called for songs and the torturers for cheerfulness,' reads the famous Psalm 137.[7] This sense animates Cohen's darkest and also his most luminous verse from first to last.

In his 'Song for Abraham Klein' (1961), Cohen wrote:

> The weary psalmist paused
> His instrument beside
> Departed was the Sabbath
> And the Sabbath Bride.

Through many wanderings among the teachers of the heart and the soul, he returned to this theme most powerfully in some of his finest lyrics of the 1980s. Two of the most beautiful and haunting are 'If It Be Your Will' and 'Hallelujah', from his 1984 album *Various Positions*. The latter opens with an invocation of King David as psalmist:

> I've heard there was a secret chord
> That David played to please the Lord
> But you don't really care for music, do you?
> It goes like this: the fourth, the fifth
> The minor fall, the major lift
> The baffled king composing hallelujah.

The Book of Mercy (1984) overflows with this redolence of the Psalter, for example in a passage headed 'Sit Down, Master', in which Cohen wrote: 'Sit down, master, on this rude chair of praises, and rule my nervous heart with your great decrees of freedom ... In utter defeat I came to you and you received me with a sweetness I had not dared to remember. Tonight I come to you again, soiled by strategies and trapped in the loneliness of my tiny domain. Establish your law in this walled place ...'

The enduring influence of Isaiah can be found, Nadel rightly remarked, in Cohen's eloquent repudiation of illusions, oppression and deceit. One of his better known late lyrics, which became a byword in Europe years ago, is 'First We Take Manhattan':

> They sentenced me to twenty years of boredom
> For trying to change the system from within
> I'm coming now, I'm coming to reward them
> First we take Manhattan, then we take Berlin.

> I'm guided by a signal in the heavens,
> I'm guided by the birth mark on my skin,
> I'm guided by the beauty of our weapons.
> First we take Manhattan, then we take Berlin.

This was little more than folksy apocalyptic for Cohen, which he described wryly as 'a sort of demented manifesto.' The undercurrent was forceful, though. It could erupt on occasions, as when he challenged an unruly German crowd at the Berlin Sportspalast in 1972, in the very words Joseph Goebbels had used there, thirty years before: *'Wollt ihr den totalen Krieg?'* – Do you want total war?

From *Stories of the Street* (1968) to *Everybody Knows* (1988), Cohen's dark outlook on the post-Holocaust world has been one of his most abiding characteristics. Yet the luminous influence of Isaiah glows like a candle even in this darkness. Nowhere is this more so than in his brilliant late song 'The Future', which makes no concessions at all to Babylonian optimism. The refrain draws the ancient Hebrew prophet into the apprehended twenty-first century:

> Things are going to slide in all directions
> Won't be nothing, nothing you can measure any more.
> The blizzard of the world has crossed the threshold
> And it has overturned the order of the soul
> When they said REPENT
> I wonder what they meant.

The third verse spells it out even more clearly for the hard of hearing among the *goyim*:

> You don't know me from the wind
> You never will, you never did
> I'm the little Jew who wrote the Bible
> I've seen the nations rise and fall
> I've heard their stories, heard them all
> But love's the only engine of survival.

Cohen participated in the folk revival of the late fifties and early sixties in America, but moved on when the merchants took it over. He has never succumbed to the dark powers of the music industry. Instead, he has sung mournful songs of exile.

'I am like the crow in the desert and the owl in the ruins. I sit awake wailing alone like a bird on the roof', says Psalm 102. The last phrase inevitably brings to mind Cohen's famous song 'Bird on the

Wire' (1969): 'Like a bird on the wire, like a drunk in a midnight choir, I have tried in my way to be free.' Cohen's way of being free, however, has been cumulatively extraordinary. His hypnotic songs call on us to own the radical freedom that is our birthright. His prayer is that the Lord of Song will bless all of us in his mercy:

> If it be your will
> If there is a choice
> Let the rivers fill
> Let the hills rejoice
> Let your mercy spill
> On all these burning hearts in hell
> If it be your will
> To make us well.

His vocation has been just what Franz Rosenzweig might have hoped for from a Jewish poet. We, in Babylon, owe him a debt of gratitude for both his weeping and his singing of the songs of Zion.

9.   *Endnotes*

1   William Shakespeare, *The Tempest*, act 4, scene 1, lines, 146-158.

2   King James Bible, Isaiah 2:2-4.

3   Leonard Cohen, *Stranger Music: Selected Poems and Songs*, Vintage Books, New York, 1993, p. 357.

4   Stephane Moses, *System and Revelation: The Philosophy of Franz Rosenzweig*, Wayne State University, Detroit, 1992.

5   Ira Nadel, *Various Positions: A Life of Leonard Cohen*, Pantheon, New York, 1996.

6   *Leonard Cohen: More Best Of*, Sony Music Entertainment, 1997.

7   Peter Levi, *The Psalms*, Penguin, 1976.

The fox knows many things, but the hedgehog knows one big thing.
-Isaiah Berlin

Hedgehogs remind one ... of Churchill's definition of a fanatic: someone who cannot change his mind and will not change the subject.
-Philip Tetlock

Insisting on anonymity was the only far-sighted thing those characters did.
-anonymous sceptic

10. On Reason and the Prediction of the Future

We rely on experts – academics, intelligence analysts, stock market analysts, strategic planners in business and finance – to forecast the future for us. It has long since been established that stock market analysts in general do no better than random.[1] There has been widespread scepticism about intelligence analysts since the 9/11 debacle and the WMD fiasco in Iraq. Yet the belief in forecasting expertise as such is tenacious. Philip Tetlock's work suggests we should start to regard expertise in political forecasting with the same scepticism with which the well-informed now regard stock market forecasting. Political pundits are attempting, he argues, to do with confidence what they demonstrably cannot do very well at all.

Tetlock is a psychologist who, after completing his PhD at Yale University in 1979, spent many years at the University of California, Berkeley and Ohio State University exploring the capacity of experts to learn, which is to say, to admit errors and alter their beliefs and assumptions in the face of evidence. His book *Expert Political Judgment* was the climax of a generation of research. His ongoing research interests are on how experts think about possible pasts (historical counterfactuals) and probable futures (conditional forecasts), how they respond to confirmation or disconfirmation of expectations, and how people in general cope with various types of accountability pressures and demands in their social world.

Tetlock gained tenure in 1984 and, whatever the downside of the tenure system in universities, he has to be accounted one of its outstanding success stories. Tenure gave him the stability needed to engage in a long-term project with remarkable tenacity and energy. He might, perhaps, have found the means to do it anyway, but twenty years of patient analysis is not something purely commercial interests will normally underwrite. This is especially so, given that it isn't clear there is a commercial pay-off for the work he has done – just a deeper understanding of how many experts are paid a good deal of money for making dubious forecasts without any ultimate accountability.

In a preface he explains the motive for his long-term study of expert political judgement. 'I have long been puzzled,' he writes, 'by why so many political disagreements – be they on national security or trade or welfare policy – are so intractable. I have long been annoyed by how rarely partisans admit error, even in the face of massive evidence that things did not work out as they once confidently predicted. And I have long wondered what we might learn, if we approached these disputes in a more aggressively scientific spirit.'[2] It is the scientific spirit with which he tackled his project that is the single most notable thing about his book, but the findings of his inquiry are important and, for both reasons, everyone seriously concerned with forecasting, political risk, strategic analysis and public policy debate would do well to read the book and ponder its lessons.

His project dates back to 1984 when, freshly tenured at Berkeley, he was involved in exploring the psychological and strategic dilemmas of the Cold War. 'I was struck,' he writes, 'by how frequently influential observers offered confident, but flatly contradictory, assessments that were impervious to the arguments that were advanced by the other side.'[3] This is, clearly, an issue of abiding importance which afflicted the conscientious centuries before the Cold War. The European wars of religion in the sixteenth and seventeenth centuries ended with both Catholics and Protestants signing up to the maxim *cuis regio, eius religio* (whose territory, his religion), signifying that intellectual exchange had been able to achieve no more than a truce.

After the end of the Cold War, despite the famous announcement of 'the End of History' by Frank Fukuyama, we plainly still had an intractable problem in this regard. Just consider the brouhaha over the war in Iraq, to say nothing of the controversies over Islam. One possible 'answer' to this problem is to embrace 'postmodernism' or some other form of relativism and simply declare that all world views are equally valid and that 'objectivity' in matters of argument, whether about the past or the future, is a chimera: there are only different perspectives, different interests and irreducible antinomies in how they represent themselves.

One of the many beauties of Tetlock's book is that, while clearly an Enlightenment man and no relativist, he did not merely dismiss this 'answer', but went out of his way to consider its merits. In an effort to get beyond relativism, he set out to design a research project that would make it possible to rigorously explore why we keep 'running into ideological impasses rooted in each other's insistence on scoring its own performance'. To do that, as he declares 'we need to start thinking more deeply about how we think. We need methods of calibrating expert performance that transcend partisan bickering and check our species' deep-rooted penchant for self-justification.'[4]

Here was a project surely daunting in its complexity and daring in its reach. It is something to make even the ghost of the legendary Stanley Milgram – originator of the controversial and disturbing obedience experiments of 1961 – stir and look on with interest.[5] But, whereas Milgram's bold and unusual experiments failed to gain him the conventional professional recognition he hoped for,[6] Tetlock was, by 2005, a man at the pinnacle of his profession – and his book demonstrates why.

'The goal', Tetlock tells us, 'was to discover how far back we could push the "doubting Thomases" of relativism by asking large numbers of experts large numbers of questions about large numbers of cases and by applying no favoritism scoring rules to their answers. We knew we could never fully escape the interpretive controversies that flourish at the case study level. But we counted on the law of large numbers to cancel out the idiosyncratic case specific causes for forecasting glitches and to reveal the invariant properties of good judgment.'[7] To that end, he and his research team identified a pool of 284 experts in world politics, from government service, think tanks, academia and international institutions, who had shown themselves 'to be remarkably thoughtful and articulate observers of the world scene' and asked them to make forecasts about world affairs looking out from 1988 to 2003.[8]

The forecasts were numerically weighted by the 284 (anonymous) forecasters, in terms of their own confidence in their predictions, carefully recorded by the research team, then monitored for accuracy, sometimes nearly two decades later. Great effort was expended to make the methodology both systematic and objective. Tetlock's meticulous explanations of how the data were gathered and interpreted, adjusted and tested is most impressive in its dispassionate lucidity. This is important at two levels. First, because it demonstrates the commitment to the scientific spirit he invokes as the inspiration for the project; second, because it addresses the numerous defences the experts mounted when the results came in and were embarrassing to them – as they often were.

Tetlock's findings are disconsoling for anyone who believes that expertise confers reliable forecasting powers. They are, however, highly enlightening for anyone seeking to understand how judgement works, where it goes astray and how tenacious experts can be in retrospectively defending their judgements – regardless of what their recorded opinions and the later evidence show to have been the case. Nor did Tetlock shrink from stating just how far off base the forecasters generally were. 'The results,' he reported, 'plunk human forecasters into an unflattering spot along the performance continuum, distressingly closer to the chimp [throwing darts at random at a board] than to the formal statistical models.'

The depth and longitudinal range of his study gave considerable weight to his findings and buttressed several trenchant and unsettling conclusions: '... it is impossible to find any domain in which humans clearly outperformed crude extrapolation algorithms, still less sophisticated statistical ones ... across all judgments, experts on their home turf made neither better calibrated nor more discriminating forecasts than did dilettante trespassers'[9] and 'it made virtually no difference whether participants had doctorates, whether they were economists, political scientists, journalists or historians, whether they had policy experience or access to classified information, or whether they had logged many or few years of experience in their chosen line of work.'[10] And, not least among his findings, 'Bad luck proved a vastly more popular explanation for forecasting failure than good luck proved for forecasting success.'[11]

Not only were field of expertise and depth of experience not correlated with either accuracy of forecasting or well-calibrated self-confidence in judgement, but neither were ideological commitments or world views. As Tetlock puts it, '*Who* experts were – professional background, status and so on – made scarcely an iota of difference to accuracy. Nor did *what* experts thought – whether they were liberals or conservatives, realists or institutionalists, optimists or pessimists. But the search bore fruit. *How* experts thought – their style of reasoning – *did* matter.'[12] And as regards styles of reasoning, he found it most useful to invoke the old analogy, coined by Isaiah Berlin, of foxes (as sceptical, circumspect thinkers) and hedgehogs (as true believers or ideologues with the 'courage' of their convictions) and then allow a continuum in between – of foxhog and hedgefox hybrids.

The truly sobering finding of his project is that, overall, none of the human experts did well at forecasting. 'Foxes,' Tetlock found, 'are not awe-inspiring forecasters: most of them should be happy to tie simple extrapolation models, and none of them can hold a candle to formal statistical models. But foxes do avoid many of the big mistakes that drive down the probability scores of hedgehogs to approximate parity with dart-throwing chimps. And this accomplishment [modest though it may be] is rooted in foxes' more balanced style of thinking about the world – a style of thought that elevates no thought above criticism. By contrast, hedgehogs dig themselves into intellectual holes. The deeper they dig, the harder it is to climb out and see what is happening outside ... Hedgehogs are thus at continual risk of becoming prisoners of their preconceptions ...'[13]

Tetlock offered no formulaic answer to the challenges his findings confront us with. He did, however, draw particular attention to the overwhelming statistical finding of his study: that experts tended not to adjust their prior beliefs when the evidence came in, but to rationalise or outright deny their errors in forecasting. A variety of defences were used – challenging the robustness of the research project itself; appealing to an 'exogenous shock' as the reason for the future not panning out as they had predicted; invoking what Tetlock called 'the close-call counterfactual defence', which is to say, claiming that they were 'almost right'; using the 'just off in the timing' defence, or the 'politics is hopelessly cloudlike' defence; or the 'I made the right mistake' defence; or blithely claiming that the 'low probability outcome just happened to happen'. Often, experts would simply deny that what had been recorded as their forecast had in fact been what they said.[14]

All of this, in a way, might be waved away as 'human nature', but the carefully assembled research data with which Tetlock presented us should, surely, give us greater pause than that. These defences, after all, have important consequences in the world. If they characterise the cognitive behaviour of experts, how can experts themselves demand anything approaching rational belief adjustment among the *hoi polloi*? In any case, as Tetlock points out, the real gravamen[15] of his findings is not merely these bad faith defences, but 'the pervasiveness of double standards: the tendency to switch on the high intensity search light for flaws only in disagreeable results ... It is telling that no-one spontaneously entertained the possibility that "I guess the methodological errors broke in my direction this time".[16]

Now, it is often suggested – and, I confess, I have been one of those who have so suggested – that scenario-based thinking can serve as a corrective to feckless forecasting. True to his scientific mission, Tetlock did not leave this stone unturned and his findings are not very reassuring. 'The need for such correctives should not be in question,' he observes, 'but the scenario experiments show that scenario exercises are not cure-alls. Indeed, the experiments give us grounds for fearing that such exercises will often fail to open the minds of the inclined-to-be-closed-minded hedgehogs, but succeed in confusing the already-inclined-to-be-open-minded foxes.'[17] Considering many possibilities and misreading their relative probability due to the psychological impact of dramatic scenarios whose aggregate plausibility is less than it seems, can make scenario-based thinking counter-productive. We are better off, on balance, simply acknowledging our uncertainty and hedging against it.

All of this raises profound epistemological questions as to the degree of certainty we are able to attain and what constitutes credibility in a forecast or even a research finding. Tetlock, to his lasting credit, fully appreciated this. To play it out, he composed a kind of Socratic dialogue toward the end of his book, between four interlocutors: an unrelenting relativist, a hard-line neo-positivist, a moderate neo-positivist and a reasonable relativist.[18] It is beautifully constructed and well worth reading in its own right for the sheer intellectual pleasure of observing a first class mind exercising itself by cross-examining the very foundations of its beliefs about knowledge, truth and reality. One is reminded of Plato's famously demanding dialogue *Parmenides* or, less strenuously, of David Hume's lucid and entertaining *Dialogues Concerning Natural Religion*.

Tetlock himself, at the end of the day – or at the end of his book – is, by his own account, a reasonable positivist. He believes that scientific methods give us our best chance of avoiding error and overcoming illusion and prejudice. He also allows that, on an everyday basis, we require something more 'user friendly' than a statistically driven scientific research program to monitor our thinking. Here, intriguingly, he refers us to Harold Bloom's reflections on Shakespeare. 'The dominant danger,' he concludes, 'remains hubris, the mostly hedgehog vice of close-mindedness, of dismissing dissonant possibilities too quickly. But there is also the danger of cognitive chaos, the mostly fox-vice of excessive open-mindedness, of seeing too much merit in too many stories. Good judgment now becomes a metacognitive skill – akin to "the art of self-overhearing".'[19]

His footnote at this point refers the reader to 'Harold Bloom *Shakespeare: The Invention of the Human*, Riverhead Books, New York, 1998', with no specific page reference, which is uncharacteristically imprecise. The key passage is, in fact, that in which Bloom wrote of Hamlet overhearing himself speak and changing with every self-overhearing.[20] Bloom, of course, regards Hamlet as the most supremely realised literary character in history and the very avatar[21] of the modern human being, if not the very paragon of animals. Yet he also, in his rather unscientific and flamboyant manner, celebrates Hamlet's 'nihilism' and traces it, with Nietzsche, to Hamlet's having thought not too much but too well. It is a little difficult to reconcile this with Tetlock's Enlightenment project, in which our self-overhearing and consequent better thinking would lead to more rational and responsible behaviour. Doubtless, that is why he recommended 'something *akin* to' the self-overhearing of the Prince of Denmark.

At the end of his book, Tetlock came close to specifying more precisely what he meant in this regard. 'Good judgment, then, is a precarious balancing act ... Executing this balancing act requires cognitive skills of a high order: the capacity to monitor our own thought processes and to strike a reflective equilibrium faithful to our conceptions of the norms of intellectual fair play. We need to cultivate the art of self-overhearing, to learn how to eavesdrop on the mental conversations we have with ourselves as we struggle to strike the right balance between preserving our existing worldview and rethinking core assumptions. This is no easy art to master. If we listen carefully to ourselves, we will often not like what we hear. And we will often be tempted to laugh off the exercise as introspective navel-gazing, as an infinite regress of homunculi spying on each other ... all the way down.'

'No doubt such exercises can be taken to excess,' the psychologist concludes philosophically. 'But if I had to bet on the best long term predictor of good judgment among the observers [studied in his project] it would be their commitment – their soul-searching Socratic commitment – to thinking about how they think.'[22] The problem here, however, is that this is a veritably monastic, or at least Pythagorean,[23] demand to make of any individual human being, given the tide of events, the pressures of the marketplace, the constitutive force of the passions, the insistent demands on us for group cohesion and loyalty, the urgencies of our mundane interests, the fears we hold of competitors and predators and of the looming unknown. That Tetlock, the psychologist, should have held to such a pure faith, after everything his study had revealed, is itself slightly unsettling.

He offers a more robust suggestion, but one which, for different reasons, as he allowed, is likely to find much resistance in the real world. 'From a broadly non-partisan perspective,' he reasons, 'the situation cries out for remedy. And from the scientific vantage offered by this project, the natural remedy is to apply our performance metrics to actual controversies; to pressure participants in debates – be they passionate partisans or dispassionate analysts – to translate vague claims into testable predictions that can be scored for empirical accuracy and logical defensibility. Of course, the resistance would be fierce, especially from those with the most to lose – those with grand reputations and humble track records.'[24] But, where the stakes are high and we cannot afford to rely on experts keeping their own scorecards, perhaps it is time to create serious research and training programs that would generate metrics for market and intelligence analysts and hold them more accountable.

One of Tetlock's consulting roles in recent years has been in critical analysis of political forecasting and risk assessment techniques for US government intelligence agencies. Such agencies are among those most commonly pilloried for their failures in forecasting, not least in the past few years, so Tetlock plainly has some good work to do. But as he himself remarks with characteristic restraint, 'The recommendations of this book are much in the spirit of Sherman Kent, after whom the CIA named its training school for intelligence analysts.' It is not, therefore, out of any malicious or self-satisfied glee at the errors of intelligence analysts that Tetlock urges new and ambitious programs in research and training, but out of a resilient belief that we can do better.

Alluding to Sherman Kent's own reflections of many years ago, he concludes, 'We can draw cumulative lessons from experience only if we are aware of gaps between what we expected and what happened, acknowledge the possibility that those gaps signal shortcomings in our understanding and test alternative interpretations of those gaps in even-handed fashion. This means doing what we did here: obtaining explicit probability estimates (not just vague verbiage), eliciting reputational bets that pit rival worldviews against each other, and assessing the consistency of the standards experts apply to evidence.'[25] Getting this done will require specific and tenacious commitment, at a time when resources are heavily committed to current intelligence and field operations, to analysing how analysis is done in intelligence work. A good place to start might be for intelligence analysts (and stock brokers and political pundits and academic experts and serious journalists) to read Tetlock closely and take his sober-mindedness, as well as his sobering findings, to heart.

10.   *Endnotes*

1   The classic study is Burton Malkiel, *A Random Walk Down Wall Street*, Norton, New York, 1999 – first published in 1973. But see also, John Allen Paulos, *A Mathematician Plays the Market,* Basic Books, New York, 2003. (Penguin 2004).

2   Philip E. Tetlock, *Expert Political Judgment: How Good Is It? How Can We Know?* Princeton University Press, Princeton and Oxford, 2005, preface, p. xi.

3   Ibid., p. xiii.

4   Ibid., p. 2.

5   Thomas Blass, *The Man Who Shocked the World: The Life and Legacy of Stanley Milgram, Creator of the Obedience Experiments and the Father of Six Degrees*, Basic Books, New York, 2004, esp. chapters 5-7 and 12.

6   This failure, according to Blass, had to do with 'the impression he created among some psychologists of a dilettante, who flitted from one newsworthy phenomenon to the next, not staying with any long enough to probe it in adequate depth.' (pp. 259-260).

7   Tetlock, *Expert Political Judgment,* p. 8.

8   Ibid., p. 44.

9   Ibid., p. 54.

10   Ibid., p. 68.

11   Ibid., p. 22.

12   Ibid., p. 20.

13   Ibid., p. 118.

14   Ibid., pp. 129-138.

15   'Gravamen: 1. grievance; memorial from Lower House of Convocation to Upper on disorders or grievances of Church. 2. Essence, worst part of, accusation. (Latin = inconvenience, from *gravare* to load, *gravis* heavy.' *Concise Oxford Dictionary: New Edition.* 7th Imp., 1978.

16   Tetlock, *Expert Political Judgment*, pp. 160-1.

17   Ibid., p. 199.

18   Ibid., pp. 219-29.

19   Ibid., p. 23.

20   Harold Bloom, *Shakespeare: The Invention of the Human*, Riverhead Books, New York, 1998, p. 423.

21   'Avatar: (Hindu myth) descent of deity to earth in incarnate form; incarnation, manifestation, phase. *Concise Oxford Dictionary: New Edition*. 7th Imp., 1978.

22   Tetlock, *Expert Political Judgment*, p. 215.

23   Arnold Hermann, *To Think Like God: Pythagoras and Parmenides – The Origins of Philosophy*, Parmenides Publishing, Las Vegas, 2004, provides an intriguing reconstruction of the world of these great pre-Socratics and their ground-breaking work in discerning the nature of proof, contradiction and truth.

24   Tetlock, *Expert Political Judgment*, p. 218.

25   Ibid., p. 238n.

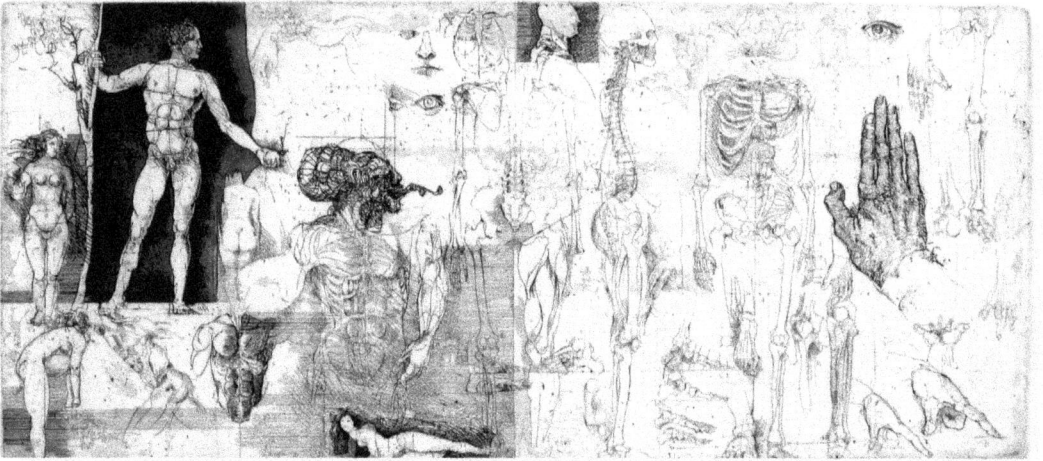

JS 110, *Sketches Also After Dürer*
Jörg Schmeisser, etching, 1974.

Acknowledgements

This book would not have come into existence had it not been for the friendship, patronage, enthusiasm and love of many people. It was Rowan Callick, then writing for the *Australian Financial Review* on Asian affairs, who put me in touch, in late 2000, with Hugh Lamberton, editor of the Friday literary review that the newspaper publishes. Over the following six years, Hugh proved to be the most professional and engaging of editors and our partnership flourished. I owe him a debt of gratitude for his openness to my writing on such a broad range of topics over such an extended period. My readers, with their feedback over those years encouraged me to believe that I was reaching people who longed for the kind of free and deep perspective on world affairs that I was seeking to offer. But it was the enthusiasm of Ian Gordon of Barrallier Books, for publishing a set of my essays to sit alongside his gorgeous edition of my *Sonnets to a Promiscuous Beauty*, that brought this project to life.

Deeper in the background and deserving of profound gratitude are three women, whose contributions to my life as a writer have been immeasurable. My mother, Brenda Monk, née Fitzgerald, though she never claims any credit for it, gave birth both to me and to my passion for the history of ideas. From my tenderest years, she nurtured in me a love of good stories and, sensing my interest in history, fostered it, first with beautifully illustrated and high quality children's books and magazines, then with ever more serious books. As she knows, my favourite such instance was the gift of Isaac Deutscher's biography of Stalin, at my own request, for my twelfth birthday. On it was (and still is, for I have kept the book ever since) the fond, if incongruous inscription 'Happy birthday to dear Paul, love from Mummy and Daddy.'

The second woman to have a vital impact was Kathleen Rose, who, as a young school teacher, in 1967, undertook, out of her own passion for enchanting stories, I'm sure, to read to our fifth grade class *The Wind in the Willows*, *The Hobbit*, several of C. S. Lewis's Narnia books and, above all, and unforgettably, *The Lord of the Rings*. Nothing else in my primary, or indeed my secondary education had anything like the profound impact on the development of my imagination and my fascination with history, world politics, language, poetry, religion and personal life that *The Lord of the Rings* had after I first heard it from her. It laid a foundation upon which I was able imaginatively to build an interest in everything from the most remote past to the gravest contemporary challenges; from simple folk ways to the most magnificent literature, music and philosophy, and to see them all as part of one grand drama.

And then there is Claudia Alvarez Ortiz, truly my own Ariadne, the golden thread of whose love for me and passion for my work as a writer has enabled me to come out of the labyrinth. She also gave me reason to believe that, somehow, I had actually slain the Minotaur. By both religion and folk song, we are taught that 'love is all you need', but I never believed this until Claudia demonstrated its truth to me. From the time that Ian Gordon suggested this project, her enthusiasm for it has been infectious and for all the right reasons: commitment to the integrity of knowledge, the raising of standards in public discourse and the sharing of insight and exploratory questions with all those who toil in the labyrinth themselves and lack an Ariadne's thread to guide them. This book is dedicated to her.

Melbourne
May 2008

Index